Proceedings of the Workshop on

Particle Astrophysics: Forefront Experimental Issues

December 8 — 10, 1988

University of California at Berkeley
Lawrence Berkeley Laboratory
Berkeley, California

Edited by
Eric B. Norman

World Scientific
Singapore • New Jersey • London • Hong Kong

Published by

World Scientific Publishing Co. Pte. Ltd.,
P O Box 128, Farrer Road, Singapore 9128

USA office: 687 Hartwell Street, Teaneck, NJ 07666

UK office: 73 Lynton Mead, Totteridge, London N20 8DH

PARTICLE ASTROPHYSICS: FOREFRONT EXPERIMENTAL ISSUES

ISBN 9971-50-835-4
9971-50-869-9 (pbk)

Printed in Singapore by JBW Printers & Binders Pte. Ltd.

Particle Astrophysics: Forefront Experimental Issues

Preface

This book contains the Proceedings of the ***Workshop on Particle Astrophysics: Forefront Experimental Issues*** held at the Berkeley Conference Center in Berkeley, California on December 8-10, 1988. The idea to hold such a meeting came about as a result of the growing recognition in both the physics and astronomy communities of the emergence of the new field of particle astrophysics. The local organizing committee consisted of M. Davis, E. B. Norman, P. J. Oddone, C. R. Pennypacker (Chair), P. B. Price, B. Sadoulet, J. Silk, and G. F. Smoot.

This 2-1/2 day workshop was attended by 311 registered participants from 13 different countries. Sessions of talks covering six major topics were put together by both local and outside organizers:

1) Dark Matter (D.N. Schramm, B. Sadoulet)
2) Large-Scale Structure (M. J. Geller, M. Davis)
3) Cosmic Microwave Background (D. Wilkinson, G. F. Smoot)
4) Supernovae (R. P. Kirshner, C. R. Pennypacker)
5) X-Rays, Gamma Rays, and Cosmic Rays (J. Grindlay, P. B. Price)
6) Neutrino Physics (F. Reines, E. B. Norman)

These topics were arranged in morning sessions of plenary overview and summary talks, while afternoons were devoted to parallel sessions of talks on each of the six major topics. At the start of the meeting Richard Nicholson, from the National Science Foundation, and John O'Fallon, from the Department of Energy, described the funding climate for this new field.

The general problem of dark matter was a major theme in much of what was discussed during the workshop. Reviews were given of both the theoretical arguments and the observational evidence that the bulk of the material in the universe is in the form of matter which does not radiate. There was a great deal of interest in the present results and future prospects for laboratory detection of dark matter particles. The possible effects of dark matter on the large scale structure of the universe and on the cosmic microwave background were also discussed. Searches for evidence of dark matter particle decays in the x-rays, gamma-rays, cosmic rays, and in neutrinos were lively topics of discussion. In addition, the wealth of new data provided by SN1987a and other recently observed supernovae provoked a great deal of interest. The general feeling during the workshop seemed to be one of great excitement over the emergence of a new multi-disciplinary field of particle astrophysics and the prospects for it to provide answers to many of our most basic questions about the nature of the universe.

The smooth running of the workshop was due in large part to the planning and technical support provided by Robert Johnson, Darlene Gonzales, William O'Connor, Peggy Little, and their staffs from LBL and by the staff of the Berkeley Conference Center. I thank them very much for the excellent jobs they all did to make the workshop a success.

Eric B. Norman
Lawrence Berkeley Laboratory
and
Center for Particle Astrophysics

This workshop was supported in part by the Director, Office of Energy Research, Office of High Energy and Nuclear Physics, Divisions of High Energy Physics and Nuclear Physics of the U. S. Department of Energy under Contract No. DE-AC03-76SF00098 and by the National Science Foundation's Science and Technology Center Research Program.

Contents

SUPERNOVAE

X RAYS, GAMMA RAYS, and COSMIC RAYS

NEUTRINO PHYSICS

DARK MATTER

DARK MATTER:
An Overview

David N. Schramm
The University of Chicago and
NASA/Fermi Astrophysics Center

ABSTRACT

The need for dark matter in Cosmology is reviewed. The needs of the halo and the requirements of $\Omega = 1$ are examined. It is pointed out that the halo problem could be satisfied with either non-baryonic cold dark matter or clumps of dark baryons ("Jupiters" or Blackholes), but that the inflationary inspired need of $\Omega = 1$ when coupled with the arguments of big bang nucleosynthesis seems to force us to non-baryonic dark matter, either of the hot or cold variety. Due to this significance of the Nucleosynthesis constraint, the possible loopholes of quark-hadron fluctuations or decaying particle scenarios are examined. It is also argued that some baryonic dark matter must exist either in the halos or in hot intergalactic gas or in failed galaxies. Astronomical arguments favoring one or another type of dark matter are shown to depend on both observations and the theory of galaxy and structure formation. The key observations include: microwave anisotropies, large scale structure (velocities, correlations and topology), gravitational lensing and the existance of objects at high redshifts. The scenarios for galaxy and structure forming "seeds" include:

(1) Quantum induced gaussion fluctuations.

(2) Cosmic strings (either accreting or superconducting) or

(3) Late time ($T \lesssim 10^3 K$) phase transition induced remnants such as domain walls.

Future laboratory experimental constraints on the various candidates are mentioned.

INTRODUCTION

Some of the most critical problems in cosmology today and also one of the best examples of the interplay between particle physics and astrophysics involve the dark matter problem and the related problem of the formation of galaxies and structure in the early universe. It is in the arena of dark matter and galaxy formation where traditional astronominal observations of cosmological relevence come face to face with elementary particle models both for predicting new and exotic types of matter and for predicting the origin of various types of seeds that eventually make the structure.

This overview will go through arguments as to why there must be dark matter in the universe and then discuss the types of dark matter and the proposed structure formation mechanisms, and finally discuss observations and experiments that will eventually determine the answers to the problem. Remember that the key reason why the cosmology/particle interface is so vital today is the close interplay between theory, observation and experiment. Unlike cosmology of past centuries or even past decades, current models and ideas are indeed testable and those observations and experiments are being carried out.

THE NEED FOR DARK MATTER

The arguments requiring some sort of dark matter fall into two separate and quite distinct areas. One is the arguments using Newtonian mechanics applied to various astronomical systems that show that there is more matter present than the amount that is shining. These arguments are summarized in the first part of Table 1. It should be noted that these arguments reliably demonstrate that galaxies have dark halos that carry at least 90% of the total mass of the galaxy. In other words, the halos seem to have a mass ~ 10 times the visible mass The arguments do not in any way imply that Ω is unity from dynamical considerations alone.

The other argument is what we choose to call the inflation paradigm. This is the argument that the only long-lived natural value for Ω is unity, and that inflation or something like it provided the early universe with the mechanism to achieve that value and thereby solve the flatness and smoothness problems. It should be remembered that it is this latter argument when confronted with the results of Big Bang Necleosynthesis (see Fig. 1 as well as Table 1.) that tells us that Ω in baryons Ω_B is ~ 0.01 and therefore that if Ω total is truly unity then the bulk of the mass of the universe must be in the form of some sort of non-baryonic matter. Thus, our need for exotica is dependent on inflation and Big Bang Nucleosythesis and not on the existance of dark galatic halos. This point is frequently forgotten, not only by some members of the popular press but occasionally by active workers in the field. Therefore, rather than spending any further time on the dynamical arguments, I will focus my attention on a brief review of the argument for the inflation paradigm and then concentrate on Big Bang Nucleosynthesis since it is really the pivotal argument that drives us to non-baryonic and therefore, exotic solutions.

I will also focus on the point that Big Bang Nucleosynthesis also requires that the bulk of the baryons in the universe are dark. In fact, locating the dark baryons maybe a very important way of discerning the nature of the non-baryonic component as we shall see. Recently some clever possible loop holes in the Big Bang Nucleosynthesis argument have been proposed; these shall be discussed.

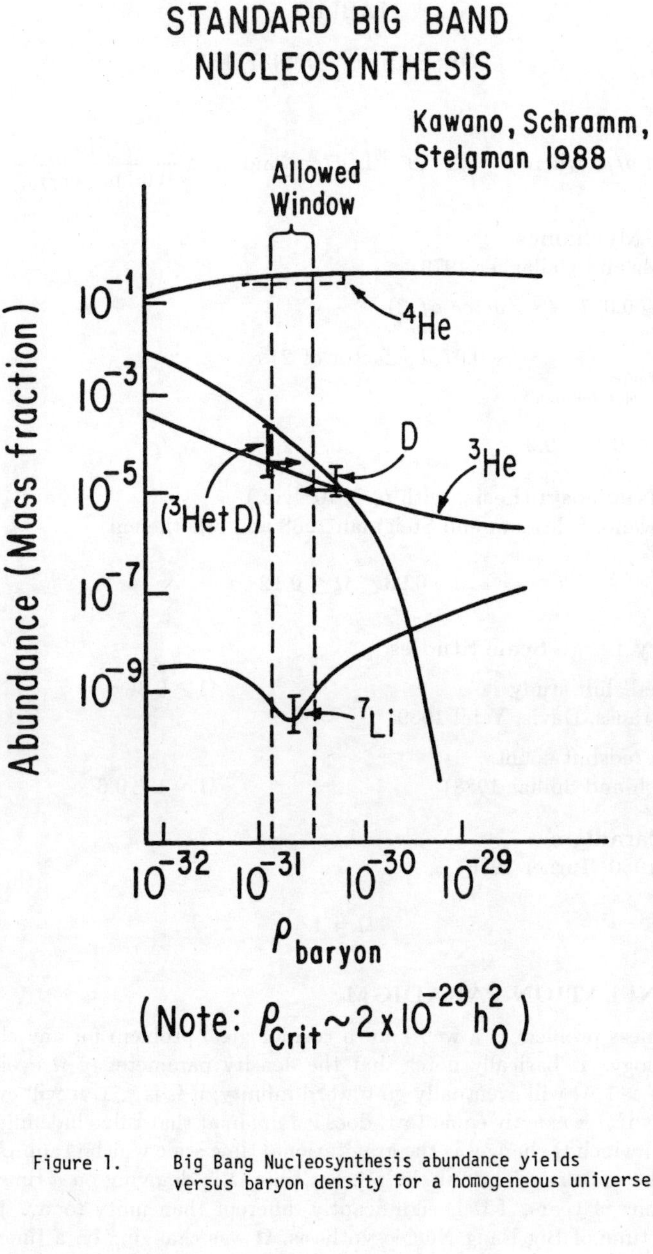

Figure 1. Big Bang Nucleosynthesis abundance yields versus baryon density for a homogeneous universe.

TABLE I

"OBSERVED" DENSITIES

$$\left[\Omega \equiv \rho/\rho_c \text{ where } \rho_c = 2 \cdot 10^{-29} h_0^2 \text{g/cm}^3 \text{and } h_o \equiv \frac{\text{Ho}}{100 \text{ km/sec/mpc}}\right]$$

Neutonian Mechanics
(cf. Faber and Gallagher 1979)

$\Omega_{\text{visible}} \sim 0.007$ ($\sim$ *factor of* 2)

Ω_{binaries} ~ 0.07 ($\sim$ factor of 2)
small groups
extended flat relation curves

$\Omega_{\text{clusters}} \sim 0.1$ *to* 0.3

Big Bang Nucleosynthesis (with $t_u \gtrsim 10^{10}$yrs.)
(c.f. Kawano, Schramm and Steigman 1988 and ref. therein)

$$0.03 \leq \Omega \leq 0.12$$

Preliminary Large Scale Studies

IRAS red shift study (Strauss, Davis, Yahil 1989) — $\Omega \gtrsim 0.3$

Density redshift counts (Loh and Spillar 1988) — $\Omega \sim 1 \pm 0.6$

Inflation Paradigm
(Guth 1980, Turner 1986)

$$\Omega = 1$$

THE INFLATION PARADIGM

The flatness problem is a well known cosmological problem for any classic big bang cosmology. It basically notes that the density parameter of Ω evolves with time. If Ω is > 1, Ω will eventually go toward infinity; if Ω is < 1, it will eventually go to 0. Only if Ω is exactly equal to 1, does it remain at that value indefinitely. The time scale on which Ω changes is the gravitational time scale which is approximately the age of the universe. Thus at the present time Ω is changing on a time scale of tens of billions of years, if Ω is significantly different than unity today. However, back at the time of Big Bang Nucleosynthesis, Ω was changing on a time scale of seconds. If we extrapolate as far as any rational person has confidence, back to the Planck time, then Ω was changing on time scales of $\sim 10^{-43}$ per second. Thus, in order for us to still be here today and talk about it, Ω had to be fine tuned to be equal to unity to nearly 60 decimal places at the Planck time. Another way of saying this is that since our existence is not compatible with Ω of 0 or infinity, the only long term value that Ω can have is unity. Thus, baring the possibility of our living at an epoch in cosmic time when Ω has just dropped below unity, for the first time but is still far from 0, we would otherwise say that Ω is unity.

These arguments went on long before *inflation* provided us the natural physical mechanism to drive Ω to unity in the first moments of the universe. Thus it did not have to rely on some arbitary fine tuned initial condition. Inflation[1] is the rapid expansion of the early universe that would take a wide range of initial conditions and convert them to conditions wher Ω was unity to a high accuracy. Although the detailed physical mechanism for driving the expansion is is not well determined and differs in different grand unified theories, a basic point is that any scalar field present in the early universe will cause inflation.[2] Inflation provides us with a plausable mechanism to set the initial conditions and avoid special fine tuning, this enables most cosmologists to believe Ω must be unity today.

BIG BANG NUCLEOSYNTHESIS

Figure 1. shows the abundances verses baryon density for standard homogeneous Big Bang Nucleosynthesis. The actual big bang calculations themselves are natural evolution from the early primitive work of Alpher, Bethe and Gamow[3] evolving through the almost complete picture used by Alpher, Follin and Herman[4] and receiving only minor physical modifications since the first post $3K$ discovery calculations with numerical reaction networks of Peebles,[5] Wagoner, Fowler and Hoyle.[6] However, it should be remembered at the time of Wagoner, Fowler and Hoyle,that the only nuclei produced in the big bang that was thought to be of significance was ^4He. Fowler, Greenstein and Hoyle[7] had argued that the other light elements where made in protostellar processes. Thus during the 1960's the abundances of deuterium, ^3He, and ^7Li produced in the big bang were merely a curiosity, and were not seriously utilized for cosmological purposes. That situation changed dramatically in the 70's, when a variety of events occurred affecting Big Bang Nucleosynthesis.

The first point of significance was a demonstration, not only that the Fowler, Greenstein and Hoyle process failed[8,9] but eventually the dramatic statement that deuterium cannot be produced in significant quanities in any astrophysical location other than the big bang due to its basic nuclear fragility[10] and with those arguments deuterium use as a powerful density constraint became cemented. This helped support, for example the arguments of Gott, Gunn, Schramm and Tinsley[11]. The development of these nuclear arguments coupled with the development of the observations, in particular the Copernicus satellite finding deuterium in the interstellar medium[12] and the implications for deuterium from solar wind observations[13] on the moon and metoritic observations [14].

Once deuterium was established as cosmological, similar but more complex proceedures were applied to establishing the cosmological relevance of ^3He and ^7Li as demonstrated by Yang, Rood, Steigman and Schramm[15] following the important ^3He work of Tinsley, Rood and Steigman [16]. In particular it was eventually shown that ^3He plus the deuterium that is part of ^3He in stars provides a strong lower bound density since ^3He is also manufactured in stars.[17] Furthermore, it was noted in the series of papers, by the Chicago Group and their collaborators [15,17,18] that the only allowed value for Li consistant with the ^3He and deuterium observations will be a value of Li near the minimum of its abundance curve. Namely Li/H of approximately 10^{-10} at the time this was first noted it seemed somewhat problematic because Li in Pop I objects and in the interstellar medium seemed to imply a value an order of magnitude higher. However, agruments were made that the Pop I abundance might have been significantly enhanced by later production processes. The definitive observation came in 1980 when the Spites[19] measured the Li in the extreme Pop II stars and found the higher surface temperature Pop II stars all had the same Li abundance and it was at the level of 10^{-10} in agreement with the minimum derived from the Big Bang Nucleosynthesis agruments.

With ^7Li as a keystone, standard Big Bang Nucleosynthesis was fitting abundances all the way from ^4He at 25% by mass down to ^7Li at 10^{-10} by number; a range of over 9 orders of magnitude. Such quanitative agreement not only seconded the 3K backgrounds establishment of the basic big bang model, but also led to the establishment of the particle/cosmology connection. It said that we understood the universe not just at the epoch of the 3 thousand background decoupling ($t \sim 10^5 yrs.$), but also at the epoch of Big Bang Nucleosynthesis ($t \sim 1sec.$). Big Bang Nucleosynthesis was used to predict the number of fundamental particle types [20], which explicitly substantiated the particle/astro connection by making a cosmological prediction about a quanitity of explicit interest to particle physicists and by enabling experimental tests in accelerator laboratories to be made regarding the big bang model.[21,22]

For dark matter the important implication of all these arguments was that Ω_B was constrainted to be between 0.03 and 0.12. Thus the universe cannot be closed with baryons, but furthermore the lower bound was greater than Ω visible. To obtain this lower bound it should be noted that one has added the additional constraints[23] that the age of the universe is greater than 10^{10} years, which thus constrains any $\Omega_{total} = 1$ model to have an H_0 of less than $70km/sec/Mpc$. It was noted by Gott et al.[11], that the Big Bang Nucleosynthesis derived Ω_B is in good agreement with the $\Omega's$ implied by the dynamics of galactic systems, thus to explain halos one is not forced to look beyond some form of baryonic dark matter. It is only if one goes to an Ω of unity or to be more specific $\Omega \geq 0.12$ for the standard homogeneous model that one is really forced to exotic matter.[24]

Before discussing such matter let us look at two possible loopholes in the argument. These loopholes are: (1) fluctuations generated at the quark-hadron transition and (2) alterations of nucleosynthesis by late decaying massive particles.

This latter model was developed most fully by Dimopoulos et al.[33] In it they noted that if some massive unstable particle exsisted, and decayed shortly after the time of normal nucleosynthesis it would regenerate nuclear synthetic results that were quite different than the standard model, and could even fit the observed abundances with somewhat different values of baryonic density, and/or numbers of neutrinos. A key point about these calculations though, was that they predicted that the bulk of the lithium coming from the big bang will be ^6Li rather then ^7Li. At the time of this writing there are no firm observations to prove that the most extreme Pop II stars contain ^7Li rather than ^6Li, those observations are now being done following the arguments of Brown and Schramm.[34] Current measurements of the Li isotopic ratio are being made by Hobbs and Pilachowski and by Beckman et al. Thus we should have a definitive test on this model in the new future.

As to the quark-hadron transition possibilities, much has been written. It was first noted by Applegate, Hogan and Sherrer[35] followed by work by Alcock, Fuller and Mathews[36] that if a quark-hadron transition were a first order phase transition, then density fluctuations produced at the phase transition could yield inhomogeneities at the time of nucleosynthesis. Furthermore the differential defusion of neutrons relative to protons out of the density inhomogeneities will yield variable a neutron/proton ratio, as well as the previously studied density fluctuations. (For previous studies of inhomogeneities see Yang et al.[17]). In the initial calculations, it was thought that one might be able to obtain an Ω_B of unity while fitting all of the light abundances with the exception of Li, which would have been over-produced by a significant amount. Later work by Fowler and Malaney[37] argued that when a more detailed treatment of the two-phase model was carried out that included back defusion of the neutrons, then the Li could be depleted to more acceptable values. However, work by Kurki-Suonio, and Matzner[39] and Alcock et al. [40] showed that back defusion also resulted in high helium abundances. Whether parameters could be found that enabled this He to be depleted is problematic among the different groups. However, they all seem to be in agreement that except for some very narrow range in parameter space. Li and He are usually over-produced for high Ω_B. In phase transitions, one usually expects some distribution of parameters, not single values. That a phase transition would exactly pick out those parameters that avoid excessive over-production of Li and He seems difficult. Especially when one realizes that the separation of the nucleation sites required is the order of hundreds of meters for a phase transition that operates on Fermi scale processes. However, until this is completely explored, there clearly remains a loophole that must be investigated further.

Ignoring this possible loophole (which, now appears far less compelling than it did in the initial papers) let us apply to the quark-hadron transition the normal abundance constraints that we have used from nucleosynthesis and not relax the Li constraint. If we apply our normal constraints we obtain [41] Figure 2. The paramenter on the vertical scale there is the separation of the nucleation sites measured in meters at the time critical for the phase transition. The horizontal parameter is again the density in baryons. Once the Pop II constraints on Li are put in, the highest baryonic density to be achieved is only slightly higher than the standard models, regardless of separation of nucleation sites. Similarly, the lower bound does not drop significantly. It should also be noted that these kind of arguments can be turned around to constrain the parameters of the phase transition, since values of over one hundred meters appear to be excluded by nucleosynthesis. This same kind of argument was also made by Reeves and Audouze.[42]

Before leaving Big Bang Nucleosynthesis it should be remembered that the quark-hadron loophole is critically dependent on Li. One point that has been raised is, if we are willing to use the Li for this argument we should understand fully how Li evolves in the galaxy. A key question for Li has been how to get from the Pop

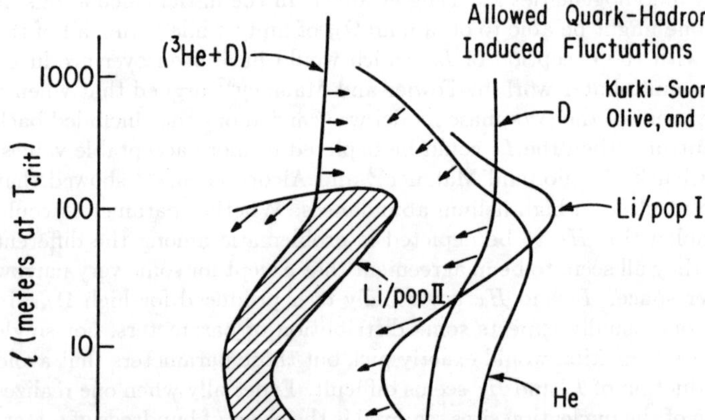

Figure 2. Allowed parameter space for quark-hadron induced fluctuations with nucleation sites separated by distance ℓ and average baryon.

II abundance of $Li/H \sim 10^{-10}$ to the Pop I abundance of $Li/H \sim 10^{-9}$. Some[40] have argued that perhaps Li is depleted from some high initial value down to both Pop I and Pop II values. However, recently Dearborn, Schramm, Steigman and Truran [43] have found that Li will be produced in type II supernova and thus will be enhanced in exactly the same objects that produce the metal abundance of the Galactic disk. With this mechanism it is easy to understand why the Pop I value is an order magnitude higher than the Pop II value and why it appears constant for old as well as young Pop I stars that have not depleted their surface Li. In fact, if Li can be proven to be made in supernova then it will be impossible to reconcile successive high Li production and $\Omega_B = 1$ universes with the galactic evolution of Li. Thus Li evolution continues to be a critical point of study.

DARK MATTER CANDIDATES

Table 2. summarizes both the baryonic and non-baryonic dark matter candidates. Some baryonic dark matter must exist since we know that the lower bound from big bang nucleosynthesis is greater than the upper limits on the amount of visible matter in the universe. However, we do not know what form this baryonic dark matter is in. It could either be in condensed objects in the halo, such as brown dwarfs, and jupiters (objects with $\lesssim 0.08$ solar masses so they are bright shining stars) or black holes, (which at the time of nucleosynthesis would have been baryons). Or, if the baryonic dark matter is not in the halo, it could be in hot intergalactic gas. Hot enough to not show absorption lines in the Gunn Peterson test, but not so hot as to be seen in the x-rays. Evidence for some hot gas is found in clusters of galaxies however, the amount of gas in clusters would not be enough to make up the entire missing baryonic matter. Helfand [44] has argued that the isotropic x-ray background may be due to hot intergalactic gas in sufficent density to account for all the dark baryons. If verified this would reduce the possibilities of halos containing condensed dark baryons since all the baryonic matter would be accounted for. Another possible hiding place for the dark baryons, would be failed galaxies. Large clumps of baryons that condense gravitationally, but did not produce stars. Such clumps are predicted in galaxy formation scenarios that include large amounts of biasing where only some fraction of the clumps shine.

Hegyi and Olive[45] have argued that dark baryonic halos are unlikely. However, they do allow for the loopholes mentioned above of low mass objects or of massive black holes. It is worth noting that these loopholes are not that unlikely. If we look at the initial mass function for stars forming with Pop I composition, we know that the mass function falls off roughly like a power law for standard size stars as was shown by Salpeter. Or even if we apply the Miller-Scalo mass function, the fall off is only a little steeper. In both cases there is some sort of lower cut off near $0.1 M_\odot$ However, we do not know the origin of this mass function and its shape. No star formation model based on fundamental physics predicts it.

TABLE II

"DARK MATTER CANDIDATES"

Baryonic (BDM)	
Brown Dwarfs and/or Jupiters	$M \lesssim 0.08 M_\odot$
Blackholes	$M \gtrsim 1 M_\odot$
Hot intergalactic gas	$M \sim 1 GeV$, $T \sim 10^6 K$
Failed galaxies	$M \gtrsim 10^5 M_\odot$
Non Baryonic	
Hot (HDM)	
Low Mass Neutrinos	$M_\nu \sim 20 \pm 10 eV$
Cold (CDM)	
Massive Neutrinos	$M_\nu \sim 3 GeV$
Wimps, Lightest Supersymmetric Particle (Photino, Gravitino,Sneutrino)	$M_{susy} \sim 4 GeV$
Axions	$M_a \sim 10^{-5} eV$
Planetary mass black holes	$M \sim 10^{15} g - 10^{-30} g$
Quark nuggets	$M \sim 10^{15} g$
Topological debris (monopoles higher dimensional knots, balls of wall, etc.)	$M \gtrsim 10^{16} GeV$

We do believe that whatever the origin of this mass function is, that it is probably related to the metalicity of the materials, since metalicity affects cooling rates, etc. it is not unreasonable to expect the initial mass function that was present in the primordial material which had no heavy elements, only the products of Big Bang Nucleosynthesis, would be peaked either much higher than the present mass funtion or much lower. Higher if the lower cooling from low metals resulted in larger clumps, or lower if some sort of rapid cooling processes ("cooling flows") were set up during the initial star formation epoch, as seems to be the case in some primative galaxies. In either case moving either higher or lower produces the bulk of the stellar population in either brown dwafts and jupiters or into massive black holes. Thus the most likely scenarios are that a first generation of condensed objects would be in a form of dark baryonic matter that could make up the halos, and could explain why there is this interesting coincidence between the implied mass in halos, and the implied amount of baryonic material. However, it should also be remembered that to follow through with this scenario one would have to have the condensation of the objects occur prior to the formation of the disk. Recent observational evidence,[46], seems to show disk formation is relatively late, occurring at red shifts Z $\lesssim$ 1. Thus, the first several billion years of a galaxy's life may have been spent prior to the formation of the disk. In fact, if the first large objects to form are less than galactic mass, as many scenarios imply, then mergers are necessary for eventual galaxy size objects. Mergers stimulate star formation while putting early objects into halos rather than disks. Thus, while making halos out of exotic material may be more exciting. It is certainly not impossible for the halos to be in the form of dark baryons. One application of William of Occum's famous razor would be to have us not invoke exotic matter until we are forced to.

Non-baryonic matter can be divided following Bond and Szalay[47] into two major categories for cosmological purposes. Hot dark matter (HDM) and cold dark matter (CDM). Hot dark matter is matter that is relativistic until just before the epoch of galaxy formation. The best example, being low mass neutrinos with $M \sim 20eV$. (Remember $\Omega_\nu \sim \sum \frac{m_\nu(eV)}{100\ h^2}$).

Cold dark matter is matter that is moving slow, at the epoch of galaxy formation because it is moving slow, it can clump on very small scales. Whereas, HDM tends to have more difficulty in being confined on small scales. Examples of CDM could be massive neutrino-like particles with masses greater than several GeV or the lightest super-symmetric particle which is presumed to be stable and might also have masses of several GeV. Axions, which while very light would also be moving very slow[48] thus would clump on small scales. Or one could also go to non-elementary particle candidates such as planetary mass blackholes [49], or quark nuggets of strange quark matter, also found at the quark-hadron transition. Another possibility would be any sort of massive toplogical remnant left over from some early phase transition.

When thinking about dark matter candidates, one should remember the basic work of Zeldovich,[50] later duplicated by Lee and Weinberg,[51] which showed for a weakly interacting particle that one can obtain closure densities, either if the particle is very light, $\sim 20eV$, or if the particles very massive, $\sim 3GeV$. That is because if the particle is much lighter than the decoupling, temperature then its number density is the number density of photons (to within spin factors and small corrections), and so the mass density is in direct proporation to the particle mass since the number density is fixed. However, if the mass of the particle is much greater than the decoupling temperature, then annihilations will deplete the particle number. Thus, the temperature of the expanding universe drops below the rest mass of the particle, thus depleating the number density via annihilations. For normal weakly interacting particles decoupling occurs at a temperature of $\sim 1MeV$ so higher mass particles are depleated.

Before leaving the discussing of DM candidates, it might be noted that in addition to the curious coincidence of the density of baryons being approximately equal to the density impied by halos that there is another coincidence which may have exactly the opposite resolution. This is the coincidence in the ratio of halos to visible matter (~ 10) is the same as the ratio of critical density to the baryonic density, (also about 10). This coincidence is nicely explained in CDM models with biasing. Since in these models there will be many clumps of baryons and CDM, but only some biased fraction would shine. Once the ratio of CDM to baryons is set in the early universe it would propogate in all objects and thus would yield the same ratio of shining to non-shining matter everywhere. Of course these "coincidences" are only good to factors of a few so as observational data improves the "coincidences" may vanish.

GALAXY FORMATION MODELS

As much a part of any DM scenario as the DM itself, are the seeds that enable the DM and the baryons to form the observed clusters of galaxies and other structures. While many statements are frequently made about the ability of one or another kind of DM to make realistic structures those statements are always made in the context of an explicit model of galaxy formation. Since we do not really know how galaxies form, all such statements need to be taken with several grains of salt. At the present time there are two basic galaxy formation scenarios. One using quantum induced Gaussian fluctuations generated at the end of inflation. The other using some topological remnant, again produced by some new fundamental physics. It should be noted that each of these mechanisms involves new fundamental physics and it should also be noted that prior to the marriage of cosmology with particle physics we had no models for generating the initial seeds, and fluctuations were merely put in by hand. Now we have models that do relate the structure formation seeds to fundamental physics, but the fundamental physics we need is not just the standard model particle physics interactions.

TABLE III

GALAXY FORMATION SEEDS

1. **Quantum Induced Gaussion Adiabatic Fluctuations with Harrison-Zeldovich Spectrum** (c.f. Guth and Pi)

2. **Cosmic Strings** (Kibble, Zeldovich, Velinkin)
   A) Accreting (Turok and Albrecht, Bennett and Buchet)
   B) Exploding Superconducting
   (Ostriker, Thompson and Witten)

3. **Late Time Phase Transitions** (Hill, Schramm and Fry)

The quantum induced Gaussian fluctuation model with the production of the Harrison-Zeldovich spectrum has been the standard model for galaxy formation in the '80's. And with that model, CDM is favored since HDM is not able to make small galaxy- like structures fast enough. However, CDM with Gaussion fluctuations, as we will see, may run into problems on the large scale side if present reported observations continue to hold up. The advantage of the CDM plus Gaussion model is it is easy to calculate and has been explored in far more detail than any of the other models, and does amazingly well considerating the rate at which new observations are being generated. The most detailed work on these models have been the numerical simulations of Frenk, White, Davis and Efstahiou.[52]

The alternative of using topological remnants as the seeds as opposed to density fluctuations in the matter is best epitomized in the cosmic string scenarios. First noted by Kibble and Zeldovich[53] and later developed by Vilenkin[54] The last few years these scenarios have divided themselves into two sub-catagories. One, the gravitationally accreting string developed most fully by Turok and Albrecht[55] with recent interesting simulation being carried out by Bennett and Buchet[56]. The other variant of this has been the exploding seed scenarios, where the strings are superconducting. This model has been put forth by Ostriker, Thompson and Witten[57] In the exploding scenario, instead of the strings being gravitational accretion points, the strings radiate and push the baryons about, thus creating a segregation between the baryonic matter and the non baryonic matter. The exploding scenario is in some way a natural outgrowth of the earlier model of Ostriker and Cowie[58] but in the earlier model they had no energy sources strong enough to push matter about on cluster scales. The superconducting strings provided them with that energy source.

A new alternative using topological remnants but instead of coupling them to a phase transition back in the primitive early moments of the universe, it ties the topological remnants to a phase transition that occurs late, after decoupling has recently been developed by Hill, Schramm and Fry.[59] In this scenario, the late phase transition produces domain walls, strings, etc. which can be the seeds of structure formation. Because the transition occurs after decoupling it produces the minimum possible fluctuations in the microwave background, for a given structure that is produced. Since there would be no fluctuations on the surface of last scattering the only induced fluctuations in the microwave background in this model are due to the propagation of the microwave protons through the potential wells in the transparent media and due to the seeds themselves changing during the propagation. If the universe was static and not expanding, the differential red shift/blue shift could cancel and there would be no net effect. However, because the universe is expanding while the protons are propagating the red shift and blue shift does not quite cancel. It can be seen that in late time transitions, larger structures, giving a larger differential between the red shift and blue shift would yield the largest microwave fluctuations. The maximum size structure that could be created in such a model can be in principle up to the horizon at the time of the phase transition, and that horizon is larger than any presently observed structure, including the giant structures noted by Tully[60] Geller and Huchra[61]. However, it should also be remembered that if the evolution of the late-time structures leads to larger and larger structures. Then this model may have the problem that will produce too much power on large scales which would be exactly the opposite problem of the quantum fluctuation scenario. However, producing larger structures and consequently $\frac{\Delta T}{T}$ depends on the details of how the walls, strings, etc. evolve with time. Simulations of the type that were carried out for cosmic strings, need to be carried out for late time phase transitions. Preliminary simulation work has been begun by Kawano,[62] and Spergel. Press and Widrow[63]. In any model for the simulations of domain wall evolution, one needs to make assumptions about the number of minima which produce the numbers of different types to domains. Futhermore, one also needs to make assumptions about the intercommutability of the domain walls which is yet to be proven, and one needs to look at the possibility that the vacuum minima are not all degenerate, but that there may be some weakly broken symmetry yielding one vacuum slightly preferable over the others. This latter possibility will mean that eventually the walls would disappear. If they were there long enough to generate structures and then disappear one would avoid having the problem of too much power on large scales. In this latter possibility one might still retain small "balls of wall", which would behave like non-topoligical solitons.[64]. Non topological solitons produced by such a late time phase transition could be very good seeds for making galaxy and structure with the seeds, distributed in some pattern depending on the evolution of the phase transition.[65]

These late-time phase transitions are the most recent of the ideas for seeds and thus the least explored. However, as we will see if present trends and observations are verified and continue this may be an extremely promising model. For example Stebbins and Turner[66] have shown that this model and variations of it might easily give large scale velocity fields.

Before leaving the discussion of this model it should also be noted that the physics that could produce such a late time phase transition is probably no more ad hoc than the physics that is invoked to enable inflation to work and still obtain sufficiently small primordial fluctions. In both cases there is some tuning and in both cases one is invoking a phase transition based on "new" physics. The toy model proposed in the initial paper[59] was actually motivated not to try to solve this problem but the solar neutrino problem where it was noted that if MFW mixing is right than neutrinos have masses of $\sim 10^{-2} eV$.If that mass is generated out of some vacuum energy then you naturally have a phase transition at the order of $\sim 10^{-2} eV$. If that phase transition is related to a GUT scale having GUT scale coupling of $\sim 10^{16} GeV$, then the compton wave length of the resulting Psuedo-Goldstone particles is $\sim 1 Mpc$. Thus yielding cosmological scales derived from particle considerations. Numerous alternative particle models that also yield late-time transitions have been proposed. For example Dimopolous [67] has noted that if one uses a running coupling constant, analogous to QCD, one can have that running coupling constant grow strong at some temperature such as $10^{-2} eV$, thus yielding a phase transition. Another alternative has been proposed by Fuller and Schramm[68], where they note that if majorons exist and are produced by a phase transition at $\lesssim leV$, the majoron induced neutrino interaction enables those regions that first undergo the phase transition to work like neutrino fly paper gobbling up any neutrinos in the vicinity, thus creating a non-linear density enhancement. It is certain that many other late-night phase transition models can exist that would have some sort of generic properties of the type needed. Thus just like the case of inflation where a variety of particle models can all inflate, we have here a variety of particle models that can all yield late-time structures that could be interesting.

OBSERVATIONAL CONSTRAINTS

Table IV. summarizes some of the constraints on different candidates. For example dark baryonic halos DBH are not compatable with CDM, since CDM would also cluster on small scales and would thus be present in enough quanity to produce the halos. Any large amount of DBH would then be unnecessary, and difficult to understand. While there are no direct observations of DBH searchers using micro-lensing may resolve this. Recent observations by Thuan, Gott and Schrader[69] argue that dwarf galaxies are distributed in the same patterns as are the brighter galaxies. If this continues to be born out it certainly will be a difficulty for any CDM plus biasing model, since such a model argues that dwarfs are more uniform then big bright galaxies.

Observations of dark halos around dwarf galaxies is inconsistent with HDM as that halo material, but HDM could well exist and only clump on much larger scales so such observations are not wide reaching in their implications. In fact DBH may be quite appropriate for dwarfs in HDM models.

TABLE IV
CONSTRAINTS

Dark Baryonic Halos (DBH) (Brown Dwarfs, Jupiters, Blackholes)
Not compatible with CDM. Halos of Dwarfs require either CDM or DBH.

CDM
requires significant BDM in failed galaxies.

HDM
requires either cosmic strings or late time phase transitions - it is not compatible with quantum Gaussian fluctuations.

CDM with Quantum Gaussian fluctuations
not compatible with
a. high cluster-cluster correlations
b. high large scale velocity flows
c. coherent structures $\gtrsim 50$ Mpc
d. dwarf galaxies being distributed like bright galaxies

Quantum Gaussian Adiabatic Fluctuations
not compatible with current limits on $\delta T/T$ and
large numbers of condensed objects at $Z \gtrsim 4$

Microwave Anisotropy Limits $\delta T/T \lesssim 5 \times 10^{-6}$
If found only compatible with late-time transitions.

Submillimeter Excess
would require energy input at $Z > 10$. Need objects to form early or decay of particles or topological defects, neither consistent with CDM plus quantum.

Peculiar velocites are certainly expected in all models on small scales (the earth goes around the sun, the sun around the galaxy, galaxy around the local group. However, for this peculiar motion to persist up to scales of the order of 50 Mpc as the preliminary observations of the 7 Samurai[70] indicate and as the recent work of Dressler and Faber[71] supports, is very problematic for the CDM Gaussian fluctuation scenario, since that scenario will build everything up from small scales. Similarly observation of structures much bigger than 50 Mpc, such as those claimed by Tully [60] and Geller and Huchra[61] are very problematic for models, building things up from small scales unless they are just a few rare accidents.

An observation that has impact on any model that starts with quantum fluctuations and the current limits on the anisotropy of the microwave background has to deal with the number of condensed objects observed at red shift $\gtrsim 4$. Any model that starts with small fluctuations and requires linear growth has difficulty in producing large numbers of objects at red shifts much greater than unity. While quasars are known to exist at high red shifts, recent reports are that some galaxy like objects may exist back then. The question is how ubiquitous are these objects? If they are rare multi-sigma fluctuations then all may still be well. However, if they are truly common, that is, if standard structures really started forming and yielded condensed objects at high Z, then we really are forced to some sort of topological model.

Furthermore, if the microwave limits are continued to be brought down and eventually are shown to yield fluctuation limits of only $5 \cdot 10^{-6}$ it would only be compatable with a late time phase transition. All other models including cosmic string models to produce structures require the microwave background to have large magnitude anisotropies. Another microwave constraint that is important is the sub-millimeter excess of the Berkeley-Nagoya group.[72] If true this requires a large amount of energy input at red shifts $\gtrsim 10$. Such energy input would either need objects to form at that enormously high red shift or have decay with emission of energy from either particles or topological defects. None of these possibilities is consistant with the standard CDM and primordial Gaussian fluctuation scenario.

One important large scale structure constraint is the correlation of clusters. Bahcall and Soneira[78] have argued that clusters are more correlated than galaxies. Szalay and Schramm[79] have pointed out that such correlations, if real, support some sort of fractal initial seed patterns as opposed to gaussian. In fact string models may naturally yield such correlations[55]. However, projection effects[80] may be responsible at least in part for the previously reported correlations, but preliminary work from other groups also finds strong correlations[61]. More work with new data is clearly needed.

For future tests, beyond what we have already discussed (see Table V.) Of course gravitational lensing is a key in our search for DBH. Gravitational lensing might also help find cosmic strings. Strings should produce pairs of images. A clump of pairs found by Cowie and Hu has recently been found to have even more pairs (S. Lilly private communication). If verified, this may indicate a cosmic string still existing in that direction.

The x-ray background may also help us find the baryonic dark matter if it is not in our halos. X-ray observations can also help tell us something about galaxy structure and formation since these early structures might inevitably produce x-rays. Thus, AXAF when it flies may help us to find the baryonic matter. COBE which will fly in the near future will be testing whether the sub-millimeter excess is real and will be pushing the limits on anisotropy. If they find something this will of course tighten the arguments and point the direction for the models.

TABLE V
Future Tests

Gravitational Microlensing -
Tests DBH

AXAF -
Tests hot x-ray gas and activity at time of galaxy and structure formation.

COBE -
Tests submillimeter excess; pushes limits on $\delta T/T$

Other Limits on $\delta T/T$ -
Could push limits to $\sim 10^{-6}$; can also check for characteristic patterns for cosmic strings and domain walls

Limits on m_ν
Tritium endpoint should soon eliminate ν_e. Limits on ν_μ and ν_τ require either accelerator mixing experiments or another supernova with a neutral current detector operating.

Supernovae
In addition to supernovae limiting neutrino masses, they also limit axions and other exotic particles with $M \lesssim 10MeV$.

Accelerator Tests
In addition to ν-mixing, also searches for supersymmetry. Searches for Higgs could reveal structure of vacuum. Identification of any new stable particle could yield the dark matter; width of Z^0 tests Big Bang Nucleosynthesis.

Antiprotons in cosmic rays
Limits constrain annihilations of CDM in galactic halo

Laboratory Searches for CDM
Axion searches using resonant cavities. Limits on ν's from annihilation in the Sun (and Earth) - using underground detectors. Direct searches (cryogenic detectors, etc.) should be able to detect WIMPs, if they exist.

As to HDM one of the best ways of finding it would be to find a mass for a neutrino. Tritium end-point measurements should soon eliminate ν_e as a candidate. Already limits from both laboratory experiments and the supernova seem to show that the mass of the ν_e is $\lesssim 20eV$. Experiments that are underway now will be able to push that limit down to $\sim 5eV$. Thus eliminating it as a candidate for the dominate matter of the universe. However, the ν_e was never a serious candidate. Most likely the neutrino that would have the most mass will be the neutrino associated with the τ. Measuring the ν_μ and ν_τ masses requires either accelerator mixing experiments or another supernova with neutral current detectors operating to pick out the distribution in time of these species. Supernova also could make a wonderful laboratory to further constrain other weakly interacting particles with masses $\lesssim 10MeV$. For example supernova 1987a constrained axions [73]. The fact that the supernova emitted neutrinos on a time scale of $\sim$ 10s second argues that there is no significant axion emission. These limits force the axion to have masses $\lesssim 10^{-3}$ electron volts, that is, the only masses of the axion are the masses that would make it (if it exists at all), an important DM candidate.

Accelerator tests in the future are also very important. In addition to the neutrino mixing mentioned above, searches for super-symmetry could enable the dark matter particle to be found. Similarly searches for the Higgs tell us something about the structure of the vacuum itself which leads to the formation of the seeds. In fact identification of any new stable particle (even one not predicated) would of course reveal the dark matter. Of course the of the mass Z from accelerators tests the neutrino counting agruments from Big Bang Nucleosynthesis and thus helps confirms our baryonic arguments.

Another observational test is the search for anti-protons in cosmic rays. These limits constrain the annihilation of CDM in The Galactic halo since massive CDM particles would produce anti-particles via annihiliation processes.

Perhaps the most exciting of all the dark matter constraining observations and experiments are the direct laboratory searches for CDM. This is a wonderful example of how new technology can be brought to play on an exciting problem. Axions searches using giant resident cavities may find this elusive particle directly. Limits on neutrino fluxes that might have been produced by the annihilation of CDM in the sun or even in the center of the earth might be found in underground detectors. Already the constraints from these kinds of experiments have eliminated and/or seriously constrained certain classes of models.[74,75] Perhaps the most exciting new detectors are the direct searches using cryogenic detectors and superconductivety. These should be able to detect any form of weakly interacting massive particle if it exist in the halo of our Galaxy. Details of the search possibilities are summarized in an excellent review by Primack, Sadoulet and Seckel[76]

SUMMARY

The dark matter problem and its related problem of large scale structure generation is one of the most exciting and vital problems in physical science today. It is being approached from many angles by particle theorists, by astrophysical theorists, by astronomical observers at many wave length regimes, and by particle experimentalists both with accelerators and with non-accelerator experiments.

While the ultimate answer to all our questions may not occur until we do experiments at the Planck scale (extrapolation of the Livingston Curve reveals that such experiments might[77] occur in the year 2150). It does seem that the important problem of finding out what the bulk of the matter of the universe is, may be resolved by the end of this century.

ACKNOWLEDGMENTS

I would like to thank my colleagues at Chicago for many interesting discussions particularly Chris Hill, Michael Turner, and Rocky Kolb. I would also like to thank my frequent collaborators, Gary Steigman, Keith Olive and David Dearborn. This work is supported in part by NASA NAGW-1321, and the NSF AST 85-15447 at the University of Chicago and by the NASA/Fermilab Astrophysics Program.

REFERENCES

1. Guth, A. (1981) *Phys. Rev.* **D23**, 374
2. Linde, A. (1983) *Phys. Lett.* **129B**, 177
3. Alpher, R., Bethe, H. and Gamow, G., (1948), *Phy. Rev.* **73**, 803
4. Alpher, R.A., Follin, J.W., and Herman R.C., (1953) *Phys. Rev.***92**, 1347
5. Peebles, P.J.E., (1966) *Phys. Rev. Lett.* **16**, 410
6. Wagoner, P., Fowler, W.A. and Hoyle, F. (1967) *Ap. J.* **148**, 3
7. Fowler, W. Greenstein, J. and Hoyle, F. (1961) *Am. J. Phys.* **29**, 393 (see also ref. 32)
8. Ryter, C., Reeves, H., Gradstajn, E. and Audouze, J. (1970) *Astron. and Astrophys.* **8**, 389
9. Reeves, H., Audouze, J., Fowler W.A., and Schramm D.N.(1973) *Ap. J.* **179**, 909
10. Epstein, R., Lattimer, J. and Schramm, D.N. (1976) *Nature* **263**, 198
11. Gott III, J.R., Gunn, J., Schramm, D.N. and Tinsley, B.M. (1974) An Unbound Universe? *Ap. J.* **194**, 543
12. Rogerson, J. and York, D. (1973) *Ap. J.* **186**, L95
13. Geiss, J. and Reeves, H.(1972), *Astron. and Astrophys.* **18**, 126
14. Black, D. (1971) *Nature* **234**, 148
15. Yang, J., Schramm, D., Steigman, G. and Rood, R. (1979) *Ap. J.* **227**, 697
16. Tinsley, B.M., Rood, R. and Steigman, G.
17. Yang, J., Turner, M.S., Steigman, G., Schramm, D.N., and Olive, K.A. *Ap. J.* **281**, 493

18. Olive, K., Turner, M., Steigman, G., Schramm, D.N. and Yang, J. (1980) *Ap. J.* **246**, 557
19. Spite, J. and Spite, F. (1981) *Astron. Astrophys.* **115**, 357
20. Steigman, G., Schramm, D.N. and Gunn, J.E. (1977) *Phys. Lett.* **66B**, 202
21. Schramm, D.N. and Steigman, G. (1983) *Phys. Lett.* **141B**, 337
22. Cline, D., Schramm, D., and Steigman, G. (1987) *Physics Lett. B* Neutrino Families. *Comments on Nuclear and Particle Physics* **A17**, 145
23. Freese K. and Schramm, D.N. (1981) Freese, K. and Schramm, D.N. (1984) *Nuclear Physics* **B233**, 167
24. Schramm, D. and Steigman, G. (1980) *Gravity Award Essay in G.R. and Grav.* **13**, 2 (1981)
25. Faber and Gallagher, J. *Annual Rev. Astron and Astrophys.* **17**, 135
26. Kawano, L., Schramm, D. and Steigman, G. (1988) *Ap.J.*, **327** 750-754
27. Strauss, M., Davis, M., Yahil, A. (1988) *Univ of Calif., Berkeley* preprint.
28. Loh, E. and Spillar, E. (1988) *Ap. J.* **329**, 24
29. Guth, A. (1980) *Phys. Rev.* **D23**, 374
30. Turner, M. (1986) *Proc. of Erice Symposium*, Reidel, Holland in press
31. Ryter, C., Reeves, H., Gradsztajn, E. and Auodouze, J. (1970) *Astron. and Astrophys.* **8**, 389
32. Fowler, W., Greenstein, J. and Hoyle. (1962) *J.R. Astron. Soc* **6**, 148
33. Dimoupolis, S., Ismailzadein, R., Hall, L., and Starkman, G. (1988)*Phys. Rev. Lett.* **60**, 7
34. Brown, L. and Schramm, D. (1988) *Ap. J.* **329** L103
35. Applegate, A., Hogan, C., Scherrer, R. (1987) *Phys. Rev. D* **35** 115
35. Alcock, C., Fuller, G. and Mathews, G. (1987) *Ap. J.* **320**,439
36. Malaney, R., Fowler, W.A. (1988) *Ap.J.* **333**, 14
39. Kurki-Suonio, R., Matzner, (1988) *Univ. of Texas preprint*
40. Alcock, C., Fuller, G., Myer, B. and Mathews, G. (1988) *Livermore preprint*
41. Kurki-Suonio, H., Matzner, R., Olive, K. and Schramm, D. (1989) *FNAL preprint*
42. Reeves, H. (1988) *Proc. Erice Symposium on Dark Matter*; Audouze, J. (1988) *Proc. Recontre de Morionde*
43. Dearborn, D., Schramm, D. , Steigman G., and Truran, T. (1989) The Shocking Development of Lithium (and Boron) in Supernovae *Ap. J.* submitted
44. Helfand, D. (1989) *Columbia Univ. preprint*
45. Hegyi, D. and Olive, K. (1986) *Ap. J.* **303**, 56
46. Gunn, J. 1988 talk at *ITP Santa Barbara*; York, D. 1988 talk at *Univ. of Chicago*
47. Bond, R. and Szalay, A. (1982) *Proc. Texas Relativistic Astrophysics Symposium* Austin, Texas
48. Turner, M.S., Wilczek, F. and Zee, A. (1984) *Physics Rev. (Lett.)*
49. Crawford, D. and Schramm, D.N. (1982) *Nature* **298**, 538
50. Zeldovich, Y. and Novikov, I. (1967) *Relativisitic Astrophysics* (Soviet edition)
51. Lee, B., and Weinberg, S. (1977) *Phys.Rev. (Lett.)*
52. Frenk, C., White, S., Davis, M. and Efstatiou, G. (1985) *Ap. J.* **292**, 371
53. Kibble, T. (1980) *Phys. Rev. (Lett. B)*; Zeldovich, Ya., (1981) *Nature*
54. Vilenkin, A. (1986) *Phys. Rep.*

55. Turok, N. and Albrecht, (1986) *Phys. Rev. (Lett.)*
56. Bennett, D. and Buchet, F. (1988) *Fermilab* preprint
57. Ostriker, J., Thompson, C. and Witten, F. (1986)*Phys. Lett.B*
58. Ostriker, J. and Cowie, C. (1981) *Ap. J.* **243** L127
59. Hill, C., Schramm, D. and Fry, J. (1989) *Comments on Nuc. Part. Physics* in press
60. Tully, B. (1988) presented at *Aspen Workshop on Large Scale Structure*
61. Geller, M. and Huchra, T. (1989) *Proc. Berkeley Part/Astro Workshop*
62. Kawano, L. (1989) *Univ. of Chicago Thesis*
63. Spergle, D., Press, W. and Rydon, B. (1989) in preparation
64. Lee, T.D. (1986) *Phys. Rev.D* L. Widrow (1989) *Harvard Univ.* preprint
65. Griest, K. and Kolb, E. (1989) *FNAL preprint*
66. Stebbins, A. and Turner, M. (1989) *FNAL preprint*
67. Dimoupolis, S. and Starkman, G. (1989) private communication
68. Fuller, G. and Schramm, D. (1989) in preparation
69. Thuan, T., Gott, J.R. and Schrader, R. (1989)*Univ. of Virginia preprint*
70. Burstein, D. et al. (1986) in Galaxy Distances and Deviations from Universal Expansion ed. B. Madore and B. Tully, **p.23** (Dordrecht Reidel)
71. Dressler, A. and Faber, S. (1989) *Proc. Texas Symposium Rel. Astro.*
72. Lang, A., (1988) *Proc. of Berkeley Part/Astro Workship*
73. Mayle, R., Wilson, J, Ellis, J., Olive, K., Schramm, D., Steigman, G. (1988) *Phys. Lett.B* in press Borrows, A., Brinkmann, R. and Turner, M. (1989) Phys. Rev. Lett. in press
74. Srednicki, M. and Olive, K. (1988) *Phys. Lett. B*
75. Gelmini, G. (1988) *Proc. Erice Symposium on Dark Matter*, in press
76. Primack, J., Sadoulet, B, Seckel, D. (1988) *Ann. Rev. Nuc. Part. Sci.* **38**, 751
77. Lederman, L. and Schramm, D.N. (1989) Understanding the Universe *Sci. Amer. Books*, W. Freeman, N.Y. in press
78. Bahcall, N. and Soniera, R. (1983) *Ap. J.* **270** 20
79. Szalay, A. and Schramm, D. (1985) *Nature Nature* **314**, 718
80. Primack, J. and Deckel, A. (1989) *Santa Cruz preprint*

Dark Matter And Particle Physics

Lawrence J. Hall
Department of Physics
University of California
and
Theoretical Physics Group
Physics Division
Lawrence Berkeley Laboratory
1 Cyclotron Road
Berkeley, California 94720

Abstract

A classification scheme is introduced for cosmic relics. Two points about direct detection of dark matter are made: first that current experiments are very close to probing techni–baryons as the dark matter, and second that it is very difficult to push the mass of a cosmic relic above 1000 TeV.

(I) Classification Scheme For Cosmic Relics

The hot big bang framework allows us, at least in principle, to calculate the contribution to Ω today from the stable objects of any particular model of particle physics. From the viewpoint of particle physics these relics can be classified as various plasma relics or phase transition relics.[1] There are several types of plasma relic. A relic which is relativistic at freezeout gives $\Omega_x \simeq 10^{-5}$ if $m_x < T_0$ and $\Omega_x \simeq m_x/30eV$ otherwise. For non–relativistic freezeout

$$\Omega_x \simeq \frac{1}{\sigma_A \beta}\left(\frac{1}{10^3 TeV}\right)^2 \tag{1}$$

if the cosmic asymmetry η_x is zero, and $\Omega_x \simeq T_0^3\,\eta_x m_x$ otherwise. Finally a plasma relic may be cosmologically unstable, in which case the stable, secondary relics survive until today, and are possibly relativistic.

There are many varieties of relics which are produced at phase transitions in the field theory vacuum as the plasma cools. Weakly coupled scalar field oscillations with damping $\Gamma \ll t_0^{-1}$, such as axions, are quite common. There are well known topologically stable defects: domain walls, strings and monopoles; and there are nuggets of false vacuum such as quark nuggets produced at the QCD phase transition. Quark nuggets are known to evaporate and hence are not the dark matter; however, their properties depend on the neutrino couplings and perhaps nuggets of some other phase transition could both be produced and survive until today.

Since non–baryonic dark matter must involve non–standard model particle physics, we should seriously consider any stable object which occurs in gauge theories.

(II) Experimental Probe of Technibaryon Dark Matter.

There are two attractive ways of breaking weak interactions and giving masses to W and Z: technicolor and supersymmetry. Both have good dark matter candidates; that of technicolor is a neutral technibaryon with a non–zero cosmic asymmetry, just like ordinary baryons.[(2)] These technibaryons have a model dependent mass of $1 - 10$ TeV, typically 3 TeV; they are expected to couple to Z's but not to strong interactions. Hence they will interact in double beta decay spectrometers in a way very similar to Dirac neutrinos. Present limits exclude 12 GeV to 1.4 TeV,[(3)] and future data can probe this extremely exciting, well–motivated and constrained idea of technibaryons as dark matter.

(III) An Upper Limit To The Mass of Elementary Cosmic Relics.

If dark matter is not directly detected will it be possible to push its mass up to arbitrarily large values making the flux at the earth arbitrarily small? If the dark matter is elementary particles this is very hard to do. This is because for a heavy particle even if its annihilation in the big bang were via a very strong interaction there is a limit $\sigma_A \beta \lesssim 1/m_x^2$. Hence from equation 1, $\Omega_x \lesssim O(1)$ implies $m_x \lesssim O(10^3 \text{ TeV})$. There may be extraordinary ways to evade this, for example the x particles could receive a large mass at a phase transition which occurs after they have frozen out. Also, it should always be kept in mind that the dark matter could be aggregations of x particles rather than single x particles. However, the rather general argument given above shows that if dark matter has to do with plasma relics the physics responsible for them cannot be pushed to scales larger than 10^3 TeV.

References

1. L.J. Hall, LBL preprint LBL–26210, Proceedings of SLAC Summer Institute (1988).

2. S. Nussinov, *Phys. Lett.* **165B**, 55 (1985).

3. D. Caldwell *et al.*, *Phys. Rev. Lett.* **61**, 510 (1988).

Horizontal branch stars as a probe of non baryonic dark matter

Pierre Salati *

Astronomy Department, University of California
Berkeley, CA 94720, U.S.A.

Lawrence Berkeley Laboratory, University of California
Berkeley, CA 94720, U.S.A.

As pointed out by Renzini [1], the weakly interacting massive particles which may solve the solar neutrino puzzle ("cosmions") [2] may also break convection inside horizontal branch (HB) stars and lead to anomalies in the stellar counts of globular clusters. However, Spergel and Faulkner [3] have shown that evaporation during the HB phase depletes the cosmion population so that convection is not affected. Even if evaporation does not occur, they claim that the heat transport by any kind of WIMP is ineffectual. We have reconsidered this exciting problem [4], taking possible variations of the dark matter (DM) environment into account.

In the core of an HB star, the total energy flux $L_{\rm core}$ is too large to be carried out by radiative transfer, and so convection develops. However, if the star has been steadily accreting cosmions since its birth, and if the maximum energy transfer L_x carried by these particles exceeds $L_{\rm core}$, the core adjusts itself to a radiative structure and convection is broken. If evaporation during the helium flash or at the beginning of the HB phase is not important, the number N_x of cosmions in an HB star is given by the number $N_{\rm MS}$ accreted during the main sequence evolution, which is found to be [5,6] :

$$N_{\rm MS} \approx 1.8 \times 10^{47}\, f_{\rm MS}\, (m_p/m_x)\, {\rm Min}(1, \sigma_{\rm eff}/\sigma_{\rm crit}). \qquad (1)$$

The critical cross section $\sigma_{\rm crit}$ above which all cosmions impinging on the star are trapped is $\sigma_{\rm crit} = 3 \times 10^{-36}\ {\rm cm}^2\ (M/0.66\, M_\odot)^{0.6}$ where M is the progenitor mass, and we have assumed that its radius varies as $M^{0.8}$. For neutrino-like WIMPs [7], the cross section $\sigma_{\rm eff}$ relevant for cosmion trapping on the main sequence may be approximated by $\sigma_{\rm eff} = 0.1\, \sigma_{\rm He}$. The "fudge factor" $f_{\rm MS}$ characterizes the progenitor and its dark matter environment by :

$$f_{\rm MS} = \left(\frac{\rho_{\rm DM}}{0.01\, M_\odot/{\rm pc}^3}\right) \left(\frac{300\,{\rm km/s}}{\overline{v}}\right) \left(\frac{M}{0.66\, M_\odot}\right)^{1.8} \left(\frac{\tau_{\rm MS}}{10^{10}\,{\rm yr}}\right), \qquad (2)$$

where $\rho_{\rm DM}$ is the dark matter density, $\overline{v}$ its velocity dispersion, and $\tau_{\rm MS}$ the main sequence lifetime. Our benchmark values for $\rho_{\rm DM}$ and $\overline{v}$ characterize the solar neighborhood. The number N_x of cosmions in the HB star will be identical with $N_{\rm MS}$ unless cosmions are ejected during the helium flash (a

*Miller Research Fellow at the University of California at Berkeley. On leave of absence from LAPP and from Université de Chambéry, 73000 Chambéry, France.

possibility which we estimate to be fairly unrealistic), or unless they evaporate from the HB star.

Fig. 1 shows the range of m_x and $\sigma_{\rm He}$ values for which convection is broken ($L_x > L_{\rm core}$) in the case of neutrino-like WIMPs. Below the "evaporation line", the cosmion population is depleted by HB evaporation and, as pointed out by Spergel and Faulkner [3], no effect on HB stars occurs. Above this line, different contours for the values $f_{\rm MS}$ = 1.5, 3, 5, and 10 delimit the regime where convection is broken. The dashed rectangle delimits the approximate range of parameters for which the solar neutrino puzzle is solved. Thus it is clear that the regime of cosmion parameters to which HB stars are sensitive depends strongly on the "environmental factor" $f_{\rm MS}$, where $f_{\rm MS} = 1$ characterizes the solar neighborhood. Number counts of HB stars in 15 well studied globular clusters clearly indicate that convection is not broken in these systems [8]. All of these globular clusters are at relatively large distances from the inner galaxy so that presumably $f_{\rm MS} \approx 1$. These observations do not preclude an interpretation of the solar neutrino deficiency as a signature of neutrino-like cosmions.

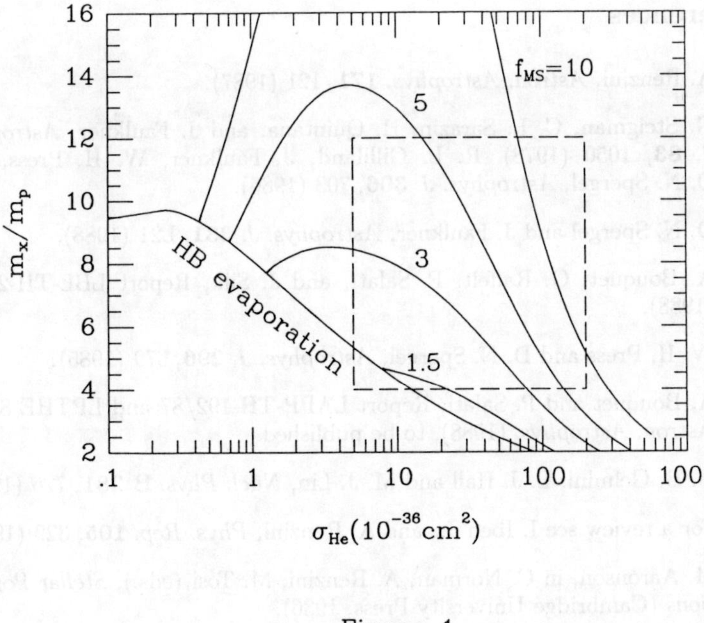

Figure 1

Moreover, one may well expect to observe an anomalous paucity of HB stars in systems where the "environmental factor" f_{MS} is only slightly enhanced over the solar values. An example would be globular clusters in the inner galaxy, where the dark matter density is enhanced by a factor of at least 2 and the one-dimensional velocity dispersion in the inner halo, inferred from the spheroidal component of the galaxy, is about 120 km/sec, whence $f_{MS} \approx 3$. An even stronger test may come from examining HB stars in nearby dwarf spheroidal galaxies. Several of these are inferred to contain substantial amounts of dark matter [9] with a density $\sim 0.1\, M_\odot\, \mathrm{pc}^{-3}$. Even though the dark matter core radius is not measured directly, the velocity dispersion in the luminous cores (around 200 pc) is $10 - 20$ km/sec, so that we infer $f_{MS} \approx 100$. While a significant population of HB stars is seen in these systems [10], the data have not been analyzed to determine the HB lifetimes. A detailed analysis of the color magnitude diagrams could either yield a signature for DM cosmions or, less interestingly, allow for the exclusion of a large range of cosmion parameters.

I greatfully acknowledge a fellowship from the Miller Institute for Basic Research in Science at the University of California at Berkeley and would like to thank Professor B. Sadoulet to have invited me to speak at this conference.

References

[1] A. Renzini, *Astron. Astrophys.* **171**, 121 (1987).

[2] G. Steigman, C. L. Sarazin, H. Quintana, and J. Faulkner, *Astrophys. J.* **83**, 1050 (1978). R. L. Gilliland, J. Faulkner, W. H. Press, and D. N. Spergel, *Astrophys. J.* **306**, 703 (1986).

[3] D. N. Spergel and J. Faulkner, *Astrophys. J.* **331**, L21 (1988).

[4] A. Bouquet, G. Raffelt, P. Salati, and J. Silk, Report LBL-TH-25215 (1988).

[5] W. H. Press and D. N. Spergel, *Astrophys. J.* **296**, 679 (1985).

[6] A. Bouquet and P. Salati, Report LAPP-TH-192/87 and LPTHE 87-20, *Astron. Astrophys.* (1988), to be published.

[7] G. B. Gelmini, L. J. Hall and M. J. Lin, *Nucl. Phys.* **B 281**, 726 (1987).

[8] For a review see I. Iben Jr. and A. Renzini, *Phys. Rep.* **105**, 329 (1984).

[9] M. Aaronson, in C. Norman, A. Renzini, M. Tosi (eds.), *Stellar Populations* (Cambridge University Press, 1986).

[10] E. W. Olszewski and M. Aaronson, *Astron. J.* **90**, 2221 (1985). G. S. DaCosta, *Astrophys. J.* **285**, 483 (1984). J. Mould and M. Aaronson, *Astrophys. J.* **273**, 530 (1983).

Tracing the Galactic Mass Distribution with Microarcsecond Astrometry

J.D. Phillips, R.W. Babcock, J.F. Chandler, and R.D. Reasenberg

Harvard-Smithsonian Center for Astrophysics
Cambridge, MA 02138

POINTS (the Precision Optical INTerferometer in Space) is a proposed space-based optical astrometric instrument now being developed in the laboratory. In a 10 minute observation of the separation of a pair of 10th magnitude stars about 90 degrees apart, the instrument will achieve an uncertainty of 5 microarcseconds (μas), three orders of magnitude better than currently achievable. POINTS will comprise two separate stellar interferometers mounted nearly at right angles, each with a pair of 25 cm telescopes 2 meters apart. The angle between the interferometer baselines will be measured by interferometric laser gauges. The principal technical challenge in building and operating the instrument is the control of systematic error, which is the central theme of the architecture. This challenge is met by careful control of structural stability, precise real-time metrology, and correction of systematic error in post-measurement data analysis. Studies to date indicate that this level of performance can be achieved within present technological limits.

A ten year observing program on a reference grid of 300 stars, with 6000 observations of star pairs per year, would result in uncertainties for the grid stars of 0.6 μas in position, 0.4 μas/yr in proper motion, and 0.4 μas in parallax. For grid stars at distances up to 10 kpc, POINTS would determine parallaxes to 0.4% and measure transverse velocities to 20 m/s. POINTS could also make ~150,000 measurements of stars with respect to that grid with 5 μas nominal accuracy, or with greater accuracy on selected objects by devoting more instrument time. POINTS and its applications are treated in detail in Reasenberg, R.D. et al., Astron. J. 96, 1731 (1988), and references therein.

The Galaxy's rotation curve seems to be flat out to about 18 kpc, so its total mass is presumably at least twice that within the orbit of the Local Standard of Rest. Distances measured by POINTS to the O and B stars that excite the H II regions used in these studies would help confirm this result. To determine the local mass density in the disk, models have been built which balance the disk's gravitational potential against the stars' velocity dispersion, using observations of 1000 stars. POINTS should be able to support a substantially larger study in both hemispheres and could, in particular, prove vital to deriving the scale height of the galaxy's third apparent component, the "thick disk".

Perhaps the best measure of the Galaxy's total mass and its dark matter component would be to measure the space velocities of the Magellanic Clouds, both of which contain stars as bright as m = 11 in spite of their 60 kpc distances. The LMC should be moving due East at 15 mas/y if the Galaxy lacks such a massive component, and even faster if it has one. Such a proper motion could have been determined by classical astrometry if there were not the problem of systematic error due to a non-inertial reference frame. POINTS, which puts all measurements in a frame tied to the quasars, would provide the required low levels of both systematic and accidental error. The POINTS frame is also tied to the "dynamical frame" of the earth's motion to ~1/2 mas.

In addition to its uses in determining the galactic mass distribution, POINTS addresses a wide variety of other astrophysical and cosmological problems. Among them are a thousandfold improvement of the accuracy of the gravitational light deflection test, a search of a large number of nearby stars for unseen companions with masses as small as 0.01 Jupiters, a direct (parallax) determination of the Cepheid distance scale to 1%, accurate measurements of individual masses of stars in binaries, and a bound on or measurement of quasar proper motion.

DESIGN FOR A PRACTICAL LABORATORY DETECTOR FOR SOLAR AXIONS

K. van Bibber*

Physics Department
Lawrence Livermore National Laboratory
University of California
Livermore, CA 94550

1. ABSTRACT

A design for a detector sensitive to solar axions or other light particles with a two-photon interaction vertex is presented. Axions would be detected by their conversion into x-rays in a strong magnetic field. The transition rate is enhanced by filling the conversion region with a gas (H_2 or He); this induces an effective photon mass which can be adjusted to be equal to that of the axion being searched for. A practical detector should be possible for the mass range 0.1-5 eV, corresponding to one of two regions not yet excluded by astrophysical constraints.

2. MOTIVATION

Combined evidence from particle physics, astrophysics and cosmology leaves only two narrow windows in mass in which axions might still exist. Axions of mass $m_a < 10^{-5}$ eV would imply overclosure of the universe. If $10^{-3} < m_a < O(1eV)$, the neutrino signal from SN1987a would have been foreshortened (axions emitted through NN bremsstrahlung would have competed with neutrino emission). The SN1987a does not preclude axions heavier than O(leV), as these axions would couple to matter sufficiently strongly to re-interact and thus be trapped. This upper limit is regarded as very uncertain. The observed evolution of Red Giants constrain m_a to be less than about 1 eV. Thus axions of mass O(leV) cannot yet be ruled out, at least for the hadronic class of axion models (no tree-level coupling to light fermions).[1]

3. PRINCIPLE OF THE EXPERIMENT

We propose a search for axions emitted by the sun in the mass range 0.1-5 eV. This search would consist of conversion of the axions to photons in a strong magnetic field, and their subsequent detection.

The axion luminosity from the sun is obtained by folding the photon-axion transition rate (including plasma screening) with the blackbody photon distribution from a standard solar model. The ratio of axion to photon luminosities is then about $3.6 \times 10^{-3}\ (m_a/1eV)^2$. For 1 eV axions, this would result in only a very minor perturbation of the sun, but a large axion flux at the earth, $7.4 \times 10^{11}\ cm^{-2} sec^{-1}$, and of average energy 4.2 keV.

The problem of converting axions back into photons is most easily dealt with in the axion-photon mixing formalism of Raffelt and Stodolsky.[2] Starting with a pure axion beam, the maximum probability of measuring a photon after a distance L in a transverse magnetic field B is given by $P(a\rightarrow\gamma) = (B/2M)^2 L^2$. Here M^{-1} is the axion-photon coupling specified by $\mathcal{L}_{a\gamma\gamma} = -(1/4M) aF_{\mu\nu}F^{\mu\nu} = aM^{-1}E\cdot B$, and for grand unified theories, $M^{-1} = 1.45 \times 10^{10}$ GeV $(m_a/1eV)$.

The difficulty is that the dispersion relationships for a photon $(k_\gamma = \omega)$ and for a massive particle $(k_a^2 = \omega^2 - m_a^2)$ are different, and in this case the axion-photon oscillation length is very much smaller than the dimensions of large magnets. Thus the photon and axion fields do not remain in phase, and the coherent limit above is not achieved.

The situation for x-rays is totally redeemed however, by recalling that the real part of the index of refraction for x-rays in a gas (far above all resonances) is less than 1. Thus its dispersion relationship is that for a massive particle, with the plasma frequency being the effective photon mass. One may then 'tune' the detector to achieve the theoretical maximum by adjusting the gas pressure of a suitable gas within the conversion volume, and the detector response will go through a sharp resonance. For reasons of both self-absorption and constancy of the atomic scattering factor throughout the relevant range of energies (1-15 keV), H_2 and He are the only viable candidates.

4. RESULTS

Figure 1 shows a conceptual design for the experiment. A wire chamber sensitive to low-energy x-rays, lines the inside surface of a large solenoidal magnet; the proportional gas (e.g. Ne, Ar, Xe) being separated from the buffer volume gas by a thin mylar window.[3] Clearly both gas pressures must be equal as the buffer gas pressure varies over a large dynamic range: for m_a = 0.1-5 eV, the pressure at 300 K goes from 0.1-300 atm. The axis of the solenoid should be on the average perpendicular to the direction of the sun, but there is no first-order diminution of the rate due to angular misalignment, if $\omega_p = m_a$.

Calculations incorporating self-absorption have been performed for the FNAL 15' bubble chamber magnet (4m diam., 3m long, 3T). Figure 2 shows the rate as a function of m_a for a plasma frequency ω_p = 1 eV. (The upper scale and dashed line are just an expansion of the lower scale and solid line.) Figure 3 shows the rate as a function of axion mass, when the detector is tuned to that mass. The detection rate goes as M^{-4} overall, and goes from O(1/sec) at 2eV to O(1/day) at 0.2 eV.

This experiment would need to be conducted deep underground to avoid backgrounds due to cosmic rays. The problems associated with backgrounds due to natural radioactivity, as well as achieving the requisite spatial uniformity of the gas density, and its monitoring, etc. are formidable but not impossible.

5. SUMMARY

It seems that a feasible scheme exists to close the 'small' window in the vicinity of 1 eV which exists for hadronic axions. The realization of this experiment nevertheless will be a significant technical enterprise.

This work was performed under the auspices of the U.S.D.O.E., under contract numbers W-7405-ENG-48 (LLNL), DE-AS05-81ER40039 (Texas A&M), DE-FG03-84ER-40161 (UC Berkeley), DE-AC03-76SF00098 (LBL), and by NASA under contract number NGR05-003-578.

* Work performed in collaboration with P.M. McIntyre, D.E. Morris and G.G. Raffelt.

6. REFERENCES

1. For a comprehensive review of all relevant astrophysical and cosmological bounds, see G.G. Raffelt, to be published.
2. G. Raffelt, L. Stodolsky, Phys. Rev. D37, 1237 (1988).
3. A detector exactly of this sort was developed and used for an entirely different application; see U. Gastaldi *et al*, Nucl. Instr. and Meth. 156, 257 (1978).

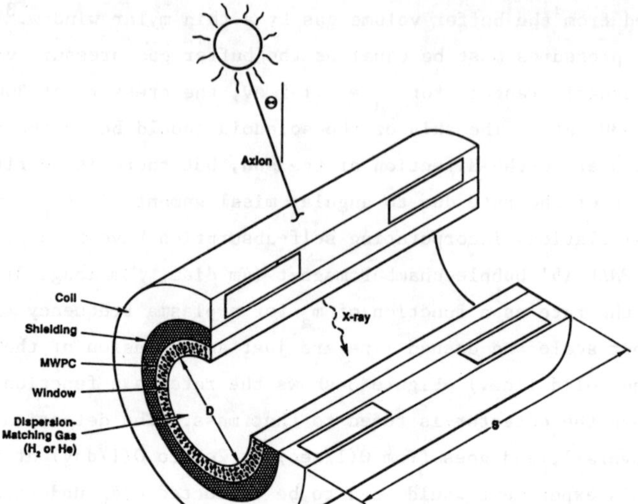

Figure 1.

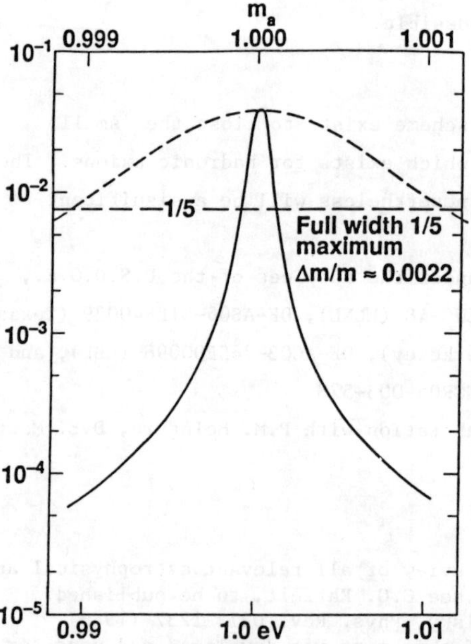

Figure 2.

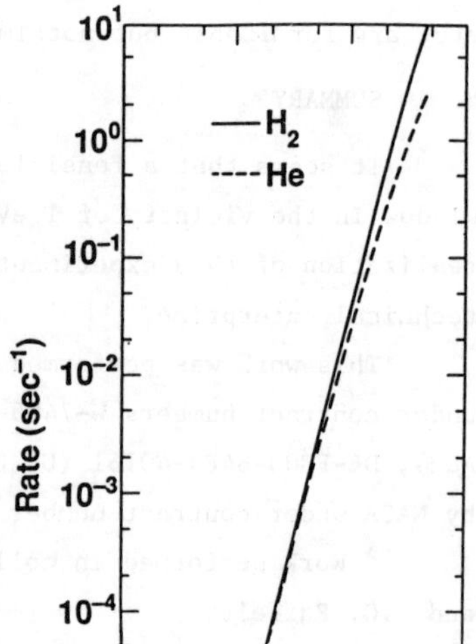

Figure 3.

Expected Rates for Direct Detection of Particle Dark Matter

Kim Griest
Astronomy and Astrophysics Center
The University of Chicago

That 90% - 99% of the mass of the universe consists of dark matter (DM) which clusters in the halos of galaxies is widely accepted, and several lines of reasoning make it plausible that this DM is not baryonic, but may consist of some as yet undiscovered elementary particle. If so, direct laboratory detection of these particles may be possible by measuring the small (order keV) recoil energy deposited when a DM particle from the halo scatters off a nucleus in a detector. Many groups are upgrading existing detectors or prototyping new detectors for this purpose, and it is therefore important to have some idea of what interaction rate to expect. General arguments lead to expectations of measurable rates almost independent of the DM particle considered, but there are large uncertainties and loopholes which should be discussed.[1]

For an isothermal halo the rate per kilogram per day in a detector can be written[2,3,4]

$$R = \sqrt{\frac{8}{3\pi}\eta_t\eta_v\eta_c\eta_e}\frac{\langle v\rangle_{halo}\,\rho_{halo}\sigma_{el}}{m_N m_x}, \tag{1}$$

where $\rho_{halo} \approx .3$ GeV cm^3 is the density of DM near the Earth (uncertain by at least a factor of two), $\langle v\rangle_{halo} \approx 270$ km sec^{-1} is the average speed of halo particles, m_x is the unknown DM particle mass ($m_x > 1$ GeV is of interest), and m_N is the mass of the detector nucleus and is under the control of the experimenter. The η factors modify the rate and introduce more uncertainties: $\eta_v \approx 1.3$ takes into account the motion of the Sun and Earth and would be smaller if the halo rotates, $\eta_e \approx 1$ takes into account the unknown galactic escape velocity, $\eta_c \leq 1$ is a coherence loss factor which is small for heavy DM particles or nuclei, and $\eta_t \leq 1$ takes into account the energy threshold of the detector.

While there is more than a factor of two total uncertainty in these astrophysical parameters, the largest uncertainty is the elastic scattering cross section, σ_{el}, which in principle could be zero. However, requiring a particle to be the DM, in general constrains σ_{el} substantially. If the DM was in thermal equilibrium with quarks and leptons in the early universe, then a "Lee-Weinberg" calculation gives a lower limit to its annihilation cross section at freeze-out, $\langle \sigma v \rangle_{ann} \gtrsim 10^{-26}\text{cm}^3\text{sec}^{-1}/(\Omega_x h^2)$, where Ω is the ratio of the DM density to the critical density and $h \approx \frac{1}{2}$ parameterizes the Hubble constant. The inequality is saturated if there is no initial particle-antiparticle asymmetry.

A typical annihilation cross section into fermions of mass m_f can be written

$$\sigma_{ann} v \simeq \sum_{f<x} \frac{{m_x}^2 C_X^2 C_f^2}{M_{ex}^4} \sqrt{1 - m_f^2/{m_x}^2} \left(a_f + b_f \frac{m_f^2}{{m_x}^2} + c_f v^2 \right), \tag{2}$$

where $M_{ex} \approx m_Z$ is the mass of the exchanged boson, C_X and C_f are the couplings of the DM particle and fermion respectively, $\langle v^2 \rangle \approx \frac{1}{3}$ is the relative velocity of the annihilating particles, and a_f, b_f, and c_f are model dependent coefficients of order unity. The sum is over all fermions with $m_f < m_x$. Particles with axial vector couplings have $a_f = 0$ which is the famous p-wave suppression. Using crossing symmetry, the annihilation matrix element can be related to the elastic matrix element and the elastic cross section can be written

$$\sigma_{el} \simeq \frac{{m_x}^2 m_N^2 C^2}{M_{ex}^4 (m_x + m_N)^2} \geq \frac{m_N^2}{(m_x + m_N)^2} \left(\frac{10^{-37}\text{cm}^2}{\Omega_x h^2} \right) \left[\frac{C^2}{\sum C_f^2 C_x^2 (a_f + \cdots)} \right]. \tag{3}$$

where C^2 is a coherence factor which comes from summing quark wave functions to make a nucleus and hides a great deal of physics. Assuming for the moment that the term in square brackets is unity, we find from eq. (1), $R \gtrsim 1$ event kg^{-1} day^{-1} which is a quite substantial rate, probably within reach of the current or next generation of DM detectors. This encouraging result has given rise to a large effort by experimenters – but it is worth going back and carefully examining the simplifications and assumptions that went into it.

First, note that we identified M_{ex} in the annihilation cross section with M_{ex} in the elastic cross section. For Z^0 exchange this is valid, but in supersymmetry, for example, it is quite possible that annihilation is mainly into electrons, via a lighter selectron exchange, while for elastic scattering off up and down quarks in a nucleus, it is only the squark exchange that is relevant. For generic supersymmetric models the scalar fermion masses are model dependent and this introduces a factor of 100 uncertainty into the rate. A very similar thing occurs if the annihilation coefficients a_f, b_f, etc. for up and down quarks are suppressed. Again, it is crossing in the up and down quark channels that is important and a suppression here will reduce the expected detection rate. Probably the biggest model dependence, however, is contained in C^2. This factor depends sensitively on the type of coupling the DM particle has. For particles such as Dirac neutrinos, which have vector couplings, $C^2 \propto$(number of nucleons)^2, which can lead to an enhancement of a factor of 1000 in the rate. This enhancement is why the strongest limits to date are on Dirac neutrinos and other particles with vector couplings.[5] For particles such as majorana neutrinos or pure photinos which have only axial vector couplings, $C^2 \propto \lambda^2 j(j+1)$, where j is the spin of the nucleus and $.1 \leq \lambda^2 \leq 1$ is a nuclear physics factor. Therefore, for common spinless elements such as oxygen, carbon and the majority of silicon and germanium, $C^2 = 0$. For favorable elements which have spin, $C^2 \approx 1$, and we recover $R \geq 1$, but we see that to find particles with only axial vector couplings the detector must be made of an element which has a spin. Also note that theoretical uncertainties in the nuclear and parton physics used introduce another factor of four uncertainty in the rates. For particles which have scalar couplings there are some subtleties, but in general, $C^2 \propto (m_N/m_W)^2$ is in the range of .1 to 10 for common elements. Example particles include the neutralino from supersymmetry and any particle when the exchange particle is a Higgs boson. Finally, particles with pseudoscalar couplings, or mixed vector-axial vector couplings have $C^2 \approx 0$; but particles of this type have not been proposed as dark matter candidates.

In conclusion, we see that while generally rates of greater than one event per kilogram per day are to be expected in DM detectors, there is a great deal of astrophysical and particle physics uncertainty. Model dependencies make rates between zero and 1000 events per kilogram per day possible. A detection, however, would give a new particle, the identity of the dark matter and valuable information about the galaxy's halo. The possible benefits outweigh the risk and so the search is well worth the effort.

References

1. K. Griest and B. Sadoulet, in preparation.
2. M. W. Goodman and E. Witten, *Phys. Rev.* **D31** (1986) 3059.
3. K. Freese, J. Frieman and A. Gould, *Phys. Rev.* **D37** (1988) 3388.
4. K. Griest, *Phys. Rev.* **D38** (1988) 2357.
5. D. O. Caldwell, *et al.*, *Phys. Rev. Lett.* **61** (1988) 510.

BACKGROUND REJECTION IN THE DETECTION OF DARK MATTER

David O. Caldwell

Physics Department, University of California, Santa Barbara, CA 93106

ABSTRACT

Ge double beta decay experiments have been converted to search for low-energy nuclear recoils from the scattering by heavy dark matter particles. Although remarkably low backgrounds had been obtained in the MeV region, new problems arose at keV energies. Some solutions and data are presented and possibilities for the future are discussed.

Ge ionization detectors have so far provided the only means by which the direct detection of heavy dark matter has been attempted.[1,2] These measurements also supply the only useful information on backgrounds at these very low energies.

The detectors were used to search for double beta decay in ^{76}Ge, which makes up 7.8% of normal Ge and would have a decay energy of 2.041 MeV. Around this energy the background has been reduced to < 1 count/keV·kg·year, so that a limit of $\sim 10^{24}$ years has been achieved for the process $^{76}\text{Ge} \rightarrow {}^{76}\text{Se} + 2e^-$.[3] Obtaining such low backgrounds requires many precautions. First, the experiment needs to be underground to exclude the cosmic ray hadronic component and enough muons to reduce the neutron rate sufficiently. However, fission neutrons from the rocks can also introduce the bothersome capture γ rays. Second, good passive shielding is required. This is usually a graded shield, such as lead and borated polyethylene, but the inside must be chosen carefully. Some have used multiply distilled mercury, many use OFHC copper, but if lead is used it needs to be very pure (generally from the St. Joseph mine in Missouri) or better, old lead so that ^{210}Pb ($T_{1/2} = 22$y) has died out. The β's from $^{210}\text{Pb} \rightarrow {}^{210}\text{Bi} \rightarrow {}^{210}\text{Po}$ provide bremsstrahlung. The shield must also exclude radon, and this is usually done by flowing the boil-off from the liquid nitrogen needed to cool the Ge detectors. Inner active shields of NaI have also been used to reduce further the effects of cosmic rays, neutrons, external radioactivity, and Compton scattering in the Ge. To be an efficient anti-Compton shield, the NaI must have very little inert material between it and the Ge, and that must be of as low a Z as possible. The NaI itself may introduce radioactivity, and its packaging must be done with great care in the choice of materials.

Since ^{40}K and the U and Th radioactivity series are ubiquitous, material near the Ge has to be minimized and carefully chosen. Generally glass, rubber, steel, and solder are to be avoided, while aluminum (most often bad), plastics, epoxy, and lead are highly variable. Lexan is permissible, with purified Ge, Si, OFHC and electroformed copper being usually good. All materials used need to be counted in large quantities, including electronic components, which are often bad.

Backgrounds in the MeV region are now limited mainly by cosmogenically induced activities, particularly in the Ge itself, produced when these materials were above ground. For example, ^{68}Ge, produced probably by ^{70}Ge(n,3n), decays by e^- capture (288 days) to ^{68}Ga, which decays by a 1.89 MeV β^+ in 68 min. If the annihilation γ's are also captured in the Ge, this gives an effect up to 2.91 MeV.

At low energies new problems arise. A half gram of indium used as a gasket had a devastating effect, despite its 4×10^{14} year lifetime, and needed to be replaced by gold. In addition to the ^{68}Ge→^{68}Ga activity which produced the large X-ray peak (mainly Ga but some Zn) shown in Figure 1, cosmogenically induced ^{3}H accounts for much of the low-energy background, as also shown in the figure. The amount of ^{3}H observed corresponds closely to that predicted by Martoff[4)] and is confirmed also by the amount of another spallation product, ^{7}Be (478 keV γ).

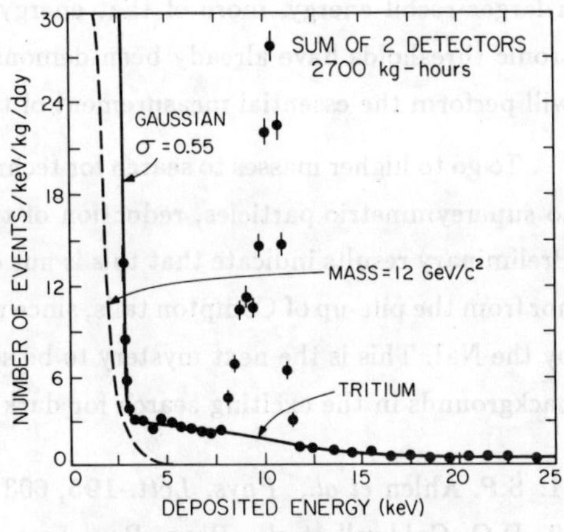

Fig. 1. Fit to ionization energy in Ge by ^{3}H from spallation, electronic noise, and a constant background.

Figure 1 shows the summed spectrum from the two of the 8 Ge detectors of about 0.9kg each of the UCSB/LBL double beta decay experiment[3] which were modified to look also at low energies. The energy plotted is ionization energy, which is much less than the recoil energy, but the ratio of the two has been determined by measurements.[5] The data are fit (solid line) by a background of $\sim 1/2$ count/keV·kg·day which (aside from a few X-ray peaks) is constant from about 20 to 200 keV, plus the ^{3}H and the gaussian ($\sigma = 0.55$ keV) electronic noise, which is consistent with the X-ray peak widths. This leaves no room for a signal, as shown, for a 12 GeV/c^2 particle having the weak interaction. Such particles, which could be Dirac neutrinos, are excluded as the sole dark matter in our galaxy for masses between 12 and 1400 GeV/c^2.

To go to lower masses it is necessary to reduce the ^{3}H and the electronic threshold. The UCSB/LBL/UCB group is trying to do this by getting a crystal with a variable impurity density underground as rapidly as possible. The density gradient shapes the electric field so that charge can be collected on a small electrode, reducing the capacitance and hence the threshold. Also, Si detectors will be installed in January 1989, and these should have three advantages:[6] the lighter nucleus gives a larger recoil energy, more of that energy goes into ionization, and lower electronic thresholds have already been demonstrated. The Rich–Spiro Saclay group will perform the essential measurement of the ionization/recoil energy ratio.

To go to higher masses to search for technicolor particles, or for more sensitivity to supersymmetric particles, reduction of the constant background is important. Preliminary results indicate that this is not due primarily to recoils from neutrons, nor from the pile-up of Compton tails, since relatively few of these events are vetoed by the NaI. This is the next mystery to be solved in the continuing quest for lower backgrounds in the exciting search for dark matter.

1. S.P. Ahlen *et al.*, *Phys. Lett.* **195**, 603 (1987).
2. D.O. Caldwell *et al.*, *Phys. Rev. Lett.* **61**, 510 (1988).
3. D.O. Caldwell *et al.*, *Phys. Rev. Lett.* **59**, 419 (1987); *Nucl. Instr. Meth.* **A264**, 106 (1988).
4. C.J. Martoff, *Science* **237**, 507 (1987).
5. C. Chasman in "Penetration of Charged Particles in Matter—A Symposium", Nat. Acad. of Sci. (1970), p. 16 and references therein.
6. B. Sadoulet *et al.*, *Ap. J.* **324**, L75 (1988).

Detection of Dark Matter Using Low Pressure Gas Detectors (TPC's)

G. Masek, K. Buckland, M. Mojaver
Physics Department, University of California, San Diego 92093

We report on experimental work to evaluate the feasibility of using low pressure gas detectors (TPC'S) as detectors of WIMP dark matter, by detecting the elastic recoil of some normal matter target atom which has interacted with a WIMP. A low pressure TPC may have several advantages over other WIMP detectors: (1) *Directionality*— Not only is the energy of the recoil measured, but the direction as well. This affords a positive signature of WIMPS, for the mean recoil direction would vary with a 24 hour period. (2)*Background Identification*— The main background may arise from Compton scattering, producing a recoil electron whose track characteristics may be quite different from those of a heavy atom recoil of comparable energy. In addition, a magnetic field can be employed to further distinguish between electrons and heavy atom recoils. A monte carlo calculation has been used to determine operating parameters for such a detector. It indicates that a practical kilogram detector could be built using light target nuclei (e.g. He or H) and at pressures ~ .025 atmosphere. Thus figure 1 shows centimeter-scale ranges(large enough to be detected) for typical recoil energies (1—15 kev). Also it should be noted that dE/dx due to atomic electrons (dashed line), which leads to ionization, decreases with decreasing energy, allowing a signature of the direction of the recoil. Implicit in these calculations is the assumption that such recoils will produce a measurable track. Below we describe experimental tests made to observe tracks produced in a TPC by kilovolt ions coming to rest in a gas.

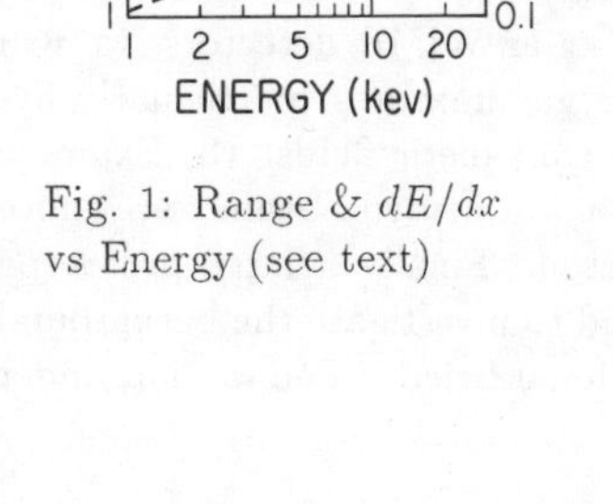

Fig. 1: Range & dE/dx vs Energy (see text)

A 0–50 kev accelerator (Fig. 2) injects Li^+ ions into a test chamber filled with the target gas (He:CH4–50:50 @.025 atmosphere). Differential pumping through two small holes allows the ions to be extracted from the accelerator into the test chamber without passing through any material windows. In the test chamber, the ions are directed into the field cage region of the TPC. The resulting ion track drifts to the bottom where it is detected by a parallel plate avalanche chamber followed by a MWPC, achieving an overall gas gain of 10^5. Figure 3 shows a typical response (pulse height vs time) of 8 sense wires (2mm spacing) to a 10 kev Li^+ ion directed parallel to the sense plane, and coming to rest between the 7th and 8th wires. Several things to note are: (1) The pulses on the first few wires represent about 35 ion pairs of primary ionization. (2) Knowing the drift velocity (4cm/μ sec), the variation of the onset time of the individual pulses gives a measure of deviation of the track from a straight line. An analysis on 200 tracks gives a spread of $\pm 20°$ about the initial direction. (3) The pulse heights decrease (wires 5,6,& 7) towards the end of the range–clearly identifying the ion going from wire 1 to wire 8, the direction of the injected beam. Thus these tests show that the necessary characteristics of these kev ion tracks are maintained and measureable over drift distances of 10 cm.

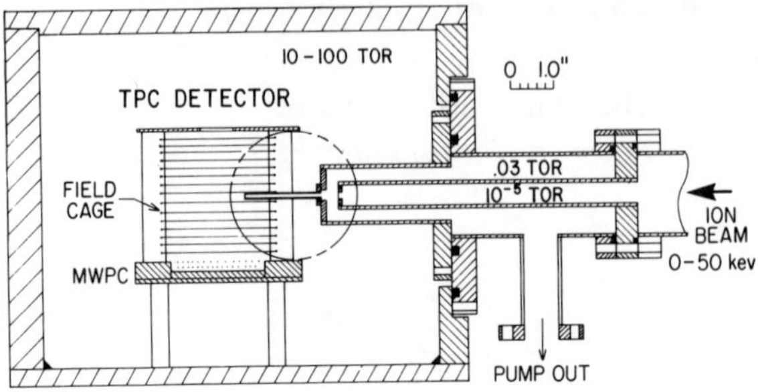

Fig. 2: Set up for observing low energy (kev) ion tracks. The circled region is the locaton of the differential pumping holes.

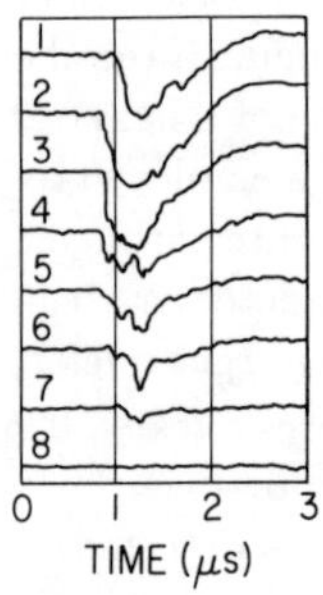

Fig. 3: Typical response to 10 kev Li^+ ions. Sense wires are numbered to the left.

To further demonstrate the feasibility of TPC's as WIMP detectors, we will use the present apparatus to: (1) Examine other gas mixtures, particularly hydrogen. (2) Run with electrons, with and without magnetic fields. (3) Examine the response to neutral atoms. Later tests would use a neutron beam to produce recoils, more closely simulating a WIMP interacton. Finally a large prototype needs to be constructed and operated underground to investigate the background problems. [This work was supported by the National Science Foundation under Grant No. PHY 85–17279.]

Dark Matter Detection with Hydrogen Proportional Counters

G. Gerbier, J.Rich, M.Spiro, and C.Tao

DPhPE, CEN Saclay, 91191 Gif-sur-Yvette, France

Most dark matter detection schemes [1] use semiconductor or cryogenic detectors [2]. In this paper we discuss the possibility of using proportional chambers [3,4] filled with a hydrogenous gas to detect protons recoiling from collisions with dark-matter particles. For dark matter trapped gravitationally in our galaxy [5], such collisions result in a roughly exponential distribution of recoil energies with an average energy near 2 keV (assuming a particle mass greater than the proton mass). Proportional chambers filled with H_2 or CH_4 have been shown in neutron scattering experiments to be sensitive to recoil energies down to 1 keV [6].

The use of protons rather than heavy nuclei as the target offers some theoretical advantages. This is especially the case for "Cosmions", particles invented to solve the solar neutrino problem [7]. In order that they be trapped in the Sun in sufficient numbers, Cosmions must have a mass in the range 4-10 GeV and a cross section on hydrogen near $10^{-36}cm^2$. An experiment using hydrogen that was sensitive to such particles could rule out the Cosmion hypothesis independent of assumptions about their cross section on heavy nuclei.

For Cosmion scattering on hydrogen the total rate is near $0.2(gram \cdot day)^{-1}$. The density of H_2 at atmospheric pressure is $90g/M^3$, but, for the background reasons discussed below, it may be advantageous to work at low pressure (.02 atm) where the H_2 density is $2g/M^3$. Hence, Cosmions can be studied with low-pressure counters [8] in the form of a time-projection-chamber (TPC) [9] of a few m^2. It would, however, require a considerable effort to construct a TPC sensitive to photinos, which have cross sections at least two orders of magnitude smaller.

The background will come from Compton scatters of gamma rays due to the ambiant radioactivity. In the keV range, the lowest rate achieved for this process is of order $10^{-3}(g \cdot day \cdot keV)^{-1}$ in the germanium diode detectors used for dark matter searches [10]. This rate is below that for Cosmion scattering, about $0.1(g \cdot day \cdot keV)^{-1}$, but these detectors have thresholds too high to be sensitive to particles with masses below 10 GeV. For proportional chambers, the lowest rates are near $10(g \cdot day \cdot keV)^{-1}$ in the $1cm^3$ counters used in radiochemical solar neutrino experiments [11]. Fortunately, the use of a low pressure TPC placed in a 0.1T magnetic field might eliminate much of the Compton background since Compton electrons will spiral in tight orbits while recoil protons of the same energy will have a much straighter trajectory. (The range for a 1 keV proton in .02 atm H_2 is near 1cm.)

The galactic origin of the recoil protons can be established because their direction is preferentially opposite the direction of the solar-system's movement through the galaxy [12]. The recoil direction can be determined in a TPC from the ionization pattern since, in the keV range, specific ionization falls with decreasing energy [6,13].

Work now in progress at Saclay and at San Diego [4] should determine if it is possible to achieve sufficiently low background with low-pressure TPC's.

Acknowlegements : We gratefully acknowledge conversations with A. Breskin, P. Mangeot, G. Masek, B. Sadoulet, P. Smith, and D.Spergel.

References

[1] M.W.Goodman and E.Witten, Phys.Rev. D31 (1985) 3059.

[2] For a reviews of dark matter detection techniques see P.F.Smith, Rutherford Lab preprint RAL-88-045; J.Rich Saclay preprint DPhPE-88-10; B.Sadoulet, LBL report LBL-24302; and B.Cabrera in these proceedings.

[3] J.Rich and M.Spiro, Saclay report DPhPE88-04 (1988).

[4] G.Masek, in these proceedings.

[5] A.K.Drukier, K.Freese and D.N.Spergel, Phys.Rev. D33 (1986) 3495; K.Freese, J.Freedman and A.Gould, SLAC preprint SLAC-PUB-4427 (1987).

[6] V.V.Verbinski and R.Giovannini, Nucl. Inst. and Meth. 114 (1974),205.

[7] J.Faulkner and R.L.Gilliland, Astrophys.J. 299 (1985)994; D.N.Spergel and W.H.Press, Astrophys.J. 294 (1985) 663; W.H.Press and D.N.Spergel, Astrophys.J. 296 (1985) 679; R.L.Gilliland et al., Astrophys.J. 306 (1986) 703; D.Spergel, in these proceedings; P.Salati, in these proceedings.

[8] A. Breskin, Nucl. Inst. Meth. 196 (1982) 11.

[9] J.A.Macdonald (editor) The Time Projection Chamber, (AIP Conf. Proc. no. 108) (American Inst. of Physics, New York,1984).

[10] D.O.Caldwell et al., Phys.Rev.Lett. 61 (1988) 510.

[11] T.Kirsten in '86 Massive Neutrinos in Astrophysics and in Particle Physics, ed. O.Fackler and J.Tran Thanh Van (Editions Frontieres,1986).

[12] D.N. Spergel, Phys.Rev. D37 (1988) 1353.

[13] H.H.Anderson and J.F.Ziegler, The Stopping and Ranges of Ions in Matter , vol. 3 (Pergamon Press,New York, 1977).

ACOUSTIC DETECTION OF WEAKLY INTERACTING PARTICLES

B. Cabrera, A. T. Lee, C. J. Martoff and B. A. Young
Physics Department, Stanford University, Stanford, California 94305

B. Neuhauser
Department of Physics and Astronomy, San Francisco State University
San Francisco, California 94132

and

J. P. McVittie
Center for Integrated Systems, Stanford University, Stanford, California 94305

1. ABSTRACT

We are developing a new class of elementary particle detectors capable of detecting weakly interacting particles such as neutrinos. These detector are based on the ballistic propagation of phonons in silicon crystals at low temperatures.

2. INTRODUCTION

Our group has proposed [1] direct sensing of the ballistic phonons produced by an event in a large insulating single crystal. Since this wavefront carries information on the event total energy and location within these silicon crystal acoustic detectors (SiCADs), substantial improvements in background rejection are possible over detection schemes which measure total energy deposition only. A threshold of 1 keV or better is important for using SiCADs as neutrino detectors and for dark matter searches.

3. EXPERIMENTS WITH TRANSITION EDGE PHONON SENSORS

Our first generation of SiCADs utilize superconducting transition edge devices as phonon senors [2]. These devices consist of thin superconducting lines deposited on the surfaces of the silicon crystals and biased with a constant current. For currents below a critical current, self-terminating voltage pulses are observed. These pulses are caused by the heating of line segments above the superconducting transition temperature. The sensors are made by depositing 40 nm of titanium on crystalline silicon wafers which are 1 mm thick. A meander pattern (Fig 1a) is produced using conventional photolithographic techniques and consists of 299 parallel lines each 2 μm wide on 5 μm centers. The active area of the pattern is 4.5 mm long and 1.5 mm wide. The normal resistance just above the superconducting transition temperature ($T_c \approx 312$ mK; 5 mK wide from 10 to 90% of resistance) is ~ 18 kΩ per line.

The two recent x-ray experiments are described here and were performed with a 24 μCi source of ^{241}Am. The decay spectrum is dominated by alpha particles at ≈ 5.5 MeV, and the source emits a nuclear gamma at 60 keV and two atomic x-rays at 14 and 18 keV. We used 125 μm thick foils of Sn or Pb as an absorber placed in front of the source. These foils readily stop all of the alphas and the 14 and 18 keV x-rays before they reach the sensor; however the more penetrating 60 keV gamma rays pass through the absorber.

Using a cryopumped ^{3}He refrigerator at 0.3 K we biased the Ti/Si devices at the foot of the resistive transition in temperature and below the latching critical current in bias current, and we observe self-terminating voltage pulses when the film is bombarded by phonons produced by the interaction of an incident x-ray (or gamma ray) within the Si crystal. These pulses occur because enough phonon energy from the

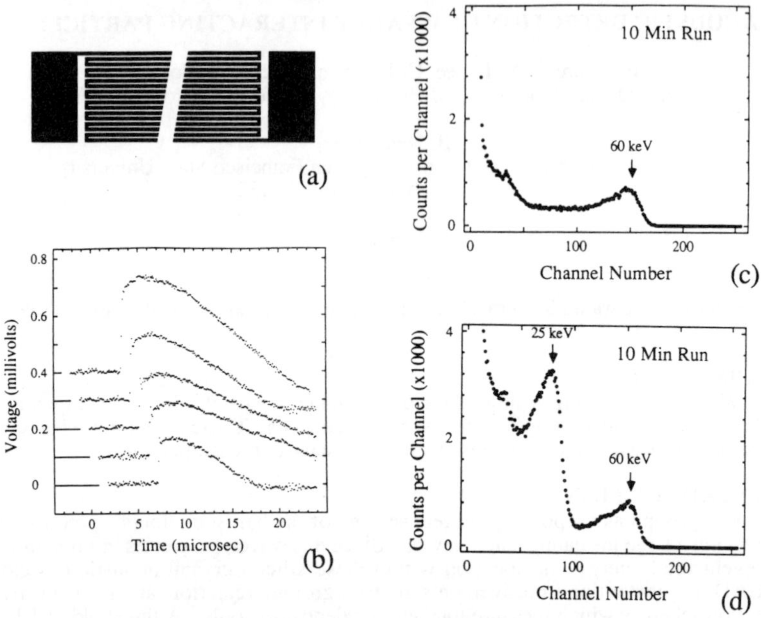

Figure 1.(a) Transition edge sensor pattern. (b) Typical pulses from 60 and 25 keV x-rays with 120 nA bias current. (c) Pulse height spectrum of 60 keV x-rays through Pb absorber and (d) with additional 25 keV x-rays from Sn absorber.

photoelectron in the silicon reaches the Ti film to drive portions of the film normal. The threshold energy density is about 0.5 eV/μm^2.

Fig 1b shows (for I ≈ 120 nA and T ≈ 286 mK) some typical single pulses with electronics-limited risetimes of ≈ 140 ns. Fig 1c is a pulse height spectrum obtained for the case of a 125μm thick Pb absorber. The prominent upper peak is the photopeak due to 60 keV gamma rays, and the sharp lower feature is consistent with the position of the Compton edge for 60 keV gamma rays at 11 keV. In Fig 1d, we show a pulse height spectrum with the Sn absorber. The features due to the 11 keV and 60 keV interactions appear as they did with the Pb absorber. Furthermore, we observe an additional peak around 25 keV due to secondary emission of Kα x-rays from the Sn. We find that the peaks in the pulse height spectrum are proportional to x-ray energy. We can qualitatively reproduce these spectra by applying Monte Carlo methods to a simple model of the detector response.

4. CONCLUSIONS

Transition edge sensors can be used for imaging low threshold detectors made of silicon crystals. These crystal acoustic detectors will set improved limits on the WIMP dark matter candidates, and can be used for neutrino experiments at reactors. Our work at Stanford is supported in part by DOE Contract # DE-AT03-82ER40-076.

5. REFERENCES

[1] B. Cabrera, et al, *Nucl. Instr. Meth.* **A275**, 97 (1989).
[2] B. A. Young, et al, *IEEE Trans. Mag.*, in press.

DARK MATTER DETECTOR DEVELOPMENT IN THE UK

P F Smith

(Rutherford Appleton Laboratory, Chilton, Oxfordshire, England)

OBJECTIVES

Studies of Galactic dark matter detection have been in progress at RAL since 1981, initially for neutrinos or light axions and subsequently including photinos and other heavy particle candidates. In 1985, an interdisciplinary collaboration (comprising physicists from a number of particle physics, astrophysics, and low temperature physics groups) was established to obtain funding for the development of low temperature dark matter detectors. Since 1987, this development work has received support from both particle physics and astronomy funding sources. The collaboration currently involves ICST (London), RHBNC (London), Birkbeck College (London), QMC (London), Sussex, Cambridge, and RAL.

The present objective is to develop a detection system capable of observing axially-coupled heavy particles (in particular photinos) at an event rate of order 1/day in a 1kg target. Current plans are to site such a detector in a 3000 mwe salt mine in the UK, where measurements have identified a low background region, and a new underground laboratory is being set up with the co-operation of the mine management. In this location, the background of muon-induced neutrons will be negligible, so that from the detector design viewpoint there will be no competing nuclear recoil events and the background will result mainly from radioactivity in the shielding and detector, as in the case of double beta decay experiments.

Development work in progress is directed towards both low temperature bolometric detectors and ionization detectors, the latter providing options either for dark matter detection or, in combination with bolometric detection, as a gamma veto [1]. Because of the basic requirement of detecting axially-coupled particles, our work is directed specifically towards crystalline target nuclei with spin - eg LiF, GaAs, CsI, in the case of bolometric detection, and GaAs, GaP, in the case of ionization. We are also interested in the possibilities of gaseous proportional chambers as detectors which could provide some discrimination between nuclear recoil and ionization events.

BOLOMETRIC DETECTOR DEVELOPMENT

We are aiming at a sensitivity in the region 1–10keV for an individual target crystal of mass ~10g. Our first tests in 1987 demonstrated thermal pulses from neutrons incident on a 1g silicon target, using a graphite layer thermometer. We have also carried out tests with carbon fibres, thin slices from carbon resistors, doped Ge and doped Si sensors, all of which have demonstrated the power sensitivity known as the "hot electron effect", and which limits the bias power typically to $< 10^{-12}$W. Nevertheless, recent results by various groups indicate that the required energy resolution/mass should still be achievable. Since a dark matter target may require perhaps ~100 individual modules, or larger modules equipped with many sensors, we have given priority to the development of a reliable and reproducible semiconducting temperature sensor which can be manufactured in large numbers at reasonable cost. Using the microelectronics fabrication facilities at RAL and Southampton University, we have manufactured arrays of doped thermistors on standard 10cm Si wafers. Each wafer contains about 150 devices, and the doping level and geometry can be varied to optimize both thermometer slope and absolute resistance value [2]. Tests with these sensors on 1–10g crystals of Si and LiF are currently in preparation.

IONIZATION DETECTOR DEVELOPMENT

Ionization measurement corresponding to a sensitivity of a few keV in 10–100g is well-known to be achievable in Si and Ge, but has not previously been attempted in large crystals of odd-A materials such as GaAs. Nevertheless, some types of commercially-available GaAs and GaP become insulating at temperatures ~ 10–20K (thus providing a large effective depletion depth) and have a sufficiently long carrier lifetime (~0.1–1 μs) to allow, in principle, separation and measurement of the carriers produced by an ionizing collision. Experiments using both capacitative and ohmic coupling are in progress, and if successful could offer either a possible odd-A detector or, with thermal sensors on the same crystal, the possibility of gamma discrimination by simultaneous measurement of ionization and thermal energy. The latter is complicated by the fact that the energy produced by ionization measurement process itself may dominate the thermal energy, although this phenomenon itself provides thermal amplification and yet another detection option (see contribution by F Goulding at this workshop).

REFERENCES

1 P F Smith, J D Lewin, preprint RAL-88-045 (1988).
2 C C Zammit et al., Proc.MRS Workshop (Strasbourg, 1988).

PARTICLE DETECTION VIA ROTONS IN HELIUM-4 SUPERFLUID

R. E. Lanou, H. J. Maris, F. S. Porter, G M. Seidel, C. A. Smith, R. H. Torii
Department of Physics, Brown University Providence, RI 02912 USA

Any experiment to detect recoils from solar neutrinos or from elementary particle dark matter faces severe constraints. The combination of flux, cross section and low kinetic energy of these particles leads to expectations of low event rates and low energy deposits in any detector. Consequently a suitable detector for such dark matter should possess a low content of intrinsic or spallation produced radioactivity, a low energy threshold, good energy resolution, good sensitivity (size vs. rate), a choice of integer or half integer spin targets, real time detection and good signal to noise discrimination. The design choices for any detector for this purpose depend on optimization of these competing requirements.

Recently, we have begun work on a new calorimetric method which is quite different from any of those previously discussed. It is based upon the properties of liquid ^{4}He and its bulk excitations at very low temperatures. Our primary motivation has been to develop a device suitable for studying the spectra of solar neutrinos down to the very lowest energies — that is, those of the p–p and ^{7}Be reactions. Such a solar detector consisting of 10 tons of ^{4}He would collect 20 events per day from those two reactions. The same properties of superfluid helium which make it useful for solar neutrinos should make it an excellent candidate for certain types of dark matter searches; but on a smaller scale — typically, less than 5 kilograms might produce 100 events per day. Also a detector of roughly 150 kg placed at a high flux reactor could test for the existence of coherent neutral current effects at a predicted rate of 700 events/day.

In general terms, ^{4}He is attractive because it is inexpensive ($\sim$ \$4 per liter), clean, quiet and simple. ^{4}He at low temperature ($<$ 100 mK) is the purest material known. All other atomic species freeze out as helium remains liquid to absolute zero unless pressurized.

Radioactive backgrounds are the principal limitations on solar neutrino and dark matter experiments. This background has three sources: a) long-lived radioactive isotopes *intrinsic* to the detector medium, b) long-lived isotopes made *cosmogenically* while the materials are above ground, and c) neutrons and gammas from surrounding material. Helium has no natural radioactive isotopes and its first excited state is 21 MeV. So the chief concern for any helium detector are gammas generated in its container and its immediate shielding. Rates from copper dewars seem to be quite manageable (0.09 decays/min/kg).

If a neutrino or dark matter candidate interacts with the atoms of a helium detector there will be a recoil — typically an electron or an entire nucleus — and it is the kinetic energy of the recoil which we must detect and measure. These energies are very small ranging from a few kilovolts to electron volts.

In liquid helium the recoiling particle deposits most of its energy into the form of bulk excitations called rotons in less than 10^{-6} sec. Only about 0.7 milli-electron volts are required to make the majority of these carriers so a one kilovolt recoil would produce $\sim 10^6$ such carriers thus leading to good statistical resolution $\left[\Delta E/E = \sqrt{10^6} \times (.7meV/1KeV) \simeq 0.1\%\right]$. These carriers propagate ballistically at about 100 m/sec which, if they can be efficiently collected, permit real time detection and measurement of deposited energy (see Fig. 1).

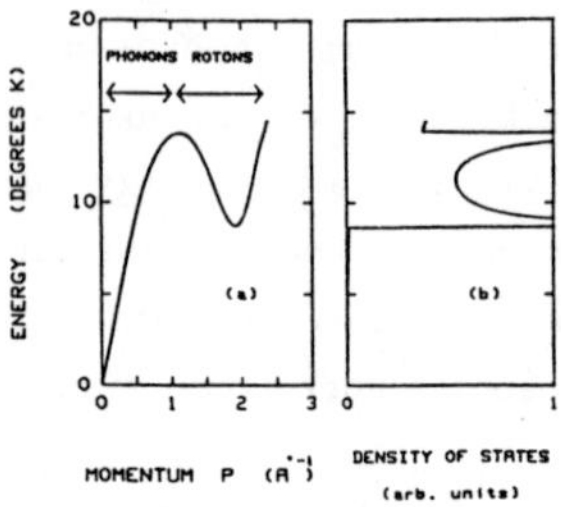

The challenge then is to collect these rotons in a realistic device. Briefly (see Ref. 1 for details) the sequence of events would be as follows: a) kinetic energy of the recoil from an interaction is converted to mechanical excitations in the helium, b) the large population of rotons propagate ballistically to the free surface, c) at the free surface $\sim 1/6$ of the rotons desorb ^{4}He atoms, d) the ^{4}He atoms, via a large van der Waals interaction, are physiosorbed onto the surface of thin silicon wafers suspended a few millimeters above the surface, e) in a few milliseconds the binding energy released into the silicon results in a measurable temperature rise (θ_D (silicon) $= 636^o K$). The ultimate energy resolution achievable by this technique will probably depend upon how well geometric effects and losses at the walls can be controlled. Other superfluid phenomena may provide mechanisms for discrimination between recoiling nuclei and Compton electrons.

We have begun construction of a 4.5 liter (0.65 kg) helium device to test the principles and obtain quantitative information needed for establishing the feasibility of a really large detector. We expect the cryogenics to be complete by this summer with cold testing to begin at that time. Many of the other facilities are already in place. (This work is supported in part by the U.S. Department of Energy.)

1. Lanou, R. E., Maris, H. J., Seidel, G. M., in Phys. Rev. Lett. 58, 2498 (1987); in "Low Temperature Detectors for Neutrinos and Dark Matter", p. 150, Editors Pretzl, K., Schmitz, N., Stodolsky, L., Springer–Verlag (1987); and in "Superfluid Helium as a Dark Matter Detector", Proceedings of XXIII Rencontre de Moriond (1988, in press).

TOWARDS THE SEARCH FOR WIMP'S USING A SUPERCONDUCTING GRANULAR DETECTOR OPERATED IN LIQUID HELIUM-4

A. K. Drukier[(1)], A. daSilva[(2)],
A. Kotlicki[(2)], M. Legros[(2)], and B. Turrell[(2)]

(1) Applied Research Corporation, 8201 Corporate Drive, Suite 920, Landover, Maryland 20785
(2) Physics Department, University of British Columbia, Vancouver, British Columbia, V6T 2A6 Canada

The use of scattering on nuclei may be the best way to detect weakly interacting particles[(1)], including the cold dark matter (CDM) candidates[(2,3)]. The technical requirements are:

- very low radioactive background;
- very good energy sensitivity,
- possibility to choose between diverse target nuclei.

The first limits[(4)] on WIMP's were obtained using ultralow radioactive background semiconducting detector, but to obtain limits on the low mass CDM candidates, cryogenic detectors should be used.

Two classes of cryogenic detectors based on dielectric crystals and superconductors are being considered. Milligram mass bolometric detectors have[(5)] much better energy sensitivity than the best semiconducting detectors. A few gram mass bolometers are being developed but still do not reach the required (sub-keV) energy sensitivity. Furthermore, they must be operated in dilution refrigerators which are "hot" from the radioactive point-of-view. For example, even the specially constructed dilution refrigerator of Milano's group has the radioactive background which is orders of magnitude higher than the expected count-rate due to WIMP's.

The Superconducting Granular Detector (SGD) depends on the change of state from superconducting to normal in small grains of few microns in diameter[(6)]. The deposition of a few hundreds of eV of energy changes the state of the grain; the readout is by magnetic means, i.e., trillions of grains are read out in parallel by a single channel of very sensitive electronics. Actually, our experiments[(7)] suggest that a Superconducting Quantum Interference Device (SQUID) can sense the change of state of a single grain in a few gram detector. This detector has been shown to be sensitive to charged particles, photons and neutrons. The extrapolation of recent experimental results suggest that even for a sub-keV energy deposition very good energy resolution and spatial resolution are possible, say $\Delta E/E \leqslant 5\%$ and $\Delta X \approx 20\text{-}50$ m. Furthermore, SGD permits excellent rejection of radioactive background[1]. WIMP's interacting with nuclei will flip <u>one and only one grain</u>, whereas radioactive background events will induce multiple flips.

The principal technological challenge of the SGD development is that both the energy threshold and the amplitude of the electronic signal due to grain flipping are proportional to the grain volume. The energy threshold is given by

$$E_{th} < 0.5\, V_{grain} * C_v(T) * \Delta T \qquad (1)$$

wherein $C_v(T)$ is specific heat at temperature T and ΔT is the temperature barrier characterizing the detector quality. Typically, $\Delta T/T \approx 2\text{-}10\%$ but recently we reached $\Delta T/T \approx 2*10^{-3}$. At very low temperature the specific heat of metals is dominated by electrons, i.e., it is proportional to T. For example, it is 30 eV/(m^3 x ^oK) and 2 keV/($m^3 * {}^oK$) at 50 mK and 1.2 K, respectively.

In the last year we have made considerable progress:

- the minimum detectable grain volume was diminished[7] by a factor of 10 to $V_{grain} \approx 10 \mu m^3$ using RF-SQUID electronics. Further order of magnitude improvement is imminent with the use of a DC-SQUID;

- the quality of detector was improved by a factor of about 20 to T<5mK.

The estimated energy threshold at T=1.2^{o}K is currently below 100 eV. Thus the <u>search for WIMP's using SGD can be performed in pumped liquid helium-4</u>. This should permit larger detector mass and much lower radioactive background.

In Figure 1 we present a typical signal due to the flip of a <u>single</u> R=8 m grain. This flip was induced by irradiation at T=3.1^{o}K of a collection of millions of grains placed within a pickup loop whose volume was about 1cm^3. The energy deposition was approximately 10keV; note the excellent signal/noise > 350.

REFERENCES

1) Drukier, A. K. and Stodolsky, L., Phys. Rev. D30, 2295 (1984).

2) Goodman, M. W. and Witten, E., Phys. Rev. D31, 3059 (1985).

3) Drukier, A. K., Freese, K. and D. Spergel, Phys. Rev. D33, 3495 (1986).

4) Ahlen, S. P. et al., Phys. Lett. B, 195, 603 (1987).

5) Moseley, S. H. et al., J. Appl. Phys. 56, 1257 (1984).

6) See papers by Drukier, A. K. et al.; Pretzel, K. P.; Bellefond, A. et al.; Gonzales-Mestres L. et al.; and Freund P. et al.; contributed to Conf. on Superconductive Particle Detectors, Oct. 87 ed. A. Barone, pub. World Scientific.

7) Kotlicki, A. et al., Proc. of the 2nd European Conference on Low Temperature Detectors..., Annency-Le-Vieux, Mai 1988, in printing.

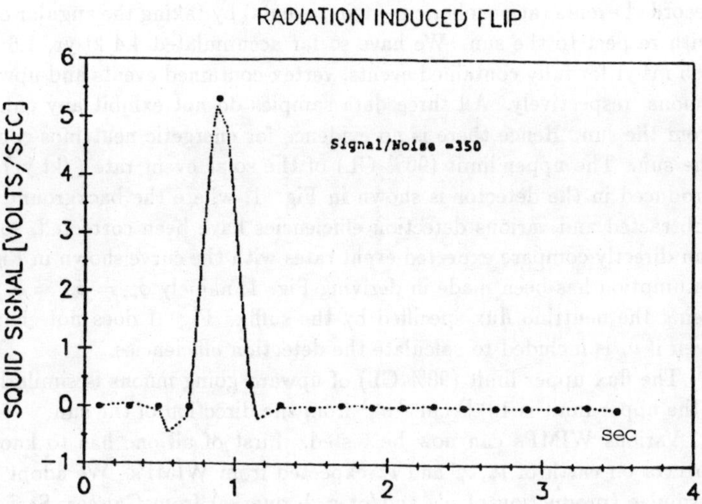

High Energy Neutrinos from the Sun : Kamiokande Results

Y. Totsuka
Institute for Cosmic Ray Research, University of Tokyo
3-2-1 Midori-cho, Tanashi, 188 Tokyo, Japan

The Kamiokande detector is of a water Cerenkov type with a sensitive volume of 2,140 t, viewed by 948 20$''\phi$ photomultipliers. It is capable of recording neutrino interactions with energies ranging from 7.5 MeV all the way to more than 100 GeV (one event with 80 GeV has been observed). They are not necessarily contained fully in the detector. Some of them have energetic secondaries that escape out of the detector so that only the lower limits of their energies are determined. These so-called vertex-contained events are equally picked up in the detector. Upward-going muons are another data sample gathered by the experiment. They are presumably produced by high energy $\nu_\mu(E_\nu \geq 100$ GeV) in the rock. They carry a new piece of information complementary to those produced within the detector.

High energy neutrinos from the sun, if any, can be picked up among the recorded events (atmospheric neutrino events) by taking the angular correlation with respect to the sun. We have so far accumulated 4.4 kt·yr, 1.6 kt·yr and 165 m^2·yr for fully contained events, vertex-contained events and upward-going muons, respectively. All three data samples do not exhibit any enhancement from the sun. Hence there is no evidence for energetic neutrinos coming from the sun. The upper limit (90% CL) of the solar event rate (/kt·yr) for those produced in the detector is shown in Fig. 1, where the background has been subtracted and various detection efficiencies have been corrected, so that one can directly compare expected event rates with the curve shown in Fig. 1. One assumption has been made in deriving Fig. 1, namely $\phi_{\nu_e} = \phi_{\bar{\nu_e}} = \phi_{\nu_\mu} = \phi_{\bar{\nu_\mu}}$, ϕ being the neutrino flux specified by the suffix. Fig. 1 does not change much even if ν_τ is included to calculate the detection efficiencies.

The flux upper limit (90% CL) of upward-going muons is similarly derived : the upper limit is 10^{-13} cm^{-2} s^{-1} from the direction of the sun.

Various WIMPs can now be tested. First of all one has to know energy spectra on earth of ν_e, ν_μ and ν_τ expected from WIMPs. We adopt the basic formulae (production of ν's and/or c, b quarks) from Gaisser, Staigman and Tilav[1] with some corrections made by Gould[2]. Decays of c and b quarks are controled by the Monte Carlo program developed by the Lund group[3]. Event rates (/kt·yr) in water and fluxes of upward-going muons are then calculated and compared with the data. More specifically one opens a window in the visible energy distribution of the data to optimize the neutrino detection and background rejection. The window is of course dependent on WIMP mass. Correcting the efficiency for the window, the data finally confront the expected curve. An example is shown in Fig.2 for massive Dirac neutrinos (ν_D).

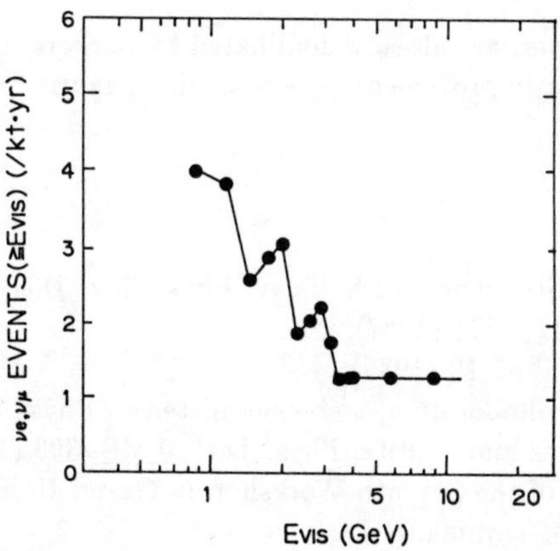

Fig.1. Upper limit (90% CL) of integrated event rate for events produced in the detector as a function of visible enegy. All relevant efficiencies are corrected and background subtracted.

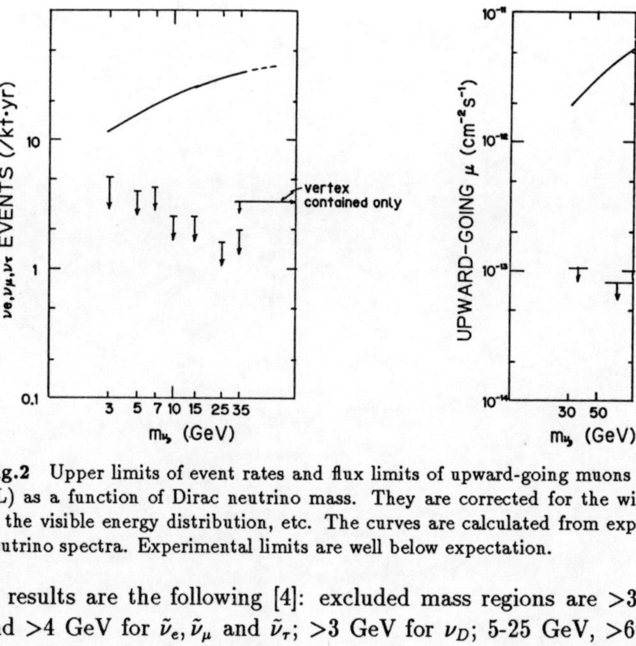

Fig.2 Upper limits of event rates and flux limits of upward-going muons (90% CL) as a function of Dirac neutrino mass. They are corrected for the window in the visible energy distribution, etc. The curves are calculated from expected neutrino spectra. Experimental limits are well below expectation.

The results are the following [4]: excluded mass regions are >3 GeV, >3 GeV and >4 GeV for $\tilde{\nu}_e, \tilde{\nu}_\mu$ and $\tilde{\nu}_\tau$; >3 GeV for ν_D; 5-25 GeV, >60 GeV for Majorana neutrinos (ν_M), no limits on $\tilde{\gamma}$, where the EMC effect[5] is taken into account for ν_M and $\tilde{\gamma}$, and the symmetry between ν_D and $\bar{\nu}_D$ is assumed which however may not be the case.

The results, however, are already dominated by background. Therefore one expects only a modest improvement even with the gigantic Super-Kamiokande detector (22,000 t)[6].

References

1. T. K. Gaisser, G. Steigman and S. Tilav, Phys. Rev. D34, 2206 (1986).
2. A. Gould, Ap. J. 321, 571 (1987).
3. T. Sjöstrand, LUTP 85-10 (1985).
4. The Kamiokande collaboration, to be submitted to Phys. Rev. D. (1989).
5. J. Ellis, R. A. Flores and S. Ritz, Phys. Lett. 198B, 393 (1987).
6. Y. Totsuka, Proc. of the Seventh Workshop in Grand Unification / ICOBAN'86, 1986, Toyama, p118.

BOUNDS ON THE LIGHTEST NEUTRALINO AS THE DARK MATTER

G. Gelmini[1,2], P. Gondolo[3], E. Roulet[1]

[1]International School for Advanced Studies (SISSA), Trieste, Italy.

[2]International Center for Theoretical Physics (ICTP), Trieste, Italy.
Associated to the INFN. On leave of absence from
Physics Department, University of Rome II, Roma, Italy.

[3]Astronomy Department, University of Trieste, Trieste, Italy.

A neutralino (supersymmetric partner of a neutral boson) is the most favoured candidate for the lightest supersymmetric particle (LSP),supersymmetry's candidate for dark matter (DM). In the Minimal Supersymmetric Extension of the Standard Model there are only four neutralinos: the photino,$\tilde{\gamma}$, the zedino, $\tilde{Z}$, and two Higgsinos, $\tilde{H}_1$, $\tilde{H}_2$. In this model the minimal number of new particles are added to the Standard Model[1]. With the usual assumption of a grand unified symmetry at some large scale, only five new parameters (in addition to those of the Standard Model) are needed to specify the model: three mass parameters, usually called M_2, μ, m, and two dimensionless numbers. The neutralino and chargino (supersymmetric partner of a charged boson) mass matrices depend on three parameters: M_2, μ, and the ratio of the two vacuum expectation values v_2/v_1.[1] The parameter M_2 is positive while μ can be positive or negative and phenomenology dictates that the ratio v_2/v_1 ranges from 1 to 20. If the parameter M_2 goes to zero the lightest neutralino, χ, is a pure $\tilde{\gamma}$, with mass of order M_2, while if μ goes to zero χ is a pure $\tilde{H}$, with mass of order μ. For general values of M_2 and μ, χ is an admixture. While a pure $\tilde{\gamma}$ or $\tilde{H}$ does not have coherent interactions with nuclei a general χ does, through squarks (Griest[2]) and Higgs exchange (Barbieri *et al* [3]). The existence of χ's coherent interactions increases the event rates in direct or indirect searches of χ particles in the halo of our galaxy. In direct searches the energy deposited by a χ in a collision with a nucleous is looked for, using existing Ge-spectrometers or future techniques. In indirect searches the productsof the annihilations of χ's are looked for. For example energetic neutrinos produced in annihilations of the halo DM particles trapped in the earth or the sun may be observed in the large underground proton decay experiments. Large signals are expected for small Higgs bosons masses, i.e. if the lightest Higgs field has a mass M_H of a few 10 GeV. Here we present bounds from indirect [4] searches on a multiparameter space: M_2, μ, v_2/v_1, M_H, and the lightest sfermion mass. Other bounds on the same parameters have been recently obtained[3] from direct DM searches with a Ge-spectrometer and from accelerators. Fixing the last three parameters within their phenomenologycally allowed values, the bounds in the μ, M_2 are obtained. As an example we present in the figures the case of $v_2/v_1 = 8$. In Fig. 1, the χ mass in GeV's is shown as a function of μ and M_2, and also the regions where $\tilde{\chi}$ is mainly $\tilde{\gamma}$, $\tilde{H}$, or $\tilde{Z}$. No indirect bounds can be obtained below the "evaporation" mass: 10 GeV for the earth and 4 GeV for the sun. The densely-hatched central region of Figure 2 is excluded by accelerator bounds because the lightest chargino mass is smaller than 30 GeV[3]. The bounds on solar and terrestrial annihilations of χ's for a Higgs mass of 10 GeV and infinite sfermion masses are also shown in Fig. 2. They exclude almost entirely the remaining area. Taking the lightest possible sfermions modifies (improves) slightly the bounds only in the region $M_2 < 30$ GeV. The limits are instead very sensitive to the Higgs boson mass (in this case, there are bounds only for $M_H < 25$ GeV). The excluded largely-hatched region corresponds to a rate of contained events from the sun in the Frejus experiment larger that 3.3 kt^{-1} yr^{-1} [5] (a similar area is excluded by the direct DM searches with a Ge-experiment[3]). The crossed region corresponds to a flux of up-going muons from the earth at IMB larger than 26.5 $10^{-14}cm^{-2}$ sec^{-1} (while the dotted line corresponds to 1 in the same units, the level necessary to get bounds comparable to those obtained from the sun)[6]. The new experimental results presented at this workshop by the Frejus and Kamiokande experiments would improved these limits.

REFERENCES

[1] Haber, H.E. and Kane, G.L., Phys. Rep. 117, 75 (1985); Gunion, J.F. and Haber, H.E., Nucl. Phys. B272, 1 (1986); Barbieri, IFUP-TH 33/87, Pisa preprint(to appear in La Rivista del Nuovo Cimento).
[2] Griest, K., Physic. Rev. D38, 2357 (1988).
[3] Barbiere, R., Frigeni, M., and Giudice, G.F., IFUP-TH 8/88, Pisa preprint.
[4] Gelmini, G., Gondolo, P., and Roulet, E., in preparation.
[5] Kuznik, B., Moriond Conference 1988.
[6] Svoboda, R. *et al* , Astrophys. J. 315, 420 (1987).

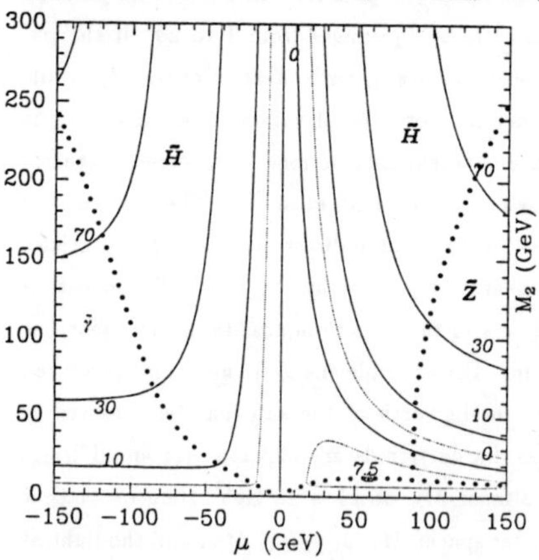

Fig. 1.

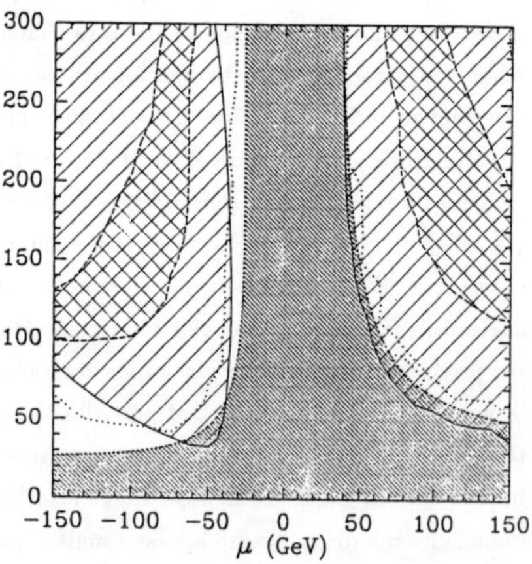

Fig. 2.

IS DARK MATTER INVISIBLE?

K. Zioutas[1], Y. Semertzidis[2)], G. Tsipolitis[3)] and G. Alberis[1]

1) University of Thessaloniki, Thessaloniki, Greece
2) University of Rochester, Rochester, NY, USA
3) McGill University, Montreal, Quebec, Canada

The origin of the missing mass in astrophysics remains enigmatic and either we have to accept another modified version for Newton's gravitational law at large scales or a tremendous mass of exotic particles must coexist in the Universe, which cannot hide everything but their gravitational force from our world. The postulated axion, with spin-parity of 0^- can appear in atomic and hadronic reactions or decays[1,2)]. Thermalized relic axions in the Universe can be the (main?) constituents of the invisible mass. The axion decays into two photons, $a \rightarrow 2\gamma$, and the associated lifetime can exceed the age of the Universe. But a non-uniform magnetic or electric field[3,4)] acts as a converter, inducing a quasi-axion decay to electromagnetic radiation. Experimental investigations[5)] are based on this reaction mechanism and the expected conversion efficiency (CE) is: CE $\approx 10^{-14}$. Such processes can also happen in different places in the Universe[4)], so that in fact we cannot claim any longer that cosmic DM particles (as for instance, the axions) are invisible, if they really exist. Therefore, the question is rather in reverse, i.e. whether at least part of the cosmic electromagnetic radiation is related to DM. This means that cosmic DM is changing in to ordinary matter and vice versa, provided T-reversal is not completely violated in such processes. Nevertheless, such a violation might be the origin of the DM dominance.

Axions can be converted to radiation if there exist: a) strong and non-uniform electric/magnetic fields[3,4)] and b) clustering of DM particles to form massive objects, in order to explain the observed luminosity.

Variable field: Pulsars, black holes, etc. can have magnetic and electric fields of enormous strength. The dipole picture of the neutron star (NS) makes them an almost perfect celestial cavity of $\sim$ 10 km radius[3-5)]. Interstellar clouds and dust consist of chemical elements with large electric dipoles having electric field strengths in their vicinity (a few Å) as strong as 10^{10} V/cm decreasing with r^{-3}. Dust particles can give rise also to ferroelectric and antiferroelectric alignment[6)]. The galactic magnetic field is considered to be homogeneous and therefore cannot contribute to the conversion of axions to photons. However, such clouds filled with electric dipoles ($n \lesssim 10^9/cm^3$) are a place where axions can change into photons. The same arguments are valid also for magnetic dipoles.

DM-clustering: A simple mechanism might trigger gravitational capture or even cause the collapse of DM particles: many reactions and atomic or nuclear transitions taking place in the stars or in their vicinity contribute to the appearance of DM particles with a velocity below the escaping velocity of the star, which is 11.2 km/sec for the earth surface and $\gtrsim 0.5$ c for a NS ($M = 1.4\ M_\odot$ and $R \simeq 15$ km). For a black hole there is no escape possibility even for the photons below the event horizon. For the case of a NS the allowed values for M ($\leq 3\ M_\odot$) and R ($\sim$ 10 km) (including also phenomena associated with the presence of strong gravitational forces) increase the escaping velocity to relativistic values. Let us consider $\sim 10^{53}$ erg as being carried away by the ν burst during a supernova (SN) explosion. If we take $m_\nu \simeq 10$ eV and assume a flat energy distribution with maximum energy $\sim$ 100 MeV, some 10^{-7} of the created neutrinos ($\sim 10^{46}$ erg) during the birth of the NS can be gravitationally captured. The capture efficiency can even be much bigger, as can be seen by mentioning only two examples: i) as long as the electron gas energy has not increased very quickly, the yield of low-energy neutrinos due to the reaction $p + e \rightarrow n + \nu_e$ is high; ii) for heavier neutrino types, the capture efficiency is much higher as well.

The same argument is valid for other DM particles, such as axions, etc. In other words *massive and dense stars can act as the seed for dark matter gravitational binding around ordinary matter and vice versa. Gravity defines the coexistence conditions for both kinds of matter!*

Luminosity: Let 10^{46} erg be bound as axions around a pulsar or black hole. With CE $= 10^{-14}$, when these dense clustered particles hit the regions with strong and locally changing field the forced decay will result to a photon luminosity of $\sim 10^{32}$ erg/s, which is not far from reality, while the amount of captured axions and their conversion yield might well be higher by orders of magnitude.

The DM processes considered here are not necessarily in contradiction with any well-known features: energy spectrum, periodicity, beaming effect of pulsars, etc. The observed γ-ray bursts with super Eddington luminosities could be explained: the 'Thomson' scattering for the accreting dark matter is many, orders of magnitude smaller, defining a new increased Eddington limit. A dense and massive DM in the form proposed in this work might well be the key to overcoming at least some of the enigmatic cosmic behaviour in those places in cosmos.

REFERENCES

1) A. De Rújula and K. Zioutas, preprint CERN-TH.5220/88.
2) K. Zioutas and Y. Semertzidis, Phys. Lett. **A130** (1988) 94.
3) P. Sikivie, Phys. Rev. Lett. **51** (1983) 1415.
4) D.E. Morris, Phys. Ref. **D34** (1986) 843.
5) B.E. Moskowith, Nucl. Instrum. Methods **A264** (1988) 98.
6) K. Zioutas, CERN-PS/87-61 (AA); and ECFA 87-110 (1987) 660.

Signals of Dark Matter Radiative Decay

John P. Ralston and Douglas W. McKay
Department of Physics and Astronomy
University of Kansas, Lawrence, Kansas 66045

ABSTRACT

Evidence for the detection of radiative dark matter decay is reviewed. Assuming only standard big bang results, dark matter neutrinos of mass $\gtrsim$ 30 eV and lifetime $\gtrsim 10^{24}$ sec can explain several observations of intergalactic ionizing photon flux. We comment on the possibility of observing dark galaxies, containing little or no ordinary luminous matter, via fluorescence from the galaxies' dark matter decay.

EVIDENCE FOR DARK MATTER DECAY

The term "dark matter" may be a misnomer. A candidate dark matter particle is a neutrino with mass in the 25 eV–100 eV range. If such neutrinos undergo radiative decay, the natural "color" of dark matter is ultraviolet,[1] with a frequency about an order of magnitude above the frequency of visible light. In this region, $E_\gamma \geq 14eV$, dark matter can become visible when decay photons ionize neutral hydrogen in space. In fact, the Gunn–Peterson limit[2] on the density of intergalactic neutral hydrogen requires an intensity of bathing UV radiation above that provided by QSO's and UV–emitting stars.[3] Adding to the Gunn–Peterson limit and several other puzzles which we do not review here,[4] Reynolds *et al.*[5] recently reported detection of H_α radiation at the $4-\sigma$ level from an intergalactic cloud of hydrogen. A flux of 10^5 ionizing photons/cm^2 sec sr is needed to explain the H_α emission.

The data can be combined with neutrino number densities to obtain the radiative neutrino lifetime $\tau_\nu \sim 10^{24}$ sec; details are given elsewhere.[4] This implies a magnetic transition moment μ_{HL} for the decay $\nu_H \rightarrow \nu_L + \gamma$, where

$\nu_{H,L}$ are heavy and light neutrinos:

$$\mu_{HL} \cong 2 \times 10^{-15} \mu_B \left(\frac{10^{24} sec}{\tau_\nu} \right)^{\frac{1}{2}} \left(\frac{30 ev}{m(\nu_H)} \right)^{\frac{3}{2}} , \qquad (1)$$

assuming ν_L is much less massive than ν_H. Such a magnetic moment implies new right–handed, lepton generation–changing physics beyond the standard model. Clearly, if the H_α observations really represent radiative neutrino decay, their particle physics implications are profound.

BRIGHT DARK GALAXIES?

Little is known about how ordinary matter becomes associated with galactic dark matter halos. Can radiative decay give a new signal for possible "dark" galaxies in which there is too little luminous matter to be visible?

Consider $10^{10} M_\odot$ in the form of 30 eV neutrinos producing decay photons at a rate of 10^{-24} sec^{-1}. The UV photon flux from such a prototype dark galaxy at a distance of 1 Mpc is approximately 3 cm^{-2} sec^{-1}. This is below the level of producing an observable signal. Even decreasing the distance by a factor of ten, and raising the mass by a factor of ten, the flux from a "dark companion" galaxy close to our own is small compared to the benchmark of $10^5 \gamma$'s $cm^{-2} s^{-1}$. This mechanism is not a viable way of testing the neutrino decay scenario. Better tests would be offered by improved H_α measurements, or by direct laboratory observation of lepton generation–changing interactions.

We thank Adrian Melott for discussions. This work was supported in part by DOE Grant No. 85ER40214.A004.

REFERENCES

1) De Rujula, A. and Glashow, S.L., Phys. Rev. Lett. 45, 942 (1980).

2) Gunn, J.E. and Peterson, B.A., Astrophys. J. 142, 1633 (1965).

3) Shapiro, P.R. and Giroux, M.L., Astrophys. J. Lett. 321, L107 (1987).

4) Ralston, J.P, McKay, D.W. and Melott, A.L., Phys. Lett. 202B, 40 (1988); Melott, A.L., McKay, D.W. and Ralston, J.P., Astrophys. J. Lett. 324, L43 (1988); McKay, D.W. and Ralston, J.P. in *Proceedings of Neutrino 88* (World Scientific, 1989, in press).

5) Reynolds, R.J., Magee, K., Roester, F.L., Scherb, F. and Harlander, J.H., Astrophys. J. Lett. 309, 9 (1986).

STATUS OF DARK MATTER SEARCHES

B. Cabrera

Physics Department, Stanford University
Stanford, California 94305

1. ABSTRACT

This paper is a summary of the current research on dark matter searches presented at the Particle Astrophysics Conference. In the first part we summarize several new observational efforts to determine the density of the universe and the distribution of dark matter. In the second part we summarize the experimental efforts aimed at laboratory detection of elementary particle candidates for dark matter in the cosmic rays.

2. INTRODUCTION

There is compelling evidence that galaxies are surrounded by low-dissipation halos and that these halos dominate their mass. This unidentified and non-luminous matter is called dark matter. So far it has only been detected through its gravitational interaction with luminous matter and it is present on all observational scales from globular clusters to galaxies to superclusters. We begin this summary by describing several new observational efforts to determine the density of the universe over this range of distance scales.

It is likely that some portion of the dark matter is baryonic and in the form of brown dwarfs or black holes. These would contribute to the mass of galactic halos, but not to the galactic luminosity. However, our current understanding of the big bang nucleosynthesis epoch [see paper by D. Schramm] together with the observational measurements on the primordial relative abundance of H, D, He, and Li limit the baryonic density to 0.2 of the critical density ($\Omega_{baryon}<0.2$). On the other hand,we know from theoretical cosmology that $\Omega=1$ is an unstable equilibrium for an expanding universe. Positive or negative deviations from $\Omega=1$ grow exponentially in time. It would be fortuitous if in our current epoch Ω were only a factor of five different from unity. Also, inflationary models of the early universe cosmology [see paper by M. Turner] provide the only suggested mechanism for diluting massive topological particles such as magnetic monopoles and domain walls below observational limits. Such topological particles arise in all unification theories. The inflation models predict in a natural way that $\Omega=1$ to high precision. Thus, if $\Omega=1$ and $\Omega_{baryon}<0.2$,there must exist a non-baryonic form of dark matter which dominates the mass density of the universe and interacts weakly with luminous matter. Such weakly interacting matter would not have clumped into condensed objects but rather would have remained as a gas of elementary particles loosely gravitationally bound to luminous matter. The density contrast would be much less than that for luminous matter and these particles would dominate the mass of dark halos around galaxies such as our own.

If these conjectures are correct and the massive halos are dominated by weakly interacting particles, then they must exist in the flux of cosmic rays which strike the surface of the earth. In the second part of this summary, we describe efforts which are currently underway to observe directly such particles with laboratory instruments.

3. OBSERVATIONAL DETERMINATIONS OF Ω

There has been much recent progress in determining the large scale structure of the universe. These proceedings contain a number of detailed reviews of these efforts [see paper by M. Davis and by M. Geller]. Here we describe three recent programs: the deep redshift survey of Loh and Spillar which has the potential for determining Ω at redshifts of $z<1$, the galaxy distribution map provided by IRAS (the NASA infrared satellite) which provides the first new spectral window on the galaxy distribution and determines Ω for redshifts of $z<.03$ and a new NASA proposal called POINTS which will provide a direct parallax distance scale and velocity map out to the Magellanic Clouds and allow a detailed mapping of the local dark matter distribution around our own galaxy.

3.1 Large Redshift Galaxy Survey

A new technique has been developed [see paper by E. Loh] for dramatically increasing the number and distance over which galaxy redshifts can be determined. Rather than obtaining direct redshift evidence from individual spectral lines which is time consuming and requires a significant photon flux, this new technique measures the galaxy luminosity in six wideband frequency ranges across the visible spectrum. The position of the sharp drop in intensity in the blue, a common feature for galaxies, is then used to measure the approximate redshift. The results to date, give the first observational evidence that Ω may approach 1 on large distance scales. However,there are large uncertainties because the galaxy luminosity function may vary with time so that more distant galaxies which are progressively younger would have different intrinsic spectra than older galaxies which are nearest to us. In addition, there is a bias towards the high intrinsic luminosity galaxies at the largest redshifts and within a given epoch it is not known accurately how the galaxy luminosity spectrum varies over the range of intrinsic galaxy luminosities. Studies are now underway to correlate the results of this new technique with the conventional redshift survey so that these questions can be answered.

3.2 Infrared Galaxy Survey

The Infrared Astronomy Satellite (IRAS) has provided a wealth of information on the sky in the infrared part of the photon spectrum [see paper by M. Strauss]. The coverage of the sky during the year and a half of data acquisition was very uniform and is missing only our own galactic plane in the Survey of distant galaxies. To date, the data have been analyzed for 2535 galaxies out to ~8000 km/s ($z<0.03$). Because essentially the whole sky is covered, these data can be converted to compute our relative velocity with respect to ever growing spheres centered about our galaxy using Newtonian physics. The results approach roughly asymptotic values for distances beyond ~4000 km/s. The value obtained for our velocity vector is consistent with the peculiar velocity determined from the dipole term in the microwave background data. The IRAS data suggests a value of $\Omega=0.7$. Remarkably, the distribution of galaxies is not as clumpy as it is in the visible redshift surveys. By preforming what is called a cluster boost which effectively increases the contrast of the IRAS to match the distribution for the visible galaxies, Ω is shifted to 0.29 with sufficient uncertainty to be consistent with the Ω determined from the visible data. However, it may be that the IRAS galaxies are a better trace of the mass distribution of the universe!

3.3 High Precision Parallax Survey

A new proposed NASA satellite called POINTS (Precision Optical INTerferometry in Space) would provide a direct distance scale out to the Magellanic Clouds [see paper by J. Phillips]. A catalog of stars would be chosen which cover the sky and the apparent angle between pairs would be determined to an accuracy of a few microarcsec. Each of the two pointing telescopes is composed of an aperture pair which form a long baseline interferometer in the optical. The absolute position of all of the stars in the survey are determined by a self-consistent fit of all the angles between pairs. Measurements will be made repeatedly on the same data set of stars throughout a several year period, so that the proper motion can be determined as well as any periodic motion caused by dark companions or planets. Interestingly, the deflection of star light by our sun which is 2 milliarcsec in a direction 90 degrees form the line of sight to the sun must be accounted for to high precision over the whole sky in analyzing the data.

If such a survey is obtained, the data would provide an absolute distance scale out to the Magellanic Clouds. In addition, a detailed map of the dark matter distribution within our own galaxy would be obtained.

4. LABORATORY SEARCHES FOR DARK MATTER

Since Ω very nearly equal to one is probable from stability arguments and since Ω less than 0.2 is probable from nucleosynthesis arguments, it is quite possible that the dark matter is in the form of weakly interacting elementary particles.

4.1 Particle Theory and Dark Matter Candidates

Given that the dark matter may be a form of elementary particle, particle theories provide a wealth of candidates [see paper by L. Hall]. All grand unification theories contain magnetic monopole solutions where the mass is the unification scale divided by the fine structure constant ($>10^{-16}$ GeV/c^2). Such massive topological particles would have been produced copiously in the early universe and would dominate the mass density of the Universe today by an unacceptable huge factor ($\sim10^{14}$) if some suppression mechanism such as inflation had not played a role. It is possible for the suppression to have left monopoles at the density observed for the dark matter, but the exponential suppression more naturally results in an unobservable small density for massive monopoles in the current universe. Nevertheless, observational limits are important.

The axion stems from theories which explain CP conservation in strong interactions by postulating the existence of this light particle. Interestingly, there is a relatively narrow window of masses for the axion which would allow it to make up the dark matter. A mass of 10^{-5} eV seems most natural and these particles would exist today as a background sea of non-relativistic particles which have adiabatically cooled as the universe expanded, in a similar fashion to the microwave background.

For a large class of hypothetical particles which were assumed to be in thermal equilibrium when they decoupled from radiation, there is a particularly elegant argument first proposed by Lee and Weinberg which connects the masses of dark matter candidates with their annihilation cross sections and thus also their interaction cross sections with ordinary matter. Any particles produced in thermal equilibrium in the early universe must today be at density which is determined by their mass and their annihilation cross section. Large annihilation cross sections would suppress the density and small cross sections would exceed Ω=1. It turns out that to obtain Ω=1 a cross section about that of the weak cross section is needed. Two mass ranges are allowed, one at low energy around 10 eV/c^2 and the other above about 2 GeV/c^2. The lower mass range suggests that neutrinos, the only nonbaryonic weakly interacting particles known to exist, may have a mass in the 10 eV/c^2 range. Most naturally, the tau neutrino would have such a mass with the mu and electron neutrinos at much smaller values.

On the other hand, the mass range above 2 GeV/c^2 is favored because it aids the process of galaxy formation. Particles which are nonrelativistic at the time of galaxy formation increase the rate of clumping in the early universe and computer simulations with such "cold" dark matter reproduce the observed galaxy distributions much better than "hot" dark matter such as neutrinos which would have been relativistic at the time of galaxy formation and thus would tend to wash out the density contrast.

These WIMPs (weakly interacting massive particles) in the GeV/c^2 mass range have become probable candidates for the dark matter in a largely model independent way. Particles which simultaneously solve the solar neutrino problem and are the dark matter are called cosmions. These are a subclass of WIMPs with a several order of magnitude larger cross section which cool the interior of the sun by transporting energy from the interior. On the one hand they must be massive enough to remain bound to the sun and on the other they must have an interaction cross section such that they undergo about one interaction per orbit. Larger cross sections are allowed in the Lee and Weinberg argument if we allow particle-antiparticle asymmetry.

Interestingly, different types of laboratory detectors can explore this whole range of possibility [see paper by K. Griest for a discussion of event rates].

4.2 Magnetic Monopoles

The first laboratory search for dark matter in the form of elementary particles passing through the earth was the search for magnetically charged particles in the cosmic rays. Grand unification theories demand the existence of massive magnetically charged particles and a density of ~1 per km^3 could account for the dark matter at our galactic location. The use of superconducting induction coils [1] provides an extremely convincing identification for magnetic particles possessing the Dirac unit of charge. In Figure 1, bounds based on the halo mass limit (if monopoles are the dark matter, they must lie on this line), the existence of the galactic magnetic field (a high density of monopoles would destroy the field), and monopole plasma oscillation models (where the monopoles are partly responsible for the galactic field in a self-consistent interaction with electric currents).

The candidate event in the prototype data set of 1982 has been largely dismissed since several generations of much larger detectors operated by a number of different groups have not observed events. The combined world limit has now ruled out the plasma oscillation models and is at the peak of the allowed region around a monopole mass of ~10^{19} GeV/c^2 (at about the Planck mass). Thus if magnetic monopoles make up the dark matter, they must be more massive than 10^{19} GeV/c^2 and have a local density below ~ 10^{-3} km^3.

4.3 Axions

As first pointed out by Sikivie [see his paper in these proceedings], detectors can be built to search for dark matter axions based on axion to photon conversion which is expected in a strong magnetic field. Two groups, one at Florida State [see paper by P. Sikivie] and the other at Brookhaven [see paper by J. Rogers] have built such detectors. These detectors consist of high-Q copper microwave cavities operated at mK temperatures in a uniform magnetic field of several tesla. The conversion process is proportional to the square of the magnetic field and ultra low temperatures are needed to suppress thermally produced photons in the cavity.

The experiments then consist of slowly sweeping the resonant frequency of the cavity over a range corresponding to that predicted by the allowed axion mass range. Since the axion rest mass is much greater than its kinetic energy, a sharp peak ($\Delta\nu/\nu \approx 10^{-7}$) should be observed as the cavity frequency passes through resonance. Figure 2 shows the current status of the BNL experiment over the interesting mass range. Improvements of about two orders of magnitude in sensitivity are needed before dark matter axions could be detected or ruled out.

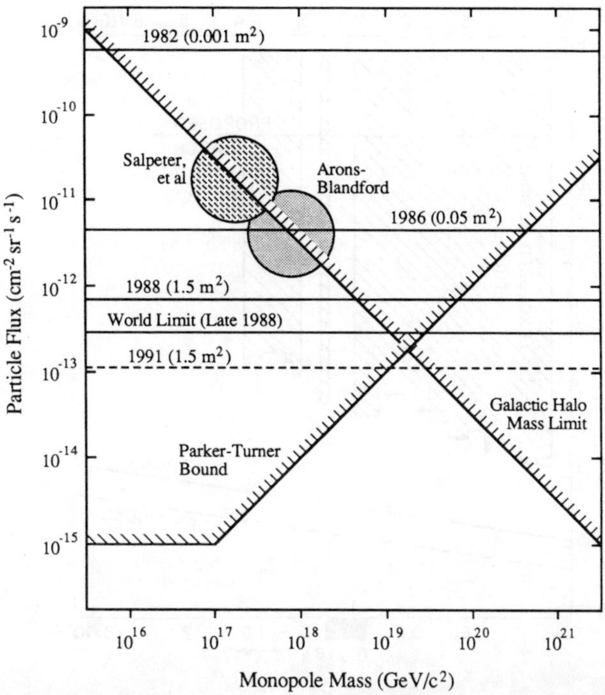

Figure 1. Particle flux limits set by superconducting inductive detectors for magnetic monopoles in the cosmic rays [from B. Cabrera].

4.4 Solar Annihilations

There remains one other allowed mass window for axions around 10^{-3} eV/c^2. In this mass range, the conversion in the core of our sun of photons to axions (now with about 1 keV of kinetic energy) should occur at a rate sufficient to observe them on earth. The Livermore group [see paper by K. van Bibber] has proposed building a multi-wire proportional counter with a high magnetic field. Then the axion conversion rate back to photons (~1 keV x-rays) would be observable.

Another experiment which takes advantage of the sun is the new interpretation of the data from the Kamiokande water Cherenkov detector [see paper by Y. Totsuka]. The thresholds are now sufficiently low (~ 7 MeV) so that the high energy tail of the ^{8}B solar neutrinos has been identified at a flux consistent with the Davis flux and a factor of two below that predicted by the standard solar model. These same data set limits on the annihilation rate to neutrino-antineutrino pairs of dark matter candidates in the solar core. Upper mass limits of about 3 GeV/c^2 are obtained for massive Dirac neutrinos and for electron, mu and tau sneutrinos. Limits are also set on the massive photino and on the Majorana neutrino. The most important caveat to these limits is the assumption that there exist equal numbers of particles and anti-particles, not necessarily true as we see from the hadron-antihadron asymmetry.

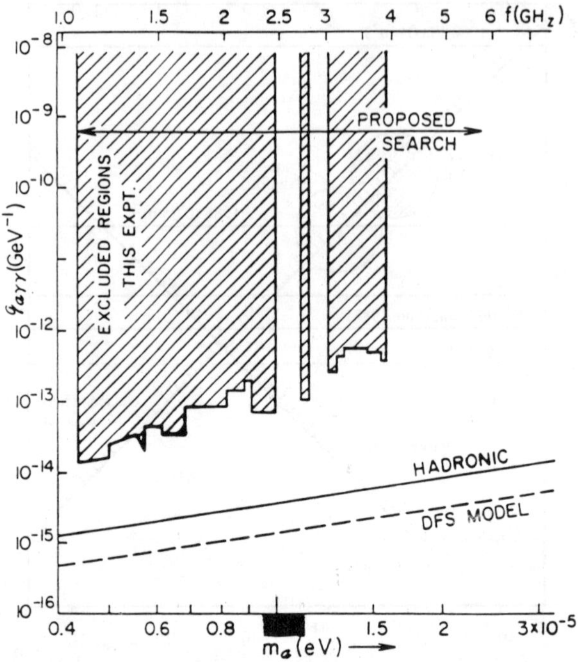

Figure 2. Experimentally excluded regions of the coupling constant versus axion mass range around 10^{-5} eV with Sikivie-type detector [from paper by J. Rogers].

5. WIMP SEARCHES USING CONVENTIONAL DETECTORS

Particularly interesting among the dark matter searches are the laboratory detectors now under construction to look for a flux of several GeV/c^2 mass neutral particles which are weakly interacting. In this section we summarize the status of the impressive early results from conventional semiconductor diodes which have already set useful limits on WIMPs and new efforts to use time projection chambers as cosmion detectors.

5.1 Semiconductor Diodes

Two groups [see paper by D. Caldwell and paper by F. Avignone], both working on neutrino mass experiments aimed at detecting the neutrinoless double beta decay of ^{76}Ge, have made dramatic improvements in the background levels of their Ge semiconductor diodes and are now able to obtain data all the way down to the electronics limit corresponding to about 3 keV in equivalent electron energy deposition.

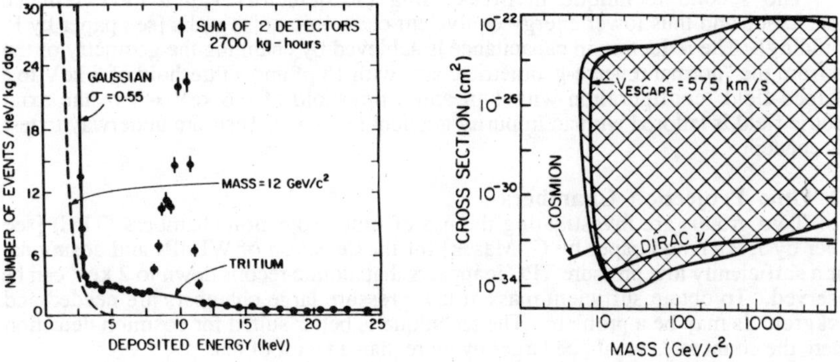

(a) (b)

Figure 3. (a) Low energy spectrum of underground Ge semiconductor diode detector. (b) Experimentally excluded regions of cross section versus WIMP and cosmion mass range [from paper by D. Caldwell].

In the case of the LBL/UCSB collaboration [see paper by D. Caldwell] a dramatic reduction in low energy background was achieved by removing indium from the apparatus, which had been used to improve the thermal contact among cryogenic components. All isotopes of In are radioactive with ^{115}In having the longest half-life at $\sim 6 \times 10^{14}$ years. Figure 3a shows data from two detectors showing the know x-ray lines for Zn and Ga around 10 keV and a beta decay distribution consistent with the tritium decay spectrum. As first predicted by J.C. Martoff [2], the rate of tritium production from spallation of Ge while the crystal is exposed to sea level cosmic rays is in excellent agreement with the observed event rate. The dotted line in the figure is the count rate expected from 12 GeV/c^2 massive Dirac neutrinos if they made up the halo dark matter. Such a signal would be observable as additional rounding at the foot of the electronics noise wall.

These data set the limits shown in Figure 3b for Dirac neutrinos (< 12 GeV/c^2), Magninos (< 11 GeV/c^2) and Cosmions (<9 GeV/c^2). The upper boundary of the excluded region corresponds to candidate particles which interact with too high a cross section and do not reach the underground detector. A very important question with regard to the interpretation of these data is the relative energy calibration of the Ge detectors for nuclear recoils versus electron recoils. The data of C. Chasman [3] studied nuclear recoils down to 10-15 keV. New experiments are now underway to check Chasman's results and to extend the energy range down to lower recoil energies.

Two efforts are currently underway to lower the thresholds of semiconductor diodes. The first utilizes high mass detectors made with intrinsic Si semiconductor diodes [see paper by D. Caldwell]. The lighter mass of Si (three times less than Ge) would allow lower energy recoils to be observed. Laboratory studies on Si detectors are now underway.

The second technique involves using Ge detectors which have reduced capacitance and thus lower energy-equivalent electronic noise walls [see paper by F. Goulding]. The reduction in capacitance is achieved by changing the geometry of the Ge from the "normal reverse-geometric coax" with 18 pf and a threshold of 3 keV to a "shaped-field" configuration with 1 pf and a threshold of 0.6 keV where the axial electric field is tailored with an impurity gradient in the Ge. Tests are underway to test this concept.

5.2 Time Projection Chambers

Two groups are investigating the use of time projection chambers (TPC) [see paper by J. Rich and paper by G. Masek] for the detection of WIMPs and cosmions. For a sufficiently low pressure TPC it appears that atomic recoils down to 2 keV can be observed. To obtain sufficient mass at low pressure large chambers are needed and backgrounds may be a problem. The technique is better suited for cosmion detection where the count rates would be larger by more than a factor of ten.

6 WIMP SEARCHES USING NEW CRYOGENIC DETECTORS

In this section we describe recent advances in new types of particle detectors designed to achieve lower thresholds and higher resolutions than semiconductor diodes. The electron-hole pairs produced in diode detectors require about 3 eV of deposited energy per pair in the detector. Several physical mechanisms exist at low temperatures where the energy of quanta is meV's (milli *not* mega) rather than eV's. These include phonons in insulating crystals, quasiparticles in superconductors and rotons in superfluid helium.

6.1 Phonons in Insulating Crystals

The most straight forward idea for using phonons is to build bolometers out of large insulating crystals such as Si, Ge, Al_2O_3, NaF and others. These follow the T^3 Debye law for heat capacity down to at least 20 mK. For example, a 100 g crystal of Si at 10 mK would show a temperature rise of 2 μK for an energy deposition of 1 keV, within the sensitivity range of doped-semiconductor thermistor technology. The thermalizaton time constants are several millisec. Microbolometers [4] of about 10^{-5} g have demonstrated an energy resolution of 17 eV (FWHM) for 6 keV x-rays from ^{55}Fe. These devices operated at 100 mK have the highest energy resolution ever obtained in this energy range.

The best possible energy resolution is $\Delta E \approx (kT\ CT)^{1/2}$ which is proportional to $T^{5/2}\ M^{1/2}$ where M is the crystal mass and the heat capacity C follows the Debye T^3 law. Thus in principle moving down 2 orders of magnitude in temperature and ten orders of magnitude up in mass leaves the energy resolution unchanged. A number of effects limit such an extrapolation, including deviations in the Debye law heat capacity from defects and impurities in the crystals and thermistor technology at low temperatures where the thermistor heat capacity dominates and where the thermistor sensitivities roll off due to poorer and poorer coupling between the phonons and the electrons.

Several groups are working on these technologies to scale up the crystal mass and scale down the temperature. The UCB/LBL group, headed by B. Sadoulet, is studying the use of neutron transmutation doped germanium as a thermistor material at 20 mK. They have demonstrated sensitivity of these devices to 60 keV x-rays absorbed directly in the thermistor, and they are investigating techniques for bonding the sensors on larger crystals of silicon and germanium.

Using commercially available germanium doped thermistors Coron, et al [5], have demonstrated a FWHM resolution of 20 keV in a 25 g crystal of Ge operated at 100 mK. This value is near the theoretical limit discussed above. Similar work by the

Milan group [see paper by Alessandrello] has achieved a resolution of ~70 keV in the MeV range and their detectors are being developed for improved double beta decay experiments.

Our group at Stanford [see paper by B. Cabrera] is pursuing a different track. Rather than waiting for the phonon thermalization process, we feel that it is advantageous to observe the first phonon wavefront which arrives at the crystal surfaces. Many of the phonons travel ballistically (i.e., without scattering) from the source to the crystal surfaces. With sufficiently sensitive phonon detectors on the surfaces of the crystal, it is then possible to image the event within the crystal and to significantly improve the background suppression. A Ti film biased at the foot of its resistive superconducting transition shows rise times of 150 ns and pulse durations of several microsec. These transition edge sensors strongly absorb phonons through the breaking of Cooper pairs and the formation of quasiparticle excitations. There exists a critical phonon energy density above which the film is driven into the normal resistive state a develops a voltage from the dc bias current. We have demonstrated a resolution of ~2 keV (FWHM) with these SiCADs (Si crystal acoustic detectors) for a sensor area of ~7 mm^2 on a crystal 1 mm thick operated at a temperature of 300 mK.

6.2 Quasiparticles in Superconductors

A number of schemes are being developed which utilize a superconductor as the absorber of radiation. The first class of these detectors utilize superconducting granules as bolometers and the second class utilize superconducting tunnel junctions to detect pulses of quasiparticles in a superconductor while the superconductor remains below its transition temperature.

The superconducting granule detectors [see paper by A. Drukier and references therein], called SSCDs (superheated superconducting colloidal detectors) consist of a colloidal suspension of spherical superconducting granules each of about 10 μm in diameter. To prepare the detector, the granules are cooled below their transition temperature (e.g., 7 K for Pb and .5 K for Cd) and then a constant magnetic field is applied which is below the superheated critical field for the type one superconductor. Type one superconductors remain in the superconducting state for fields above the thermodynamic critical field and below the superheated critical field, because there exists a metastable state separated from the ground state by an energy barrier. When a normal region is formed anywhere within a granule, the whole sphere quickly decays to the lower energy normal state. In the superconducting state, each granule is a perfect diamagnet and posses a magnetic dipole moment which is proportional to its volume and to the applied field. When a granule is driven normal by deposition of energy from a particle interaction, the magnetic moment disappears. The resulting change in magnetization of the sample is detected with a magnetometer such as a SQUID (superconducting quantum interference device). Much recent progress has been made on these devices in the US, Canada and Europe [see paper by A. Drukier].

The second class of detectors utilize superconducting tunnel junctions to collect the quasiparticle excitations (broken Cooper pairs) generated by an elementary particle interaction. Since superconductors have zero resistance, in order to obtain an electrical signal it is necessary to isolate an electrode (also superconducting) using a thin insulating barrier. If the barrier is thin enough (~ 2 nm), then with an applied bias voltage, it is possible for the quasiparticle excitations to tunnel across and provide a detectable current. The junction forms a capacitor and particles are detected by the appearance of an injected charge across the junction.

The group of A. Zehnder at S.I.N. [6] has demonstrated a resolution of 48 eV FWHM for 6 keV x-rays absorbed directly by Sn-Sn oxide-Sn tunnel junctions operated at 350 mK. The junctions are operated about a factor of ten below their transition temperatures to suppress the thermally generated quasiparticles which generate the equivalent of a detector dark current. This resolution for astrophysical x-ray spectroscopy exceeds that of the best semiconductor diodes and has only been surpassed by the microbolometer of Moseley and McCammon [4].

The Oxford group [see paper by N. Booth] is developing much larger detectors utilizing superconducting tunnel junctions for future solar neutrino experiments. One electrode of their junctions is a several gram crystal of superconductor, which provides a large cross section for particle interactions. However, since the tunneling times are proportional to the thickness of the superconductors such large junctions would not have good sensitivities. To circumvent this problem, Booth has proposed a quasiparticle trap (shown schematically in Figure 4). A superconducting film with a lower transition temperature (and lower gap) is inserted between the insulating barrier and the thick superconducting crystal. Quasiparticles which stray into the lower gap superconductor loose energy by emitting phonons and are trapped in the thin film where their probability for tunneling is high since they have a high attempt frequency. Typical tunneling probabilities are $\sim 10^{-6}$ per attempt. Early test indicate that such devices are promising.

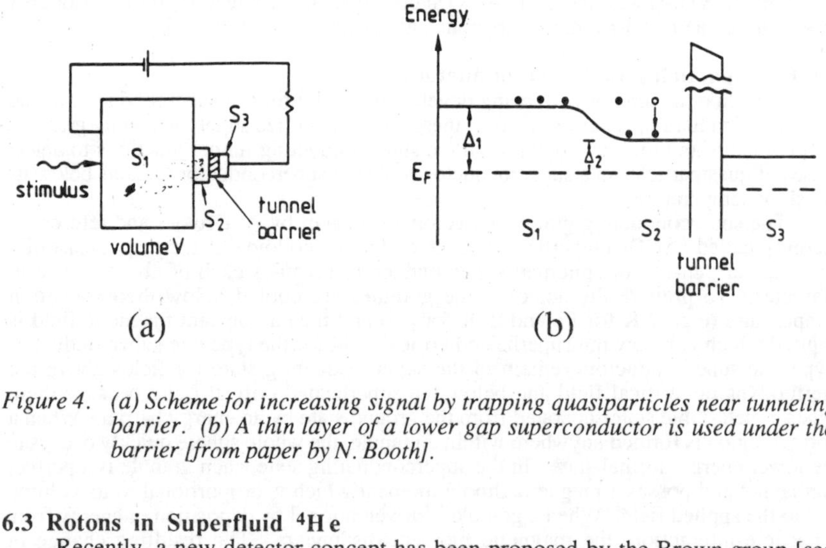

Figure 4. (a) Scheme for increasing signal by trapping quasiparticles near tunneling barrier. (b) A thin layer of a lower gap superconductor is used under the barrier [from paper by N. Booth].

6.3 Rotons in Superfluid ^{4}He

Recently, a new detector concept has been proposed by the Brown group [see paper by R. Lanou] and is based on the use of rotons in superfluid ^{4}He. Their motivation is also for a future solar neutrino experiment. Rotons are elementary excitations in superfluid ^{4}He and have characteristic energies of a few meV. At a temperature of about 20 mK, rotons formed by a recoiling helium nucleus or an electron, will have a long mean free path and propagate with little loss to the walls of the vessel where they scatter and to the surface of the He where they eject a He atom with an efficiency of about 30%. At 20 mK, the vapor pressure of He is very low (good vacuum above liquid surface) so that the ejected atoms also have a long mean free path. In their scheme, these atoms are detected with silicon bolometers which have been carefully "baked" to drive off all surface helium. The binding energy of a He atom on a clean Si surface is about 10 meV, so that an energy gain of ~10 is obtained.

One significant advantage of He is that it can be highly purified because all other gases are frozen out at these low temperatures and because only superfluid ^{4}He will pass through a superfluid plug. The Brown group is building a 600 gram prototype detector and will soon conduct tests to investigate the feasibility of their method.

7. CONCLUSIONS

There is compelling evidence that dark matter exists on all scales within our universe. The most exciting and pressing challenge for astrophysics is to identify the major constituent of the dark matter which dominates the energy density of our universe. Combining the theoretical reasoning that Ω=1 the the nucleosynthesis limit of Ω_{baryon}<0.2 leads to the conclusion that dark matter may be a new form of elementary particle. Great advances have been made in observational studies to better determine the distribution of dark matter, and new proposals will provide future improvements.

Should dark matter be in the form of a new elementary particle, laboratory detection in the cosmic rays is possible. Significant limits have been set for magnetic monopoles using superconducting inductive detectors. In the search for neutral weakly interacting particles, a number of conventional detector technologies have been used with the best current limits set by Ge semiconductor diodes. To improve on these limits, a number of new detector concepts based on meV excitations at low temperatures are being studied and show great promise. Perhaps through the efforts of these research groups and the newly formed Center for Particle Astrophysics, we will soon known the composition of the dark matter.

I thank all of the participants in the dark matter sessions for valuable discussions and for the use of several of figures in this summary paper. I also thank Bernard Sadoulet and his staff for a lively and well run conference. Our work at Stanford is supported in part by DOE Grant DE-AT03-82ER40-076.

8. REFERENCES

[1] B. Cabrera, *Phys. Rev. Lett.* **48**, 1378 (1982); for theory review see: J. Preskill, *Ann. Rev. Nucl. Part. Sci.* **34**, 461 (1984); for experiment review see: D. E. Groom, *Phys. Rep.* **140**, 323 (1986).

[2] C. J. Martoff, *Science* **237**, 507 (1987).

[3] C. Chasman, *Phys. Rev. Lett.* **15**, 245 (1965).

[4] S. H. Moseley, J. C. Mather and D. McCammon, *J. Appl. Phys.* **56**, 1257 (1984); D. McCammon, et al, *J. Appl. Phys.* **56**, 1263 (1984).

[5] N. Coron, et al, **Superconducting and Low-Temperature Particle Detectors**, eds. G. Waysand and G. Chardin, Elsevier Sci. Publ., Amsterdam, in press.

[6] D. Twerenbold and A. Zehnder, J. Appl. Phys. **61**, 1 (1987).

LARGE SCALE STRUCTURE

EXPERIMENTAL ISSUES IN RESEARCH ON THE LARGE-SCALE STRUCTURE OF THE UNIVERSE

Margaret J. Geller

Harvard-Smithsonian Center for Astrophysics

60 Garden St.

Cambridge, MA 02138 USA

ABSTRACT

Three of the issues in large-scale structure research are (1) the existence of large-scale coherent flows, (2) the nature of the distribution of galaxies on scales $\gtrsim 10h^{-1}$ Mpc (where the Hubble constant $H_o = 100h$ km/s/Mpc) and (3) the relationship between individual galaxies and clusters of galaxies as tracers of the large-scale galaxy distribution. For each of these issues, I briefly review some of the challenges presented by the data and I suggest some approaches to clarifying some of the experimental issues.

I. INTRODUCTION

It sometimes seems as though we *understand* less about the large-scale structure of the universe now than we did ten or fifteen years ago. The reason for this paradoxical situation is that we now have some data!

Here I will discuss three of the nagging issues in large-scale structure research: (1) large-scale flows, (2) the large-scale distribution of galaxies, and (3) the distribution of rich clusters of galaxies. In each case there are fundamental issues which can be resolved only by acquisition of well-designed sets of uniformly high-quality data. The technology for the resolution of these issues is available now: the resource in short supply is telescope time.

II. LARGE-SCALE FLOWS

There is now accumulating evidence that on scales $\gtrsim 20h^{-1}$ Mpc, the expansion of the universe is non-uniform. The subject is controversial at least in part because of the difficulty of the measurements. In fact, one of the greatest challenges of extragalactic astronomy is to find a reliable method of measuring distances (rather than redshifts).

Three years ago Burstein *et al.* (1986) measured the distances to a sample of elliptical galaxies. They obtained a remarkable result: their data revealed apparently coherent motions on a scale of $50h^{-1}$ Mpc. They attributed the motion to a mass concentration of some 10^{16} $M_{\odot}$ which they dubbed "The Great Attractor". The amplitude of the coherent flow is ~ 600 km s^{-1} .

The detection of a large-scale coherent flow flew in the face of cherished belief and many have rushed to meet the challenges of deciding whether the motions are real and, if so, what drives them. Burstein *et al.* (1988) have continued to work on the problem and have provided their data to others who have reanalyzed them. Recently, Groth *et al.* (1988) completed one such reanalysis. They conclude that the data demand large-scale flows and that there is a net translation driven by some mass concentration beyond the limits of the survey. This analysis is more or less consistent with the conclusions drawn by Lynden-Bell *et al.* (1988) themselves.

One can also measure the distances to galaxies using the infra-red Tully-Fisher technique (Aaronson, Huchra and Mould 1979). More recently Aaronson *et al.* (1989) have measured distance to a set of spiral galaxies. They conclude that there is a motion but that it is driven by the well-known Hydra-Centaurus supercluster. The complex dynamics of the Centaurus cluster and of the entire supercluster make interpretation of the data difficult. Because there is little overlap between the samples of spirals and ellipticals, it is hard to isolate the reasons for the discrepancies between the two results. Where the data do overlap, there are some significant disagreements (Mould *et al.* 1989).

Recently Gorski *et al.* (1989) derived a velocity field from a sample of IRAS galaxies. This velocity field is consistent with the results for the spiral sample. In other words, Hydra-Centaurus is driving the motion. In a reanalysis of the data of Burstein *et al.* (1988), Gorski *et al.* (1989) argue that the elliptical data may also be consistent with the IRAS and spiral results. The debate continues elsewhere in this volume.

II. THE LARGE-SCALE DISTRIBUTION OF GALAXIES

The idea of redshift surveys is not new. Humason, Mayall, and Sandage (1956) carried out one of the first. More recently, larger surveys have been done by Peterson *et al.* (1986), da Costa *et al.* (1988), Giovanelli *et al.* (1986), Kirshner *et al.* (1987), and Koo *et al.* (1986). The more recent of these surveys generally include more objects. I will focus on the Center for Astrophysics (CfA) survey (Huchra *et al.* 1992) which will include redshifts for 15,000 galaxies in the Zwicky-Nilson merged catalog with $m_{B(0)} \leq 15.5$.

The result of a specially designed redshift survey by Kirshner *et al.* (1981) was important in awakening interest in large structures in the universe. This survey consists (1987) of widely separated probes, each of small angular scale. This kind of survey is good for discovering large low density regions or voids — and the first one did just that. The survey uncovered the void in Boötes. Kirshner *et al.* (1981) started with surveys in three widely separated directions. They found, astoundingly enough, that the distribution of redshifts in the three directions is correlated. They found a low density region in the velocity range from 12000 to 18000 km s^{-1}, much larger than any previously discovered low density region.

Many (including me) were suspicious of the Kirshner *et al.* (1981) result. We argued that such voids must be rare and that the region of the survey was somehow specially selected. We were all destined to change our minds. Kirshner *et al.* (1987) continued to survey the region and confirmed the presence of the void. Their survey of the region now included 239 galaxies, but the survey covers only about 10% of the volume of the void.

Surveys which are complete to some magnitude (flux) limit reveal the relationship between the structure in the galaxy distribution and the voids. The more dense the survey, the higher the effective "signal-to-noise". The completed regions of the CfA survey demonstrate this point.

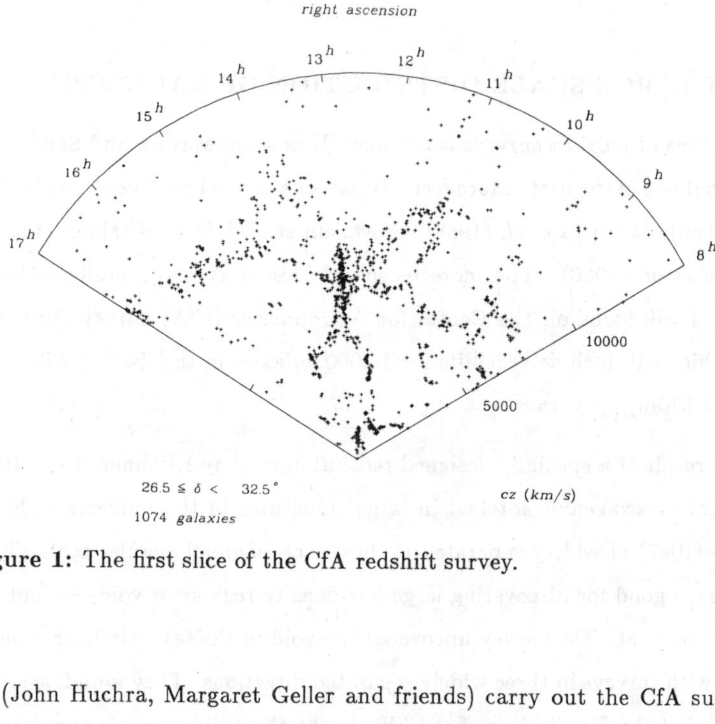

Figure 1: The first slice of the CfA redshift survey.

We (John Huchra, Margaret Geller and friends) carry out the CfA survey in 6°-wide strips at constant declination. These strips extend for as much as 9 hours in right ascension. For a survey of fixed depth, this strategy is an efficient method of constraining the scale of large coherent structures (see Geller 1987; de Lapparent, Geller, and Huchra 1988). The approach is also observationally convenient. So far, six of these slices are complete and a seventh is nearly so. Figure 1 is a cone diagram which shows the distribution of the 1100 galaxies in the first of the complete strips which cross the north pole of the Galaxy (de Lapparent, Geller, and Huchra 1986). This strip covers the declination range $26.5° \leq \delta < 32.5°$ and the right ascension range $8^h \leq \alpha \leq 17^h$. The density of points falls off toward large velocity because the sample is magnitude limited; only the intrinsically most luminous galaxies are visible at the outer reaches of the sample.

The structure visible in Figure 1 is striking: the galaxies are in thin structures which surround or nearly surround voids (low density regions). About half of the galaxies in this slice lie in the obvious structure which runs across the survey at a velocity of ~ 10000 km s^{-1}. The voids are underdense by a factor of ~ 5 relative to the rather poorly determined mean for the sample. The dense structures are overdense by about the same factor. The elongated "finger" at $\sim 13^h$ which points toward the origin (observer) is the rich Coma cluster of galaxies.

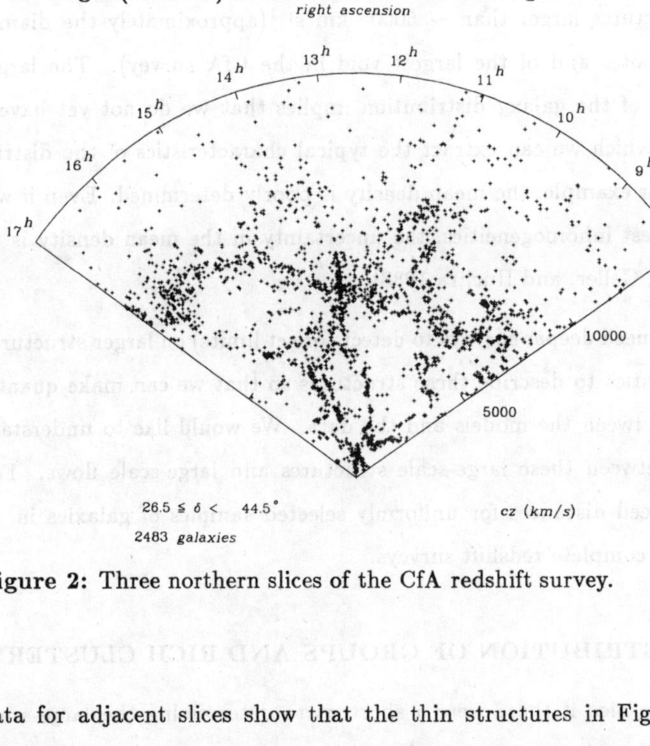

Figure 2: Three northern slices of the CfA redshift survey.

Data for adjacent slices show that the thin structures in Figure 1 are cuts through thin sheets. Figure 2 shows an 18°-wide wedge with $26.5° \leq \delta < 44.5°$ and covering the same right ascension range as Figure 1. The same general structure is visible. The dense regions are thicker than in Figure 1 because the sheets curve over this wider angle. We are now completing a strip which covers the declination range $8.5° \leq \delta < 14.5°$ and the right ascension range $8^h \leq \alpha \leq 17^h$. In this slice, we again see the sheet at $\sim$ 10,000 km s^{-1}. In other words, this thin sheet (about 500 km s^{-1}thick) is $\sim 50h^{-1}$ Mpc $\times \sim 100h^{-1}$ Mpc. It is not at all clear that either cold or hot dark matter models can produce such sheets. In the case of the cold dark matter model, the volume covered by the currently availablesimulations is smaller than the region we have observed (see, for example, White *et al.* 1987).

There are many important lessons to be drawn from this survey. One is that the inhomogeneities in the galaxy distribution are comparable with the size of the sample. Because we do not have similarly dense deeper surveys, we have no direct limits on structures larger than $\sim$ 5000 km s^{-1}(approximately the diameter of the void in Boötes and of the largest void in the CfA survey). The large-scale inhomogeneity of the galaxy distribution implies that we do not yet have "fair" samples from which we can extract the typical characteristics of the distribution of galaxies. For example, the mean density is poorly determined. Even if we have found the largest inhomogeneities, the uncertainty in the mean density is $\sim$ 25% (de Lapparent, Geller, and Huchra 1988).

We clearly need deeper surveys to detect (or set limits) on larger structures. We also need statistics to describe these structures so that we can make quantitative comparisons between the models and the data. We would like to understand the relationship between these large-scale structures and large-scale flows. For this problem, we need distances for uniformly selected samples of galaxies in regions where we have complete redshift surveys.

III. THE DISTRIBUTION OF GROUPS AND RICH CLUSTERS

It would be nice if there were a shortcut to determining the large-scale distribution of matter (or even galaxies) in the universe. It once appeared that rich clusters of galaxies might provide such a route. However, a problem has emerged. The amplitude of the correlation function for the distribution of rich clusters is a factor of 10–20 larger than the amplitude of the correlation function for individual galaxies (Bahcall and Soneira 1983; Klypin and Kopylov 1983; Postman *et al.* 1986; Shectman 1985; Huchra *et al.* 1989). Clusters thus trace the galaxy distribution in a *biased* way.

How seriously should we take the high amplitude of the cluster correlation function? Does the high amplitude reflect a problem of physics or is it a problem with catalogs of rich clusters? Just what are clusters and how are they identified?

Most of the clusters we observe today were picked out by examination of astronomical plates (Abell 1958; Zwicky *et al.* 1961–1968). Both observers identified regions on the sky where the surface density of galaxies was high relative to the surroundings. In Zwicky's cluster catalog, there are variations in the efficiency of cluster selection form one plate of the Sky Survey to the next (see Postman *et al.* 1986); the Abell catalog is not dense enough to make such variations obvious, but they are undoubtedly present.

Analyses of cluster catalogs all lead to a high amplitude for the cluster correlation function

$$\xi(r) = (r_o/r)^{1.8}$$

where

$$14h^{-1} \leq r_o \leq 24h^{-1}\mathrm{Mpc}$$

and the exponent is very uncertain (Generally the data are inadequate to fit both the amplitude and slope reliably. The slope is often fixed at the value 1.8 determined for the galaxy correlation function.) The cluster correlation function is detectable to a scale of 70h^{-1} Mpc (Bahcall and Soneira 1983; Huchra *et al.* 1989: Postman *et al.* 1986)

There are many problems with the catalogs which are the basis for the correlation function analysis. First, the catalogs contain "clusters" which should not be in them. These ringers are included because they may appear dense on the sky although they are not so dense in redshift space. In fact, cluster catalogs are biased toward systems which are artificially enriched by galaxies and/or systems superposed along the line-of-sight (see contributions by Primack (1989) and by Frenk (1989) in this volume). Then, there are clusters which are missing from the catalog. The CfA redshift survey contains well-populated "fingers" which are missing from the Abell catalog. These systems have the same properties as clusters in the survey which *were* identified by Abell.

Further problems enter when a cluster redshift is based on one or two galaxies. Often the single redshift is for the brightest galaxy in the field which may well be foreground. In fact, for clusters at redshifts $z \gtrsim 0.1$, there are ~ 10 foreground galaxies superposed within $1.5h^{-1}$Mpc of the cluster center. The recently completed deep survey of Huchra *et al.* (1989) provides a good example of this problem. In this survey of 145 clusters, there were nine previous redsjift measurements. Of these, three were substantially in error. Based on a single galaxy, the redshifts were 0.09, 0.1, and 0.12. With 3–5 redshifts, the correct valuse are, respectively, 0.22, 0.23, and 0.29. Such foreground redshifts artificially enhance the clustering amplitude.

A final problem is the small size of cluster redshift surveys. The largest so far include ~ 100 clusters (compared with several thousand redshifts for a galaxy survey). A large system — a supercluster composed of half a dozen or so clusters — included in the sample may significantly affect the determination of the correlation function (Postman *et al.* 1986; Soltan 1988). For proper application of correlation function techniques, the sample must be large enough to include a number of "rich" systems.

Some of these problems may be solved by extracting cluster/group catalogs from complete redshift surveys. A number of objective percolation (or friends-of-friends) algorithms (Huchra and Geller 1982; Tully 1980; Materne 1979) are available for identifying "fingers" in redshift space. We applied the Huchra and Geller (1982) algorithm to the first two slices of the CfA survey (Ramella, Geller, and Huchra

1989). The algorithm identifies the seven Abell clusters in the slices along with 29 other "rich" groups which contain five or more members. The properties of the rich groups are frequently indistinguishable from the properties of the Abell clusters. Furthermore, the entire sample of 128 groups in these slices trace the large-scale structure in the region. The amplitude of the correlation function for these groups is indistinguishable from the amplitude of the galaxy correlation function in the same region (Ramella, Geller, and Huchra 1989). Unlike the cluster correlation function, the group correlation function appears to be unbiased.

IV. CONCLUSIONS

Now, what data do we need to sort out some of the issues discussed above? Obviously, there is lots of argument about large-scale flows. To resolve the issues, we need distances (as well as redshifts) for large numbers of galaxies. We need a distance estimator good to 10% for an individual galaxy. We would like to have these distance estimates and redshifts for complete magnitude limited samples of galaxies. It is much more difficult to make the distance measurements than it is to do the redshift survey.

In carrying out redshift surveys there is an important issue which rarely comes to the fore (probably because it is such hard work to solve the problem). The galaxy catalogs which are the starting point for redshift surveys do not have accurate photometry. The r.m.s. error in the Zwicky *et al.* (1961-1968) magnitudes is $\sim 0.3^m$, and the systematic errors are probably comparable with the random error. Errors of this size are large enough to cause visible problems in the data; in the CfA survey there are clear systematic variations in the density of the survey which are probably caused by systematic variations in the effective magnitude limit. A digital sky survey is the ultimate solution to this problem.

In my opinion, the most profound issue raised by the redshift survey is that of large-scale homogeneity. What is the scale (if any) over which the galaxy distribution may be regarded as homogeneous? Another way of stating the question is "Do we have a fair sample of the universe?" So far the largest high contrast structures we observe in the distribution of galaxies are limited only by the size of our surveys. The structures we observe are comparable in scale with the depth of the surveys. To address this issue, we need deeper redshift surveys. A number of groups are preparing to undertake such projects.

Finally, the debate over clusters of galaxies and individual galaxies as tracers of the large-scale matter distribution may, in my opinion, be more a problem of methods of cluster identification than of physics. The optimal way to define clusters is to identify fingers in complete redshift surveys. Because of the rarity of the richest systems, surveys of large volumes of the universe are necessary for a complete and direct approach to this problem. However, the extension of the CfA survey will provide interesting limits for poorer systems (see Ramella *et al.* 1989). Another promising approach is selection of clusters from x-ray surveys (see Lahav *et al.* 1989).

Although it now seems clear how to approach some of the issues of large-scale structure, the amount of data required is large. One evening, John Huchra and I tried to guess what we would know about large-scale structure when we retire about twenty-five years from now. We plotted the number of redshift measurements as a function of time. By the time we retire, if we stay on the same exponential curve, there will be a few million redshift measurements. The number may sound large, but out to a redshift of 0.1, there are already millions of galaxies. Perhaps clever sampling will enable us to understand without having to "see" all.

REFERENCES

Aaronson, M., Huchra, J.P., and Mould, J. 1979, *Ap.J.*, **229**, 1.

Aaronson, M., Bothun, G.D., Cornell, M.E., Dawe, J.A., Dickens, R.J., Hall, P.J., Sheng, H.M., Huchra, J.P., Lucey, J.R., Mould, J.R., Murray, J.D., Schommer, R.A., and Wright, A.E. 1989, *Ap.J.*, **337**, ???.

Abell, G.O. 1958, *Ap.J. Suppl.*, **3**, 211.

Bahcall, N.A. and Soneira, R.M. 1983, *Ap.J.*, **270**, 20.

Burstein, D, Davies, R.L., Dressler, A., Faber, S.M., Stone, R.P.S., Lynden-Bell, D., Terlevich, R.J., and Wegner, G. 1986 in *Galaxy Distances and Deviations from the Hubble Expansion* ed. B. Madore and R.B. Tully (Boston: Reidel) p. 123.

Burstein, D., Davies, R.L., Dressler, A., Faber, S.M., Stone, R.P.S., Lynden-Bell, D., Terlevich, R.J., and Wegner, G. 1988, *Ap.J. Suppl.*, **64**, 601.

da Costa, L.N., Pellegrini, P.S., Sargent, W.L.W., Tonry, J., Davis, M., Meiksin, A., and Latham, D.W. 1988, *Ap.J.*, **327**, 544.

Davis, M. Efstathiou, G., Frenk, C. and White, S.D.M. 1985, *Ap.J.*, **292**, 371.

Frenk, C. 1989, this conference.

Geller, M.J. 1987 in *Large Scale Structures in the Universe*, eds. L. Martinet and M. Mayor, (Sauverny: Geneva Observatory), p.71.

Giovanelli, R., Haynes, M.P., Myers, S.T., and Roth, J. 1986, *A.J.*, **92**, 250.

Gorski, K., Davis, M., Strauss, M., White, S.D.M., and Yahil, A. 1989, preprint.

Groth, E.J., Juszkiewicz, R.,and Ostriker, J.P. 1989, preprint.

Haynes, M.P. and Giovanelli, R. 1986, *Ap. J. (Letters)*, **306**, L55.

Huchra *et al.* 1992, in preparation.

Huchra, J.P. and Geller, M.J. 1982, *Ap.J*, **257**, 423.

Huchra, J.P., Henry, J.P., Postman, M., and Geller, M.J. 1989, in preparation.

Humason, M.L., Mayall, N.U., and Sandage, A.R. 1956, *A.J.*, **61**, 97.

Kirshner, R.P., Oemler, A. Jr., Schechter, P.L., and Shectman, S.A. 1987, *Ap.J.*, **314**, 493.

Klypin, A.A., and Kopylov, A.I. 1983,*Sov. Astr. Lett.*, **9**, 41.

Koo, D. Kron, R., and Szalay, A. 1986 in *13th Texas Symposium on Relativistic Astrophysics*, M. Ulmer, ed. (World Scientific: Singapore).

Lahav, O., Edge, A.C., Fabian, A.C., and Putney, A. 1989, *M.N.R.A.S.*, in press.

de Lapparent, V., Geller, M.J., and Huchra, J.P. 1986, *Ap.J. (Letters)*, **202**, L1.

de Lapparent, V., Geller, M.J., and Huchra, J.P. 1988, *Ap.J.*, **332**, 44.

Lynden-Bell, D., Faber, S.M., Burstein, D., Davies, R.L., Dressler, A., Terlevich, R.J., and Wegner, G. 1988, *Ap.J..* **326**, 19.

Materne, J. 1979, *Astron. Astrophys.*, **74**, 235.

Mould, J.R. *et al.* 1989, this conference.

Nilson, P. 1973 *Uppsala General Catalog of Galaxies, Uppsala Astr. Obs. Ann.*, **6**.

Peterson, B.A., Ellis, R.S., Efstathiou, G., Shanks, T., Bean, A.J., Fong, R. and Zen-Long, Z. 1986, *M.N.R.A.S.*, **221**, 233.

Postman, M., Geller, M.J., and Huchra, J.P. 1986, *A.J.*, **91**, 1267.

Primack, J. 1989, this conference.

Ramella, M., Geller, M.J., and Huchra, J.P. 1988, in preparation.

Shectman, S. 1985, *Ap.J. Suppl.*, **57**, 77.

Soltan, A. 1988, *M.N.R.A.S.*, **231**, 309.

Tully, R.B. 1980 *Ap.J.*, **237**, 390.

White, S.D.M., Frenk, C.S., Davis, M., and Efstathiou, G., 1987, *Ap. J.*, **313**, 505.

Zwicky, F.,Herzog, W., Wild, P, Karpowicz, M. and Kowal, C. 1961-1968, *Catalog of Galaxies and of Clusters of Galaxies*, (Pasadena: California Institute of Technology)

LARGE SCALE STRUCTURE IN REAL-REDSHIFT SPACE IN COLD DARK MATTER

Jens Verner Villumsen

Department of Astronomy
The Ohio State University

ABSTRACT

I have studied the large scale structure of a standard Cold Dark Matter (CDM) model in real space and in redshift space with an N-body hierarchical particle-mesh code (HPM). The largest void in a cube of 30000 km/sec is 5000 km/sec across with a central density 1/4 of the mean density. The void is larger and significantly deeper in redshift space. The two-point correlation function in redshift space is enhanced by a factor 1.92 relative to real space in good agreement with linear theory.

A standard CDM model has been evolved to study the large scale structure both in real space and in redshift space. The N-body code (HPM)[1] allows you to follow the large scale structure of a simulation while singling out smaller volumes of space than can be followed with higher resolution in both length and in mass.

A problem for CDM is the existence of large voids such as the Bootes void [2], 15000 km/sec distant with a radius of 3000 km/sec,

Fig. 1

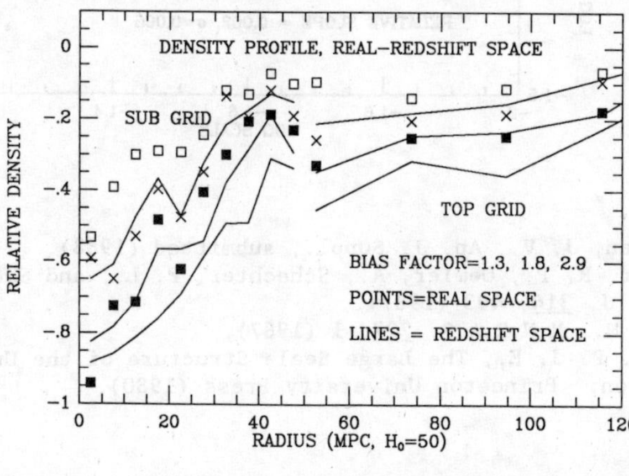

which is a scale on which CDM has little power. The underdensity in galaxies in the void is well established but the underdensity in mass is not known. We generate initial conditions for a standard CDM model 30000 km/sec across and find the largest underdensity on the scale of the Bootes void. In this region we generate initial conditions and evolve the system at higher resolution than elsewhere. The "present" is fixed using the velocity correlation function compared to linear theory with various bias factors. In Fig. 1 is shown the spherically averaged density profile of the simulated void for various bias factors in real and redshift space. The small scale structure is resolved in the sub grid and the large scale structure is resolved in the top grid. The void is approximately 5000 km/sec across nearly independent of bias factor but the depth changes dramatically with bias factor. The void is significantly deeper in redshift space but not much larger.

Kaiser [3] has calculated in linear theory for an isotropic universe the relative amplitude of the correlation functions in real ξ_r and redshift space ξ_s and finds that the correlation function ξ_s is enhanced by a factor $28/15 = 1.87$ relative to real space ξ_r for a flat universe. Fig. 2 shows the correlation functions at a time in the simulation where the density field is still linear. If we fit a powerlaw $\log(\xi_s/\xi_r)(r) = \log b + a \log r$ we find $a = 0.002 \pm 0.005$, $b = 1.92 \pm 0.04$ in good agreement with [2] and in disagreement with Peebles [4] who predicts a ratio 1.67.

Fig. 2

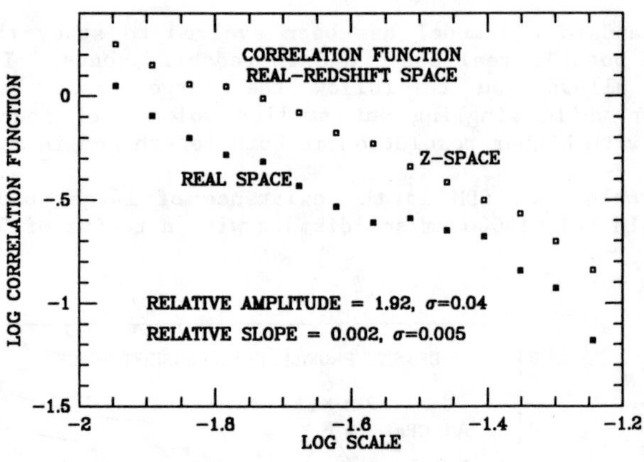

References:

[1] Villumsen, J. V., Ap. J. Suppl., submitted (1988).
[2] Kirshner, R. P., Oemler, A., Schechter, P. L., and Shectman, S. A., Ap. J. 314, 493 (1987).
[3] Kaiser, N., M.N.R.A.S. 227, 1 (1987).
[4] Peebles, P. J. E., The Large Scale Structure of the Universe, Princeton: Princeton University Press (1980).

LARGE SCALE STRUCTURE WITH COSMIC STRINGS AND NEUTRINOS

ROBERT H. BRANDENBERGER

Department of Physics,
Brown University, Providence, Rhode Island, 02912

ABSTRACT

A cosmological model in which the energy density perturbations are seeded by cosmic strings and in which the dark matter is hot (e.g. neutrinos with a mass $25h_{50}^2$ eV) produces a significant amount of nonlinear structure on large scales. Velocity perturbations due to fast moving long strings produce a concentration of galaxies in sheets with a thickness of about 1 Mpc and an average length of about $50h_{50}^{-2}$ Mpc.

This is a preliminary report on some recent work performed in collaboration with Leandros Perivolaropoulos and Albert Stebbins[1)] exploring the large scale structure which results in a model with cosmic strings and neutrinos.

Any cosmological model must specify the nature of the dark matter which makes up at least 90% of the mass of the universe. It must also specify a source of energy density perturbations. In the "standard cold dark matter model" the dark matter has negligible thermal velocity and the source of density perturbations are quantum fluctuations during a period of inflation in the very early universe. In the model considered here, cosmic strings[2)], hypothetical remnants of a phase transition in the very early universe, are taken to be the source of density perturbation.

Galaxy formation in a model with cosmic strings and hot dark matter has been investigated in Ref. 3. Since cosmic string loops survive neutrino free streaming, perturbations on small scales survive (in contrast to models with adiabatic perturbations). By studying the accretion of neutrinos about a cosmic string loop (treated as a point source) it was found[3)] that this model leads to flat halo rotation curves and to a mass function with a slope of -1.5. The differences compared to the model with cosmic strings and cold dark matter come from the fact that neutrino perturbations on a scale λ can only start to develop once the neutrino free streaming length $\lambda_\gamma(t)$ drops below λ. This leads to a flatter density profile about an individual loop, and to less total mass accreted about small loops. With the cosmic string parameters from Ref. 4, only a fraction $\Omega_e \sim 0.1$ of the mass in a flat universe goes nonlinear about any loop.

In the cosmic string model there are two ways in which nonlinear structures may form. The first is spherical accretion onto loops, the second is wakes - velocity perturbations produced behind rapidly moving long strings. Wakes in a cosmic string model with cold dark matter have been discussed in Ref. 5. It was found that at early times the ratio of mass in wakes to mass in loops is about 10^{-2} and, again using the cosmic string parameters of Ref. 4, even at the present time not more than 10% of the mass of the universe will be in wakes.

With hot dark matter, however, the relative importance of wakes compared to loops is - independent of the precise value of the cosmic string parameters - much greater. We[1)] have used the Gilbert equation approach to analyze the growth of neutrino density perturbations in the wake of a moving string. If $G\mu > 2\cdot 10^{-6}$ (μ is the mass per unit length of string) then nonlinear structures from behind long strings which are present at t_{eq} (the time of equal matter and radiation). These structures will be sheet-like with a length of about $50h_{50}^{-2}$ Mpc (the comoving horizon at t_{eq}) and a thickness of a few h_{50}^{-2} Mpc. The total mass accreted in wakes is comparable with hot and cold dark matter. Hence, since accretion about loops is suppressed, wakes are more important. We find[1)] $\Omega_w \sim \Omega_e$, where Ω_w is the fraction of critical density in wakes. With the cosmic string parameters of Ref. 4, $\Omega_w \sim \Omega_e \sim 0.1$. Thus, in this model no extra biasing is needed to reconcile $\Omega = 1$ with the dynamical mass estimates from clusters.

The most serious problem for the cosmic string model with hot dark matter comes from small scale observations: the halos of dwarf spheroidals cannot be made out of neutrinos. Also, the values of Ω_w and Ω_e depend sensitively on the precise values of the cosmic string parameters (see Ref. 6 for recent results). The model is falsifiable. The value of $G\mu$ required to get nonlinear structures in the wakes of moving strings will soon be testable by millisecond pulsar timing.

ACKNOWLEDGEMENTS

I wish to thank Leandros Perivolaropoulos and Albert Stebbins for an enjoyable collaboration, the ITP in Santa Barbara for hospitality while this work was in progress, and the A.P. Sloan Foundation for a fellowship. This research was supported in part by DOE Grant No. DE-AC02-76ER03130.026 - Tasks K & A to Brown University, and in Santa Barbara by NSF Grant No. PHY·82-17853, supplemented by funds from NASA.

REFERENCES

1. R. Brandenberger, L. Perivolaropoulos & A. Stebbins, Brown Univ. preprint (in preparation).
2. A. Vilenkin, Phys. Rep. **121**, 263 (1985);
   R. Brandenberger, Int. J. Mod. Phys. **A2**, 77 (1987).
3. R. Brandenberger, N. Kaiser, D. Schramm & N. Turok, Phys. Rev. Lett. **59**, 2371 (1987);
   R. Brandenberger, N. Kaiser & N. Turok, Phys. Rev. **D36**, 2242 (1987);
   E. Bertschinger & P. Watts, Ap. J. **328**, 23 (1988).
4. A. Albrecht & N. Turok, Phys. Rev. Lett. **54**, 1868 (1985).
5. T. Vachaspati, Phys. Rev. Lett. **57**, 1655 (1986);
   A. Stebbins et al., Ap. J. **322**, 1 (1987).
6. D. Bennett & F. Bouchet, Phys. Rev. Lett. **60**, 257 (1988).

COLD DARK MATTER WITH GAS

Dongsu Ryu
NASA/Fermilab Astrophysics Center, Fermilab
Batavia, IL 60510

Ethan T. Vishniac and Wei-Hwan Chiang
Department of Astronomy, University of Texas
Austin, TX 78712

ABSTRACT

We have studied numerically the evolution and distribution of galaxies and the intergalactic medium (IGM), as well as collisionless dark matter in a universe dominated by cold dark matter (CDM).

1. THE MODEL

Until now, most calculations of the CDM universe models have concentrated only on the evolution of dark matter distribution, and ignored the complicated hydrodynamic evolution of baryonic component. As a result, their comparisons of the mass and velocity distributions of dark matter with observations of galaxies always had large uncertainties. As a first attempt to solve the complete problem including both dark matter and baryonic matter, we have followed explicitly the evolution of galaxies and the IGM, meaning the baryonic matter between galaxies, as well as dark matter. In our calculations, the Einstein-deSitter universe with $\Omega_o = 1$ and $h = 0.5$ has been considered (here $h = H_o/100\ kms^{-1}Mpc^{-1}$ and H_o is the present value of the Hubble constant). We have assumed that initially dark matter composes 90% and baryonic matter composes 10% of total mass, and that the primordial baryonic matter is comprised of H and He, with the abundance of He equal to 10% of H by number. In our model, galaxies are allowed to form out of the IGM, if the total density and baryonic density satisfy an overdensity criterion. Subsequently, the newly formed galaxies release 10^{60} ergs of energy into the IGM over a period of 10^8 yrs. The inclusion of substantial energy injection from young galaxies makes our calculations a preliminary test of the possibility of explosive galaxy formation. Calculations have been performed with 32^3 dark matter particles and 32^3 cells in a cube with comoving side length $L = 9.6h^{-1}$ Mpc. Dark matter and galaxies have been followed by an *N*-body code using the PM method with the CIC mass assignment scheme. The IGM has been followed by a fluid code using the FCT method.

2. RESULTS

The main results from our model are as follows: 1) In our model, galaxies start to form when $z \sim 20$, and reach a maximum formation rate at $z \sim 9$. After that, the rate decreases until $z \sim 2$ and increases again up to the present. The pattern of galaxy formation depends critically on the way to form galaxies. 2) The dark matter and galaxies are distributed along connected filaments and clumps which encompass large void regions, while the IGM tends to reside mainly in the regions around the filaments and clumps as well as in the void regions. 3) The two-point correlation function of the IGM show a suppressed growth due to the energy injected from galaxies and the mass subtraction to form galaxies, showing the IGM is much less clustered. The two-point correlation function of galaxies is very similar to that of the dark matter but ~1.3 to 1.7 times larger over all separations, indicating a biased galaxy distribution (an estimated factor of the biasing between ~ 1.15 and 1.3). The cross correlation function of the dark matter and IGM has a peak at the comoving separation $x \sim 1.8h^{-1}$ Mpc, indicating a characteristic scale of the separation of the two components. 4) The mean and rms of the relative peculiar velocities of galaxy pairs increase on small separations but decrease on larger separations. The mean of the relative peculiar velocities follows the Hubble line on small scales up to the comoving separation $x \sim 1h^{-1}$ Mpc, showing galaxies are gravitationally bound with each other on these scales. The rms of the relative peculiar velocities are radially biased. 5) The high temperature regions of the IGM correspond to the low density regions and the low temperature regions correspond to the high density regions. The temperature of the IGM increases from $z = 1$ to $z = 0$, while the pressure decreases. The present temperature distribution shows a peak at ~ 10^7 K. The mass fraction of the IGM with the temperature below 10^5 K is negligible, indicating almost all the hydrogen is ionized. Pressure variation is smaller than those of the density and temperature, even though pressure equilibrium is not achieved. 6) Temperature fluctuations in the cosmic microwave background radiation due to the scattering electrons in the hot IGM gas creates temperature anisotropies(~ 4.30×10^{-6}) which are a factor of 3 smaller than the observational limit. The amount of X-ray radiation emitted by the hot IGM plasma is also expected to be much smaller than that observed in the X-ray background radiation. 7) Some of these results depend on our assumptions and the models we have used. Future work with larger number of particles and cells and/or with different assumptions and models is desirable.

A New Family of Distance-Indicator Relations for Galaxies

R. R. de Carvalho and S. Djorgovski
Division of Physics, Mathematics, and Astronomy
California Institute of Technology
Pasadena, CA 91125, USA

ABSTRACT. Measurements of large-scale peculiar velocity fields are of a considerable importance in constraining the cosmogonic models. In addition to the redshifts, relative galaxy distances are required. We describe a new family of distance-indicator relations for early-type galaxies. Distance-dependent quantities, i.e., radius or luminosity, are expressed as distance-independent combination of the mean surface brightness and a color (or any other metallicity indicator), both of which can be obtained from photometry. The importance of the new relations is that the necessary data can be obtained relatively quickly for large numbers of galaxies, and thus the velocity fields can be mapped efficiently in some detail. Our preliminary results indicate that the typical relative distance errors to individual galaxies derived from new relations are $\sim$ 25%, and the accuracy of $\sim$ 10% may be achivable with more homogeneous data; this is comparable to the accuracy obtained from the Tully-Fisher and $D_n - \sigma$ relations.

Recent work on the properties of elliptical galaxies has shown that these objects form a two-parameter family, which can be represented by an inclined plane in the space defined, e.g., by radius, velocity dispersion, and mean surface brightness (Dressler *et al.* 1987, Djorgovski and Davis 1987, Djorgovski 1987). The physical explanation behind this "fundamental plane" of early-type galaxies is simply that (*a*) galaxies are bound by the Newtonian gravity, and (*b*) they have a similar dynamical structure (Djorgovski, de Carvalho, and Han 1988). The relation is very tight, and the relative error of predicted radii is $\sim$ 20% per galaxy. This makes the fundamental plane an improved distance indicator for early-type galaxies, much better than the Faber-Jackson (1976) relation, and as good as the best versions of the Tully-Fisher relation for spirals.

Lynden-Bell *et al.* (1988) applied this distance-indicator relation (in its $D_n - \sigma$ form, proposed by Dressler *et al.* 1987) to map the local velocity field. They found large peculiar motions on the scales of tens of Mpc, and a bulk motion relative to the microwave background in the direction of the Hydra-Centaurus supercluster (the "Great Attractor" model). The amplitude of the motions claimed by Lynden-Bell *et al.* (1988) is probably incompatible with the Cold Dark Matter scenario (Vittorio *et al.* 1986). If confirmed, their result would have important implications for theories of the formation and evolution of galaxies and the large-scale structure. Using the infra-red Tully-Fisher relation, Aaronson *et al.* (1989) also find large peculiar motions in the direction of the Hydra-Centaurus, but there are important differences in their peculiar velocities of individual clusters, compared to the measurements by Lynden-Bell *et al.* (1988). Strauss and Davis (1988) point out some other possible difficulties. There is thus no consensus at this time on the exact nature of the large-scale peculiar velocity field on the scales of ~ 100 Mpc around us. An independent set of measurements for a large number of galaxies would be very valuable in clarifying the situation.

We discovered a new family of distance indicator relations for early-type galaxies, in which a distance-dependent quantity (radius or luminosity) is correlated with a linear combination of the mean surface brightness, and a metallicity term, e.g., a color or the Mg index (de Carvalho & Djorgovski 1989). The new radius – metallicity indicator – surface brightness relations can be understood as a generalization of the fundamental plane of elliptical galaxies: The velocity dispersion term there basically measures the depth of the galaxy's potential well. Since the metallicity of an elliptical galaxy also depends on the depth of the potential well, one can substitute a metallicity indicator in place of the velocity dispersion. Some examples of the new relations are shown in Figure 1.

Using the effective radii (r_e) and mean (red) surface brightness ($\langle \mu_e \rangle$) from the surface photometry of field ellipticals by Djorgovski and Davis (1987), and the $(B-V)$ colors from de Vaucouleurs and Olson (1982), we derive the relation: $\log r_e/\mathrm{pc} = 2.55\,(B-V) + 0.35\,\langle \mu_e \rangle - 5.5$, which has the scatter in predicted distances of only 28% (this includes the measurement errors and the peculiar velocities of individual galaxies). Using the colors from Burstein *et al.* (1987), we obtain: $\log r_e/\mathrm{pc} = 2.50\,(B-V) + 0.31\,\langle \mu_e \rangle - 4.9$, with the scatter in predicted distances of 36%. Using the $(V-K)$ colors from Persson *et al.* (1979), we obtain: $\log r_e/\mathrm{pc} = 1.35\,(V-K) + 0.30\,\langle \mu_e \rangle - 4.4$, with the scatter in predicted distances of only 27%. The coefficients in these examples are preliminary, and are uncertain by about 10%. We assume $H_0 = 100$ km/s/Mpc.

The new relations work almost as well as the $D_n - \sigma$ relation of Dressler *et al.* and Lynden-Bell *et al.* The typical scatter (relative errors of predicted distances) is about 25 – 30% per galaxy, if the colors are used, as in the examples above. The scatter reduces to $\sim 15-20\%$ if the Mg index from Davies *et al.* (1987) is used, demonstrating that the metallicity is indeed the driving factor. However, the Mg index requires spectroscopy, and is thus more expensive to measure and calibrate in terms of observing time. Using the data from Dressler *et al.*, we obtain in the Virgo and Coma clusters the scatter of 34% and 27% respectively, for the relations using $(B-V)$ colors, and 15% and 18%, for the relations using the Mg index. From the same photometry, the scatter of the $D_n - \sigma$ relation as quoted by Dressler *et al.* (1987) and Lynden-Bell *et al.* (1988) is 22 – 25%. The scatter appears to be limited by the accuracy and homogeneity of the available data. Our present results are based simply on the data available in the literature, and the $(B-V)$ color is not an ideal choice. One could improve the results by choosing a more metallicity-sensitive color, e.g., $(V-K)$, or $(U-R)$.

The importance of the new distance-indicator relations is in the fact that colors are much easier to measure and calibrate than the velocity dispersions, and thus large numbers of galaxies can be surveyed relatively quickly. For that reason, it is likely that the new distance-indicator relations will become very popular, and may soon lead to valuable new results in the mapping of the large-scale structure and velocity field. We have already initiated surveys to obtain large amounts of the necessary photometric data. The Mg index does work better, but it requires a time-consuming spectroscopy, and thus forefits the principal operational advantage of the new relations.

An important implication of the radius (or luminosity) – surface brightness – metallicity relation is that there is a "second parameter" in the mass-metallicity relation for elliptical galaxies, which we identify as the luminosity density. Casting the new relations in that form may provide valuable constraints for the theories of dissipative galaxy formation and the initial enrichment (Djorgovski and de Carvalho 1989, in preparation).

One possible remaining problem is the possibility that *all* of the known distance-indicator relations, including the ones presented here, may vary in different large-scale environments (e.g., field galaxies, groups, and rich clusters), or from one cluster to another. These relations between galaxian properties may well be modified by the processes of galaxy formation and transformation, many of which are environment-dependent: mergers, gas sweeping, tidal interactions, etc. (cf. Djorgovski, de Carvalho, and Han 1988). Such hypothetic variations could masquerade as false peculiar velocities. Reassuringly, the coefficients of our new distance-indicator relations evaluated separately in different clusters are the same, within the nominal errors. Still, a more detailed study is needed to explore the question of universality of distance indicators, and their relation to the formative and evolutionary processes of galaxies and clusters.

The work presented here is supported in part by California Institute of Technology. R.R.dC. is on leave from Observatorio Nacional, CNPq, Rio de Janeiro, Brasil. S.D. acknowledges a partial support from Alfred P. Sloan Foundation.

References:

Aaronson, M., *et al.* 1989, *Ap. J.* in press.
Burstein, D., *et al.* 1987, *Ap. J. Suppl.* **64**, 601.
de Carvalho, R., and Djorgovski, S. 1989, *Ap. J. Letters*, submitted.
Davies, R., *et al.* 1987, *Ap. J. Suppl.* **64**, 581.
Djorgovski, S., and Davis, M. 1987, *Ap. J.* **313**, 59.
Djorgovski, S. 1987, in the proc. of IAU Symposium #127, "Structure and Dynamics of Elliptical Galaxies", ed. T. de Zeeuw, p. 79. Dordrect: D. Reidel.
Djorgovski, S., de Carvalho, R., and Han, M.-S. 1988, in "Extragalactic Distance Scale", S. van den Bergh (ed.), *P.A.S.P. Conf. Suppl.* in press.
Dressler, A., *et al.* 1987, *Ap. J.* **313**, 42.
Faber, S., and Jackson, R. 1976, *Ap. J.* **204**, 668.
Lynden-Bell, D., *et al.* 1988, *Ap. J.* **326**, 16.
Persson, E., Frogel, J., and Aaronson, M. 1979 *Ap. J. Suppl.* **39**, 61.
Strauss, M., and Davis, M. 1988, in the proc. of the Vatican Study Week on Large-Scale Motions in the Universe, in press.
de Vaucouleurs, G., and Olson, D. 1982, *Ap. J.* **256**, 346.
Vittorio, N., Juszkiewicz, R., and Davis, M. 1986, *Nature* **323**, 132.

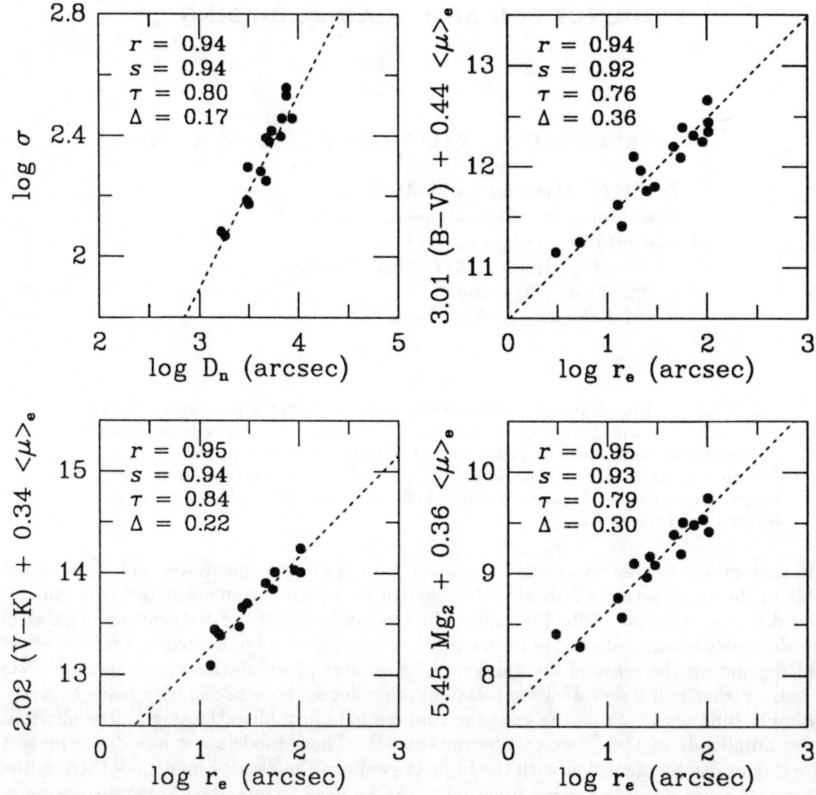

Figure 1: Examples of distance-indicator relations, for a set of the Virgo cluster ellipticals. *Top left:* the $D_n - \sigma$ relation. *Top right:* the new distance-indicator relation, with the $(B - V)$ color as the metallicity indicator. *Bottom left:* similar, but with the $(V - K)$ color. *Bottom right:* similar, but with the Mg index. The fits shown are for the data as plotted, and do not necessarily represent the "best", universal slopes. In each panel, we list the correlation coefficients, linear regression (r), Spearman rank (s), and Kendall rank (τ), and the r.m.s. scatter of the log of radius (Δ), which is also equal to the scatter of the log of predicted distances. The data are from the references listed in the text.

EVIDENCE FOR AND AGAINST BIASING...

A. BLANCHARD[1], D. VALLS–GABAUD[2], J.–M. ALIMI[1,3]

[1] *D.A.E.C., Observatoire de Meudon*
Place Janssen, 92190 Meudon, France
[2] *Institut d'Astrophysique de Paris*
98 bis, Bld. Arago, 75014 Paris, France
[3] *C.H.T Ecole polytecnique*
91128 Palaiseau Cedex, France

ABSTRACT. Biased theories of galaxy formation predict a luminosity segregation effect in the clustering of galaxies. We briefly review the confrontation of theoretical amplitudes of such an effect with the observed limits. From the fact that galaxies fainter than L_* do not present such a segregation on scales larger than a few h^{-1}Mpc, we deduce that a bias parameter of bright galaxies larger than 1.6 is unlikely.

Biased galaxy formation scenarios rely on the hypothesis that mass and light do not cluster in the same way [1)]. This already constitutes a segregation effect but it is difficult to test directly. A testable effect is expected from the fact that the distribution of galaxies must also show a segregation due to biasing[2)]. Hierarchical models predict that the effect might depend on the mass of the galaxies[3)], as numerical simulations have shown[4)]. Via the Tully-Fisher and Faber-Jackson relations, this dependence necessarily leads to a segregation in luminosity. From the gaussian random field formalism[5)] analytical predictions for the amplitude of the effect can be obtained[2)]. These models are based on the fact that galaxies can be identified with the highest peaks of the density field smoothed on the appropriate scale R_S. A selection function t_q can be used to introduce some dispersion in the correspondence between peaks and galaxies. Finally, the smoothing scale R_S may be a function of the luminosity L of the corresponding galaxy $R_S \propto L^a$. This dependence is likely if the mass of galaxies is close to the mass of the peaks. The range of a allowed by M/L observations is 0 to 1/2.

This luminosity segregation effect has recently been observed by Hamilton[6)] for galaxies brigther than L_* (see also Ref. 7). The comparison of the models with these observations lead to a rather good agreement (Fig. 1a) when the bias parameter for bright galaxies ($M \leq -18.5 + 5\log h$) $b_B \sim 1.8$. Only models with a large dispersion in the peak-galaxy relation fail to match the observed segregation at high luminosities. At luminosities lower than L_* the distribution of galaxies is virtually independent of luminosity[6)8)]. In Ref. 8 we set stringent limits on the ratio α of the J_3 integral of the cross-correlation function between galaxies differing in luminosity by one magnitude (*i.e.*, $\alpha = b(M+1)/b(M)$). This statistic is very robust against the systematic errors that arise in the determination of the correlation function. We will therefore use this quantity rather than those defined by Hamilton, although the limits that can be set from both methods are very similar. The comparison of above models with the observed limits on α at $L < L_*$ is shown in Fig. 1b. Even the most extreme models are clearly in disagreement with the data. From a larger analysis[2)], it appears that models with similar values of b_B lead to similar results on α. The selection function can noticeably reduce the discrepancy, but only for models with $b_B < 1.6$, and even these models are marginally consistent with the data. This is a

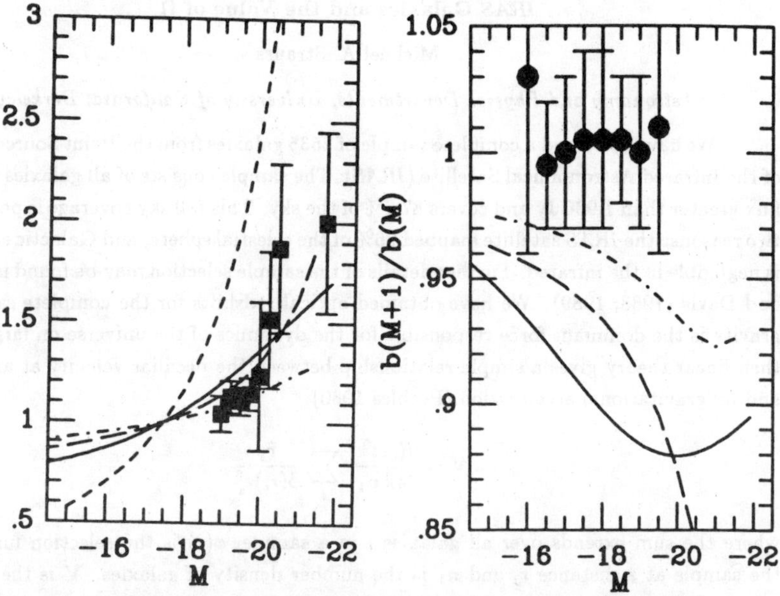

Figure 1 : Comparison of the expected luminosity segregation in biased models with predictions from the observations[6)8)]. The full line corresponds to the results from numerical simulations. The short–dashed line correspond to the prediction of the Schaffer's models with $\nu = 1$), the long dashed line and dotted-dashed line correspond to BBKS models respectively with $b_B = 1.8$ and $b_B = 1.6$

major problem for biased theories, and in fact it seems very difficult to avoid any luminosity segregation when the bias hypothesis is used. However, if the dark matter is non-baryonic, the nucleosynthesis constraint implies that about 90 % of the mass is unseen, leaving an open door for modifying biased theories. On the other hand one can wonder if the samples used are large enough to give a convincing answer to the problem of luminosity segregation. The segregation analysis used here makes full use of the whole information present in the sample, so that in order to reduce the statistical uncertainty need samples containing a much larger number of galaxies. It remains that the limits that can be set on the present catalogues are in strong conflict with the present form of biased models.

REFERENCES

[1)] Bardeen, J.M., *Inner Space/Outer Space Conference* , Chicago Univ. Press p. 212 (1986) and Kaiser, N. *idem* p. 258.

[2)] Valls–Gabaud, D., Alimi, J.–M., Blanchard, A., *preprint* (1988).

[3)] Schaeffer R., *Astron. Astrophys.* **180**, L5 (1987).

[4)] White, S.D.M., Davis, M., Efstathiou, G., Frenk, C.S., *Nature* **330**, 451 (1987).

[5)] Bardeen, J.M., Bond, J.R., Kaiser, N., Szalay, A. *Astrophys. J.* **304**, 15 (1986).

[6)] Hamilton, A.J.S., *Astrophys. J.* **331**, L59 (1988).

[7)] Davis, M., Meiksin, A., Strauss, M.A., da Costa, L.N. and Yahil, A. *Astrophys. J.* **333**, L9 (1988).

[8)] Alimi, J.–M., Valls–Gabaud, D., Blanchard, A., *Astron. Astrophys.* **206**, L11 (1988).

IRAS Galaxies and the Value of Ω

Michael A. Strauss

Astronomy and Physics Departments, University of California, Berkeley

We have extracted a complete sample of 2535 galaxies from the Point Source Catalog of the Infrared Astronomical Satellite (*IRAS*). The sample consists of all galaxies with 60μ flux greater than 1.936 Jy and covers 87.7% of the sky. This full sky coverage is possible for two reasons: the *IRAS* satellite mapped 96% of the celestial sphere, and Galactic extinction is negligible in the infrared. Further details of the sample selection may be found in Strauss and Davis (1988; 1989). We have obtained optical redshifts for the complete sample. If gravity is the dominant force responsible for the dynamics of the universe on large scales, then linear theory gives a simple relationship between the peculiar velocity at any point, and its gravitational acceleration (Peebles 1980):

$$\mathbf{V} = \frac{H_0 \Omega^{0.6}}{4\pi n_1} \sum_i \frac{\hat{\mathbf{r}}_i}{\phi(r_i) r_i^2}, \tag{1}$$

where the sum extends over all galaxies i in a sample, $\phi(r)$ is the selection function of the sample at a distance r, and n_1 is the number density of galaxies. $\mathbf{V}$ is the peculiar velocity of the Local Group, known to high accuracy from measurements of the dipole anisotropy of the Cosmic Microwave Background. A number of groups (Yahil *et al.* 1986; Villumsen and Strauss 1987; Lahav *et al.* 1988) have measured the flux-weighted *angular* dipole of the *IRAS* galaxies; the fact that flux and gravity both follow an inverse square law allows one to use eqn. (1) to calculate Ω. They find values ranging from 0.8 to 1.2. We have calculated the number-weighted acceleration on the Local Group using our redshifts, and find a similar value, $\Omega \approx 0.7$. The acceleration on the Local Group converges within $4000\,\mathrm{km\,s^{-1}}$ of us. The acceleration vector points in the direction $l = 249°$, $b = +46°$, only 23° away from our peculiar velocity vector.

It has been suggested (Kaiser 1988) that peculiar velocities can severely bias the estimation of Ω using eqn. (1). We can in fact use eqn. (1) to make predictions for the peculiar velocities of the galaxies in the sample, and correct them self-consistently (Strauss and Davis 1989; Yahil 1989); when we do this, we find that the acceleration of the Local Group is essentially unchanged in direction or amplitude. We have compared the *IRAS* predictions for the peculiar velocities of galaxies with measured values using the Tully-Fisher (Aaronson *et al.* 1982) or $D_n - \sigma$ relations (Dressler *et al.* 1987); this comparison offers an independent check on our value for Ω. We find excellent agreement with spirals in the Local Supercluster, but systematically underestimate the peculiar velocities of ellipticals on somewhat larger scales (Strauss and Davis 1989). The situation may be improved by adding a large mass concentration beyond the Hydra-Centaurus supercluster (the "Great Attractor") to our model; work is ongoing to quantify this.

A major drawback of the *IRAS* galaxy sample is the fact that elliptical galaxies, which dominate in the cores of rich clusters, are weak emitters of infrared radiation. Comparison with number counts in redshift surveys of optically selected samples in fact indicate that the *IRAS* sample systematically underrepresents the overdensities of clusters by a factor of two. In order to compensate for this effect, we have double-counted all galaxies within a redshift distance of 600 km s^{-1} of the centers of 78 clusters and rich groups within 7500 km s^{-1}, a list kindly supplied by J. Huchra. This increases the gravity on the Local Group, and the inferred value of Ω drops to 0.29. The *direction* of the acceleration, however, changes very little, to $l = 257°$, $b = +47°$, 20° from the velocity vector, and the convergence of the acceleration within 4000 km s^{-1} becomes even more pronounced. The velocity field beyond the Local Group also does not change qualitatively, and the comments above on the comparison to the spiral and elliptical samples still hold. This implies that galaxies and clusters both qualitatively trace the same structures, although one may be biased with respect to the other.

Dynamical measures of Ω using optically selected galaxies have typically found values of 0.3 and less (Davis and Peebles 1983). This may be understood as due to bias (Kaiser 1984); if we define a bias factor b such that $(\delta\rho/\rho)_{galaxies} = b\,(\delta\rho/\rho)_{matter}$, then *IRAS* galaxies have a lower value of b than do optically-selected galaxies. If this linear bias holds, relative values of b can be measured in two ways. The value of Ω found by eqn. (1) varies as $b^{-5/3}$. Alternatively, the amplitude of the galaxy two-point correlation function $\xi(r)$ scales as b^2. We have measured ξ for the *IRAS* sample and for the CfA optically selected sample using volume-limited subsets to 4000 km s^{-1}; the relative value of b is small, only about 1.3. The comparison of derived Ω values implies a larger value of $b \approx 1.7$. The relative b between the *IRAS* boosted sample and the CfA sample is close to unity by both techniques. The discrepancy in correlation function amplitudes and values of Ω probably implies that simple linear biasing is a poor approximation to the real universe. How either the *IRAS* or CfA samples are biased with respect to the true matter distribution remains an unknown.

I thank my colleagues on this project: Marc Davis, John Huchra, Amos Yahil, and John Tonry, for permission to use our results prior to publication. The support of a National Science Foundation Graduate Fellowship is gratefully acknowledged.

Aaronson, M., *et al.* 1982, *Ap. J. Suppl.*, **50**, 241.
Davis, M., and Peebles, P.J.E. 1983, *Ann. Rev. Astr. Ap.*, **21**, 109.
Dressler, A., *et al.* 1987, *Ap. J.*, **313**, 42.
Kaiser, N. 1984, *Ap. J. (Letters)*, **284**, L49.
Kaiser, N. 1988, *M.N.R.A.S.*, **231**, 149.
Lahav, O., Rowan-Robinson, M., and Lynden-Bell, D. 1988, *MNRAS*, **234**, 677.
Peebles, P.J.E. 1980, *The Large Scale Structure of the Universe*, (Princeton: Princeton University Press).
Strauss, M.A., and Davis, M. 1988, in Proc. IAU Symp. # 130, *Large Scale Structures of the Universe*, ed. J. Audouze *et al.* (Dordrecht: Reidel), p. 191.
Strauss, M.A., and Davis, M. 1989, in *Large Scale Motions in the Universe*, Pont. Acad. Sci. Study Week #27, in press.
Villumsen, J.V., and Strauss, M.A. 1987, *Ap. J.*, **322**, 37.
Yahil, A. 1988, in *Large Scale Motions in the Universe*, in press.
Yahil, A., Walker, D., and Rowan-Robinson, M. 1986, *Ap. J. (Letters)*, **301**, L1.

SUMMARY OF WORKSHOP ON LARGE SCALE STRUCTURE

Marc Davis
Depts. of Astronomy and Physics
University of California, Berkeley

INTRODUCTION

The workshop on large scale structure was small and intimate. In the two afternoon sessions and in Margaret Geller's introductory lecture, attention was focused primarily on current controversies. In particular we discussed (1) correlation properties of rich clusters, (2) new determinations of Ω, (3) the nature of the large scale density field, and (4) large scale velocity fields. In what follows, I shall discuss each of these topics in turn.

1. RICH CLUSTERS

Carlos Frenk summarized his recent work on projection contamination in the definition of Abell like clusters. Using N–body simulations he was able to show that selection of clusters by means of surface brightness enhancements will unavoidably lead to serious contamination of a large fraction of the clusters. The clusters selected by two dimensional concentration are very different from those selected in three dimensional isolation. His graphic examples were very telling. Even if full redshift information is available, cluster selection is ambiguous and can lead to serious systematic errors.

Joel Primack discussed his recent work with Avishai Dekel, George Blumenthal, and Scott Olivier on studies of projection contamination of the Abell catalog. They argue rather convincingly that outlying galaxies from foreground clusters increase the probability that a background cluster is above the threshold of selection for the Abell list. This in turn biases the estimation of the cluster–cluster correlation ξ_{cc} function upward. The published values of ξ_{cc} are a factor of 25 larger than ξ_{gg}, which has been a long standing crisis for cold dark matter models that predict an amplification of only 2–3. Primack et al have demonstrated that once the catalog is decontaminated, the estimate of ξ_{cc} drops substantially, and presents no conflict with CDM models. Similar conclusions had been reached previously by Sutherland.

The bottom line seems to be that we should heed George Abell's warning, and not use his catalog for statistical purposes. New catalogs of clusters are on the horizon. The ROSAT satellite will scan the entire sky in soft X-rays and is expected to produce a list of thousands of clusters, but this list will contain both rich and poor groups so much followup optical observation will be required. The ongoing APM–COSMOS plate scanning projects in England should soon produce a new cluster catalog over one steradian in the Southern sky. Within the next several years the new data will hopefully yield good constraints on models of large scale structure

2. NEW MEASUREMENTS OF Ω

We heard two talks concerning promising techniques for measurement of the density parameter. The first, by Ed Loh, involves counting $N(z)dz$, the number density of galaxies per unit redshift, which is sensitive the the global curvature of the Universe and so measures all the mass density, not simply the clustered component of the mass density. To perform this test Loh and Spillar use 5 color CCD data of approximately 1000 galaxies to infer individual redshifts. Their sample extends to approximately $z = 1$, and they have concluded $\Omega \sim 1.1 \pm 0.3$. This type of test has considerable promise for the future. As multi–fiber optic spectroscopy systems are developed it will be possible to obtain real redshifts for faint samples of this size, which will allay the fears of the astronomers, and help clarify the evolutionary effects that could be biasing this conclusion. This work is important and hopefully will continue.

Michael Strauss spoke about using the IRAS redshift survey to measure the local density field, from which one can infer the local gravity field. Linear perturbation theory argues that the gravity field should be aligned with the velocity field. Applying this analysis to our own motion, as given by the microwave dipole anisotropy, allows one to solve for Ω, assuming the IRAS galaxies trace the mass. The problem with this test is the systematic uncertainty concerning how well *any* class of galaxies trace the underlying mass distribution. Nevertheless, the IRAS results are intriguing for several studies in large scale structure.

3. LARGE SCALE MATTER FIELDS

Margaret Geller, in her introductory talk, showed beautiful projections of the galaxy distribution from her ongoing survey work. As she emphasized, large sheets are observed in the CfA survey, with a length comparable to the dimensions of the survey itself. What can we conclude from these large coherent structures? A fundamental question is whether these observations of large structures can be used to rule out any models of structure in the Universe. For example, in the CDM model, one would naively expect the typical coherence length of structures to be of order $10 - 20h^{-2}$ Mpc.

The answer to questions of this nature cannot be decided on the basis of beauty contests that pictorially compare observations to numerical simulations; we must be more quantitative. Correlation functions remain the major tool for quantitative analysis of density fields, and such an analysis applied to the recent CfA data does not show new features. One should caution however that the meaurement of $\xi(r)$ is not stable within subvolumes of optically selected catalogs. These volumes are too small to be fair samples of the Universe. The only known exception to this is the IRAS redshift survey discussed by Strauss, which is a dilute sample of 2500 galaxies covering 87% of the sky to a distance of approximately 8000 km/s. The volume surveyed is 5 times that of the earlier surveys, and it does have correlation behavior that is stable for different subsets of the data. A project just completed by a British collaboration was a redshift survey of every sixth galaxy in the IRAS database to a depth of approximately 16000 km/s. This is similar to the depth but with 10 times the solid angle coverage of the CfA surveys, but is very dilute. It will be extremely interesting to learn whether the $\xi(r)$ for this sample matches the shallower surveys, or shows evidence for more large scale power.

Additional statistical techniques are desperately needed to supplement $\xi(r)$ which gives no information about phase coherence in the distribution. Toward this end, perhaps the recent work on topological studies of the galaxy distribution will provide useful new constraints.

4. LARGE SCALE FLOWS

This is a topic that has been the center of attention for the past several years, and it was fitting that the most heated discussions of the workshop involved the nature of these flows. The observations are difficult, and only recently has it been possible to assemble the large data sets necessary to attack the problem. All the studies use an empirical relation between the absolute luminosity or diameter of a galaxy versus the observed rotational width or central velocity dispersion, which yields a distance estimate with a precision of order 15–20% for individual galaxies.

Alan Dressler presented new data on elliptical and lenticular galaxies in the Centaurus direction, which is consistent with the older data and supports the previous claim of the seven samurai collaboration that this region has a peculiar velocity (velocity in excess of the Hubble flow at that distance) of approximately 1000 km/s. The new data is equally precise to the original data, but goes to a fainter, and therefore deeper limit. It appears the flow of material beyond Centaurus is smaller than for Centaurus itself, but it is not clear whether inflow from the back side is observed or not. It is also not clear how the new work fits with the model of the "great attractor" favored by the seven samurai, and we eagerly await the results of detailed model studies including this enlarged database.

Jeremy Mould reviewed the data for velocity flows of spiral galaxies using the Infrared–Tully Fisher technique. He and his collaborators have also mapped the flow field around in the Centaurus direction and find a lower outflow (500 km/s) for the bulk of the cluster. Other observations have claimed the Centaurus cluster is essentially at rest in the comoving frame! It is apparent that the situation is confused, as the observational results seem contradictory. It is extremely important that studies continue in order to clarify the situation, because the implications for theoretical models are quite significant.

Kris Gorski presented new results on velocity correlation statistics, and compared the statistics derived from the elliptical galaxy data of the seven samurai, from the local supercluster spirals studied by Aaronson, Huchra, and Mould, and from the velocity field inferred from the IRAS *density* field. The results again are not consistent; the elliptical galaxy velocity correlation is four times that of the spiral or IRAS data. Perhaps this is the result of differing depths of the samples, but the situation is worrisome. Gorski and collaborators have also computed velocity correlations for CDM and Peebles baryonic isocurvature (PBI) models and conclude, as have others, that the CDM model is not consistent with the velocity field of the elliptical galaxies, but is consistent with the spiral and IRAS velocity fields, if the CDM bias parameter is $b \leq 2$. He also showed that the effect is rather sensitive to the adopted outflow in the Centaurus region, again emphasizing the importance of clarification of the observations.

Ed Bertschinger presented a beautiful new technique devised in collaboration with Avishai Dekel for inferring the full velocity field given incomplete data. They assume the field is irrotational and therefore derivable from a potential, which should be valid on large scales. Using only observed radial velocities, they can solve for the full vector velocity field and the potential field, and therefore the density field. This will permit a direct comparison to the IRAS density field, and will permit a direct map of the bias of the galaxy distribution $b(\mathbf{r})$. Perhaps by the time of the next Berkeley workshop, results from this study will be available.

In essentially every topic discussed above, new data is on the horizon. Progress in the field of Large Scale Structure has been phenomenal in the last decade. We are now arguing about details that were inconceivable 10 years ago, and the field has moved well beyond the search for two numbers. Many topics remain controversial, which is good because it keeps interest high. I think the next 10 years will see continued steady progress, particularly with the launch of the Hubble Space Telescope and ROSAT.

COSMIC MICROWAVE BACKGROUND

MEASUREMENTS OF THE COSMIC MICROWAVE RADIATION

David Wilkinson
Physics Department
Princeton University
Princeton, NJ 08544 U.S.A.

ABSTRACT

While participants munch their lunch the author attempts to review the status of measurements of the cosmic microwave radiation. Recent results on the spectrum and anisotropy of the CMR are discussed, with emphasis on evidence for spectral distortion and anisotropy.

SPECTRUM MEASUREMENTS

Some of you know, having come to Princeton to visit our research group, that we have brown bag lunches where people get up and give short talks, so I feel right at home. The crunching isn't too loud up here, but let me know in the back if the signal drops below the noise level.

This is a difficult audience because you're ranging all the way from experts in the field, the observers who are going out and getting the data, to particle theorists. I'll do my best to strike a mean somewhere in there. I'm talking about the cosmic microwave background radiation -- the radiation remnant left over from the big bang. There are two key experimental handles on this radiation. One is the spectrum. Is it really a thermal spectrum as expected in the standard cosmological model? Well, maybe. Here's a plot of radiation temperature vs. frequency, a blackbody spectrum is represented by a horizontal line. About half of the available measurements are shown on the plot; these

This work was supported by the National Science Foundation and the National Aeronautics and Space Administration.

are the most precise ones, not necessarily the most accurate ones.

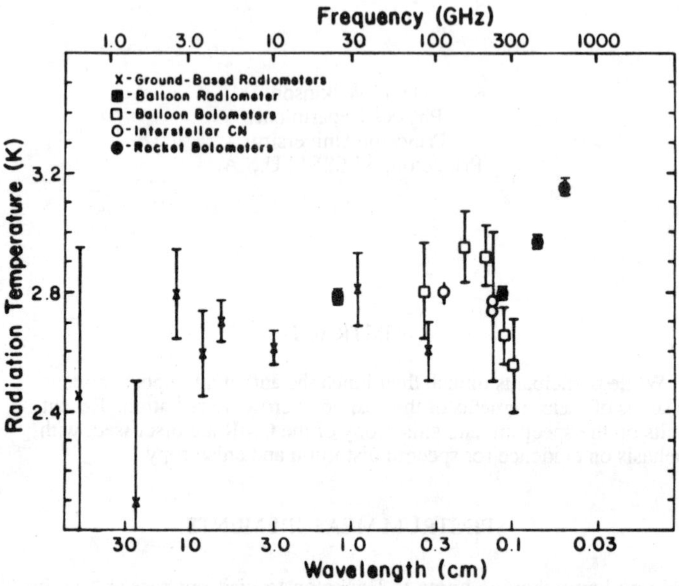

Fig. 1. Measurements of CMR temperature over 3 decades of frequency. Symbols indicate measuring technique.

This work has been going on for 25 years, so the experiments are getting pretty sophisticated. For a moment, to guide your eye, I'm going to cover up those two small points on the right. I'll come back to those. Looking at the rest of the points, you could draw a straight horizontal line which hits most of the measurements over two orders of magnitude in frequency. Remember as you do this that small is good, so your eye should be guided by the points with small error bars, not the big ones.

It's important to notice that five techniques, with differing systematic errors are represented on the plot. The long wavelength measurements tend to be ground-based, and most of those that you're looking at, were done at White Mountain, only 100 miles or so from where we're eating. Those were done by the Berkeley-Haverford-Italian collaboration.[1] You see that error bars range in size. The error bars at shorter wavelengths are dominated by the atmosphere. Measuring the atmospheric emission is a problem, and you need to subtract it from the total sky flux to get the CMR temperature .

The solid square point is a result from a balloon experiment using a Dicke radiometer[2], and the open square points are from balloon-borne bolometric radiometers.[3] So, the square points have been obtained from balloons flying out of Texas. The open ones are were done in Paul Richards group here, and the solid square one is a result obtained by David Johnson and me using heterodyne techniques. You'll notice that there are five points in the plot that, because of their small error bars, dominate a weighted mean temperature. But for the moment I'm ignoring two of the rocket points and keeping the third. It's interesting that the most precise three points were obtained by different techniques. The round, open circles are from optical absorption of starlight by cyanogen molecules floating around in interstellar space. It turns out that cyanogen has two rotational excitation frequencies that are right around the peak of the spectrum of this background radiation. It's nature's perfect cosmic radiation thermometer. One measures the populations in those three levels and infers a radiation temperature at 2.64 mm and another one at 1.32 mm. The precise point at 2.64 mm is a recent one from Crane, et al.[4] This field was rejuvenated a few years ago by Meyer and Jura[5], and Crane's group has pushed down the systematic errors due to possible collisional excitation and self absorption in the cloud.

Now I want to do something considered quite outrageous by some people (with fruit) in this room. I calculate weighted means and chi-squared values for various subsets of the data to check for internal consistency. The reason that this is a shaky procedure is that many of the error bars are dominated by systematic errors, estimated by the experimenters at a confidence level approximately equivalent to one-sigma statistics. If nothing else, the chi-squared values are a measure of how well they've done at the art of estimating systematic errors.

First of all, if you calcualte the weighted mean (leaving out the two points on the right) you get $T = 2.777 \pm 0.012$ K with a chi-square of 26.8 for 16 degrees of freedom -- not bad. The eight ground-based results give $T = 2.655 \pm 0.036$ K with chi-square = 6.1/7. The nine measurements from balloons, rockets and CN (dominated by the three with small errors) give $T = 2.794 \pm 0.013$ K with chi-square = 8.0/8. The mean for ground-based measurements is somewhat lower. Having done several such measurements, 20 years ago, I worry about whether ground radiation is getting into antenna sidelobes when the beams are tipped to measure the atmospheric component. This makes the atmosphere appear too warm, and too much is subtracted from the temperature measured at the zenith. Such an effect is hard to rule out at the 50 mk level. But I'm being very picky. Overall, the 17 results shown here are remarkably consistent, given the wide variety of techniques used and the diversity of associated systematic effects. I would remind you, however, of the history of the measured value of c. "Knowing the answer ahead of time" is a strong effect in experimental physics, the main problem being the tendency to stop trying when you get the "right" answer.

Now let me talk a little bit about these intriguing solid points at short wavelengths which obviously indicate higher temperature. These are from a rocket flight which happened about a year and a half ago -- a Nagoya-Berkeley collaboration.[6] We will be hearing more about it from Andrew Lange in our CMR workshop. The instrument was a He-cooled telescope with six bolometers, each observing a different wavelength band. The three long wavelength channels are shown in Figure 1. The point at 1 mm agrees with the long-wavelength results, but the points at 500 and 700 microns are higher. One important fact about the measurement is that the rocket axis precessed, and no time variation was seen for the flux in those channels. That's arguing that the excess is not from our own galaxy, because a dust signal (for example) would vary with galactic latitude. So, whatever they're seeing in those two channels is fairly isotropic. The lack of modulation in the signal also rules out earthshine in the antenna's sidelobes. The signal is either indicating a very important cosmological effect, not seen before, or it's coming from near the rocket. There is a long history of absolute flux measurements from rockets and balloons having systematic errors. It's hard to check everything about a remote instrument, and in rockets you have only a very short time. This kind of an excess, incidentally, had been indicated in an earlier rocket result from Gush[7], but that flight had known problems with the rocket and the rocket exhaust. The Nagoya-Berkeley rocket achieved a cleaner separation, but the exhaust fumes are still a very difficult problem. For example, one worries about icing of the cold optics. When the cover blows off the 1 kelvin telescope surface is exposed to exhaust gas and residual atmosphere which can condense on the optics and radiate at 500 or 700 microns. That certainly gives an isotropic signal. The exhaust cloud itself is worrisome, as is the residual atmosphere above the rocket. I don't mean to sound too negative. This is a very exciting result. But with this many theorists in the room, I just want to be sure to express some caution. A successful COBE mission will resolve these doubts.

If the flux is cosmological, what is it? One important fact is that the excess (if cosmological) has 10% of the total energy in the cosmic background radiation; so this is not due to some insignificant little effect. It's a major perturbation in our radiation environment. In the discovery paper, the authors suggested two possible causes. One possibility, sometimes called Compton diffusion, postulates a hot innergalactic plasma scattering long wavelength photons up into the Wien side of the spectrum, weakening the Rayleigh-Jeans side of the spectrum. Another possibility would be emission from dust in galaxies at very high red shifts. These are the two popular models at the moment.

Lacey and Field[8], and others, have examined these two models, asking how much energy is needed to ionize the innergalactic medium or to make enough dust and then heat it up in very early galaxies. They argue that if that energy comes from stars that have an approximately standard initial mass function, then there's an energy problem with each of the models. There isn't enough energy available from stars to produce either of these effects and give the observed excess radiation. One way out is to postulate a different initial mass distribution for the first stars. As we've already heard today, that wouldn't be too surprising.

ANISOTROPY MEASUREMENTS

Let's turn to the other major issue in this field, the anisotropy of the radiation. It was realized very early that if the CMR is coming out of the big bang, there should be some interesting processes in the past that might create temperature fluctuations on the sky, and that we might learn something about the early universe from mapping those bumps. The other interesting thing is that we can measure our velocity with respect to the reference frame set by this radiation by measuring the amplitude of the resulting dipole anisotropy. Current results are summarized in Figure 2.

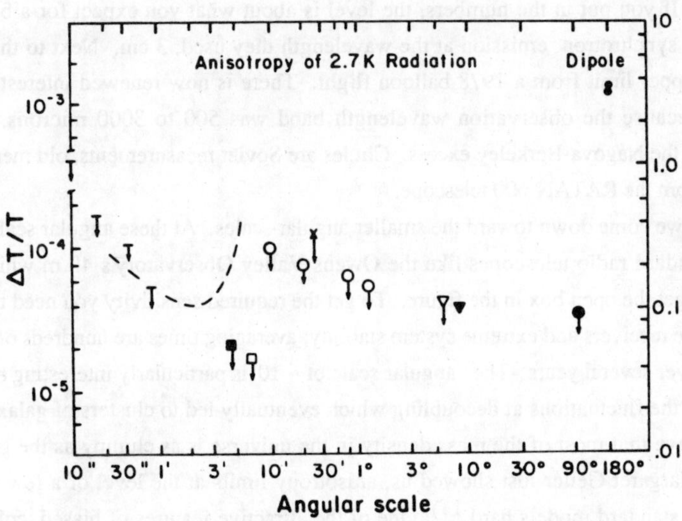

Fig. 2. Measurements of the anisotropy of the CMR. The filled circles are from balloons and a Soviet satellite; most of the others are from ground-based experiments.

Four different groups get about 3.3 millikelvins for the amplitude of the dipole effect, giving about 360 km/sec for the solar speed through the CMR.[9] When our velocity due to solar rotation is subtracted from this (it points roughly opposite to the dipole velocity), the speed of our galaxy through the CMR is about 600 km/sec. Initially, everyone was surprised that it is so large. At this conference we're hearing a lot about very large-scale structure and flow patterns which might explain such a large Galactic speed.

Let's look at the rest of the results, mostly upper limits. At 90° you see the quadrupole upper limit. Data on scales above ~ 30° are obtainted using balloons or satellites. There's no hope of looking at large angular scales from the ground; the atmosphere simply won't let you do it. The balloon-measurement art has now been perfected and in our workshop David Cottingham will show a beautiful sky map which he, Ed Cheng (M.I.T.) and Steve Boughn (Haverford) have obtained with a refurbished Princeton maser. At smaller angular scales, less than 10 degrees in the sky, you can do work from the ground on clear, cold days. Notice that one of the points does not have an error bar on it. That is a measurement[10] of an effect at the level of a few parts in 10^5. The fluctuations have not yet been shown to have the unique spectral signature of cosmic background anisotropy. One worries about whether it's our own galaxy. If you put in the numbers, the level is about what you expect for a bump in Galactic synchrotron emission at the wavelength they used, 3 cm. Next to that is an Italian upper limit from a 1978 balloon flight. There is now renewed interest in this result because the observation wavelength band was 500 to 3000 microns, which includes the Nagoya-Berkeley excess. Circles are Soviet measurements, old measurements from the RATAN-600 telescope.

Now we come down toward the smaller angular scales. At these angular scales one uses standard radio telescopes like the Owens Valley Observatory's 40 m which was used to get the open box in the figure. To get the required sensitivity you need the best low-noise receivers and extreme system stability; averaging times are hundreds of hours, spread over several years. The angular scale of ~ 10' is particularly interesting because it senses the fluctuations at decoupling which eventually led to clusters of galaxies. If you believe that most of the mass density in the universe is as clumpy as the galaxies which Margaret Geller just showed us, anisotropy limits at the level of a few x 10^{-5} push the standard models hard.[11] One of the attractive features of biased, cold-dark matter models is that most of the mass density is smoothly distributed, and the predicted CMR anisotropy levels are lower. However, nearly everyone agrees that if structure is not seen at a few x 10^{-6}, the standard model is in serious trouble.

Since factors of two in the experimental upper limits are now being taken seriously by the theorists, I should mention (especially to this audience) that there is some freedom in how one interprets anisotropy data, which are basically means and standard deviations for ΔT at a small (~ 12) number of spots in the sky. The square upper limits in Fig. 2 are based on a simple statistical test, which was standard in the early work, but suffers from instability. More recent work is being analyzed by a technique that the particle experimentalists would call a Monte Carlo. The experiment is modelled in the computer and then "run" on simulated sky fluctuations, generated from a given theoretical model to be tested. So, the limit depends on the model, which it should. The filled square shows the naive limit in a particularly unlikely case, but one that happened.[12)] The same data run on a model generating Gaussian fluctuations gives the dotted curve. This limit is higher by a factor of 2 and moves the angular scale of maximum sensitivity to somewhere between the beam size (1.5') and the beam throw angle (3.5'). The data leading to the open square is interpreted in the new way in a very lucid paper by the Caltech group.[13)]

Finally, the upper limits to the left in Figure 2 are from using the NRAO's Very Large Array in an extremely difficult mode. They are essentially trying to take pictures of the sky rather than mapping sources, which is what the VLA generally does. They look right down to the noise of the instrument for sky fluctuations at these small angular scales, and the data reduction techniques are formidable. Here one is probing things like the idea that perhaps most of the large scale structure in the universe was caused by explosive events.[14)] If so, you pile a lot of ionized matter up in bubble walls and ridges; those would show up on this scale through the Sunyaev-Zeldovich effect.

The experts in this audience will notice that Figure 2 has not changed much since the 13th Texas Meeting (in Chicago), three years ago. Isotropy measurements have been going on for over 20 years, so by now we've pretty much exploited the standard instruments, techniques and locations (or platforms). Ground-based work is now in a period of development of more sensitive detectors and finding new ways to reduce the effects of atmospheric noise. There are three groups at the South Pole right now making anisotropy measurements of the microwave background. The South Pole may turn out to be an exceedingly good place. You can look straight up and stare at the same spot all day long, and the atmosphere is cold, dry, and thin. Northern Canada and Siberia are also probably good; we've[15)] been working at Saskatoon in the winter, and the atmosphere is not a problem at 40 GHz for about 30% of the time, at the $\Delta T/T \sim 10^{-5}$ level. Better, faster beam switching techniques will help reduce atmospheric noise which has a very red spatial spectrum. Array detectors will enable a large reduction of sky noise effects, because one can then do two-dimensional beam chopping on several angular scales, simultaneously. So, the experimentalists haven't given up, it's just that we're nearing the end of the period where more integration time pays off.

There are instruments being readied to fly in balloons, rockets, and satellites which will make excellent spectral and large scale anisotropy measurements. But, because of the requirement for a large antenna, most medium and small scale anisotropy work will continue from the ground. (The Large Deployable Reflector seems a long way off.)

The Cosmic Background Explorer (COBE) is undergoing final tests at Goddard Space Flight Center for a launch in 1989. John Mather will be telling us about it in the workshop. Basically, it carries a Far Infrared Spectrometer which will measure the CMR spectrum from 300 microns to 8 mm with accuracy exceeding any of the points on Figure 1. Three Differential Microwave Radiometers at wavelengths of 8 mm, 6 mm, and 3 mm will measure anisotropy from 7° to 180° over the whole sky. The several wavelengths will allow Galactic emission to be modelled and removed, and will provide a spectral test for CMR anisotropy. A Diffuse Infrared Background Experiment will measure sky background from 1 micron to 250 microns to look for cosmologically interesting effects, such as the residual radiation from a possible epoch of bright galaxy formation. If COBE works as planned, it will certainly stimulate a lot of ideas and activity over the next few years.

REFERENCES

1. Smoot, , G., Levin, S. M., Witebsky, C., and De Amici, G., Ap. J. 331, 653 (1988) and references therein.
2. Johnson, D. G. and Wilkinson, D. T., Ap. J. Letters 313, L1 (1987).
3. Peterson, J. B., Richards, P. L., and Timusk, T., Phys. Rev. Letters 55, 332 (1985).
4. Crane, P., Hegyi, D. J., Mandolesi, N., and Dauks, A. C., Ap. J. 309, 822 (1986).
5. Meyer, D. M. and Jura, M., Ap. J. 297, 119 (1985).
6. Matsumoto, T., Hayakawa, S., Matsu, H., Murakami, H., Sato, S., Lange, A. E., and Richards, P. L., Ap. J. 329, 567 (1988).
7. Gush, H. P., Phys. Rev. Letters 47, 745 (1981).
8. Lacey, C. G. and Field, G. B., Ap. J. Letters 330, L1 (1988).

9. Fixsen, D. J., Cheng, E. S., and Wilkinson, D. T., Phys. Rev. Letters 50, 620 (1983); Lubin, P. M., Epstein, G. L., and Smoot, G. F., Phys. Rev. Letters 50, 616 (1983); Strukov, I. A., Skulachev, D. P., and Klypin, A. A., Proceedings IAU Symposium #130, Balatonfured (1987); Halpern, M., Benford, R., Meyer, S., Muehlner, D., and Weiss, R., Ap. J. 332, 596 (1988).

10. Davies, R. D., Watson, R., Daintree, E. J., Hopkins, J., Lasenby, A. N., Beckman, J., Sanchez-Almeida, J., and Rebolo, R., Nature 326, 462 (1987).

11. Vittorio, N. and Silk, J., Ap. J. Letters 285, L39 (1984); Bond, J. and Efstatiou, G., Ap. J. Letters 285, L45 (1984).

12. Uson, J. M. and Wilkinson, D. T. Ap. J. 283, 471 (1984); Nature 312, 427 (1985).

13. Readhead, A. C. S., Lawrence, C. R., Myers, S. T., Sargent, W. L. W., Hardebeck, H. E., and Moffet, A. T., Accepted Ap. J. (1989).

14. Ostriker, J. P. and Vishniac, E. T., Ap. J. Letters 306, L51 (1986); Ostriker, J. P. and Thompson, C., Ap. J. Letters 323, L97 (1987).

15. Timbie, P. and Wilkinson, D. T., Accepted Ap. J. Letters (1989).

A rocket borne Fourier transform spectrometer for measurement of the CBR spectrum at short wavelengths.

Herbert Gush and Mark Halpern

Department of Physics, University of British Columbia

Vancouver, Canada V6T 2A6

Measurements of the spectrum of the Cosmic Background Radiation (CBR) provide information about the evolution of the early universe. In particular, departures from a Planckian shape would indicate either unequilibrated energy injection between the epoch when positron-electron pairs were abundant and the epoch of production of neutral hydrogen or very large energy releases since that epoch. (Sunyaev and Zel'dovich (1980), Smoot *et al.*(1988)).

Recent measurements made at wavelengths longer than that of the peak intensity of the CBR spectrum are consistent with a Planck spectrum at T=2.74±.02 K. However at shorter wavelengths observations by Gush(1981,1986), recently substantiated by Matsumoto *et al.*(1988), indicate a non-Planckian spectrum whose brightness rises above 3K for $\nu \geq 15\text{cm}^{-1}$. (For a graph of recent data see, for example, Halpern

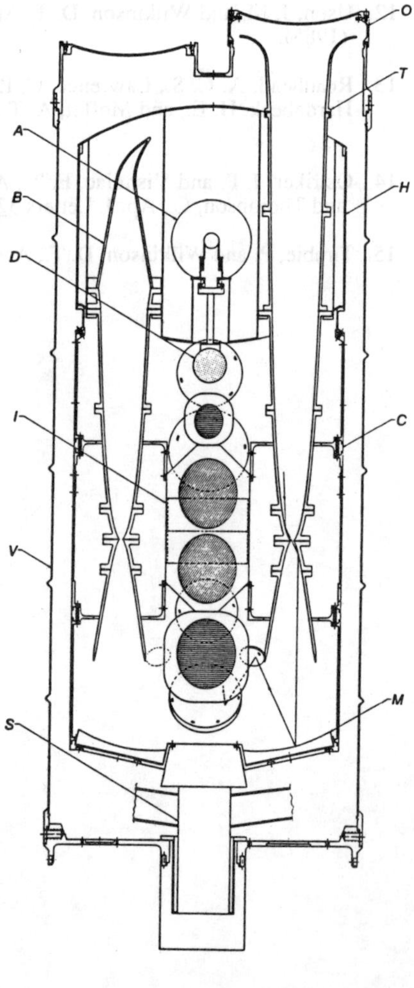

Figure 1. A cross section of the UBC rocket borne spectrometer. The plane of the cut contains the input sky and reference telescopes and the beamsplitter and polarizers of the interferometer. Components shown include: **A**, the ^{3}He refrigerator for cooling the detectors; **B**, the internal blackbody reference made of Pyrex externally coated with indium to shorten its thermal response time; **C**, the interferometer housing, which is cooled to 2K in operation; **D**, ^{3}He cooled bolometric radiation detectors; **H**, the 25 liter liquid Helium reservoir; **I**, the complex of polarizers and mirrors forming the interferometer; **M**, mirrors for directing radiation from the telescope through the interferometer. All the optical elements of this instrument including the input telescopes are made of diamond point machined aluminum, chosen for its low emissivity; **O**, the observation port; **S**, the mechanical support structure containing heat exchangers and vapour cooled radiation shields; **T**, the sky telescope with its diffraction control horn; and **V**, the outer vacuum jacket. The instrument overall diameter is 16″.

et al.) If these measurements are confirmed and the Big Bang scenario is correct one must postulate processes which reheat the CBR. Many theoretical studies are being pursued taking this approach.

Short wavelength measurements are very difficult because they must be carried out above the earth's atmosphere, either by satellites or rockets, using instruments cooled with liquid helium. Checking and eliminating sources of systematic errors is difficult and the 20 year history of this field contains several reports of spectral distortions which have not been confirmed by subsequent measurements. Even though two independent groups have obtained similar results for measurements of a high frequency distortion repeated independent measurements are crucial.

A new rocket borne instrument for short wavelength measurements has been designed and constructed at U.B.C. It is is capable of measuring the emission of a 2.7K thermal source in a 3 to 25cm^{-1} spectral region (λ=3 to .4mm) at a resolution of 1/2 cm^{-1} with a signal to noise ratio sufficient to establish the source temperature to within 10mK in 60 seconds.

The instrument, shown in figure 1, is a cooled Fourier transform spectrometer. It consists basically of a 25 liter liquid helium cryostat operating at 2K to which is attached: 1) a two beam polarization interferometer (see figure 2); 2) two bolometer detectors cooled with He-3 to 0.28K, one at each output port of the interferometer; 3) a cavity simulating a blackbody near 2.7K; and 4) two identical horn type cooled telescopes of 6° field which direct radiation from the sky and from the cavity to the two input ports. The two detectors each measure the difference in brightness between these two input ports.

The optical path difference of the interferometer is repeatedly varied in a triangular ramp fashion between ±1cm, yielding an interferogram from each detector every 1.7s, with the interferogram amplitude proportional to the difference in brightness between the sky and the calibrator. In flight the calibrator temperature will be cycled through three values, say 2.6K, 2.75K and 2.9K. If the sky has a Planck spectrum at 2.75K one would obtain a null interferogram when the internal cavity has that temperature.

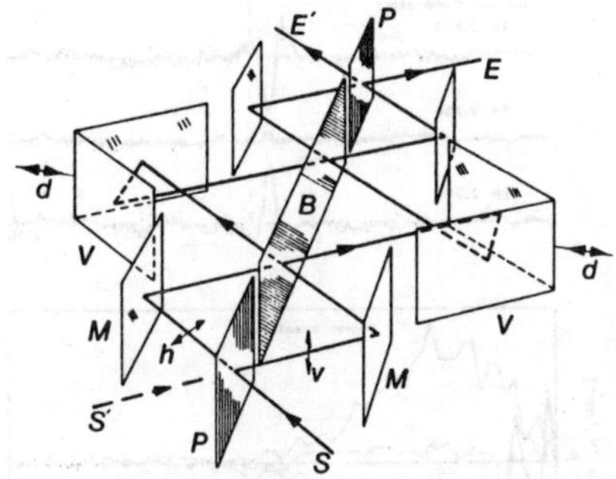

Figure 2. A schematic diagram of the polarization interferometer. The light paths shown lie in a plane perpendicular to that of figure 1. Radiation from a source at **S** is split by the wire grid polarizer, **P**, into vertically and horizontally polarized beams (**v** and **h**). Mirrors, **M**, direct these beams through the interferometer itself, which consists of the oblique wire grid beamsplitter, **B**, and the two 60°dihedral mirror pairs, **V**. Optical delay in the interferometer is varied by simultaneous displacements, **d**, of both dihedral pairs along an axis perpendicular to the plane which contains the beamsplitter. The exit beams, **E** and **E'**, are formed by directing the interferometer outputs through a vertical analyzing polarizer, **P**. The path from the second source, at **S'**, entirely overlaps the radiation path shown here after the input analyzing polarizer. The modulated signals at **E** and **E'** contain information about the brightness difference between **S** and **S'** and are independent of source polarization.

Data taken in the laboratory with the sky telescope terminated with an external liquid helium cooled blackbody radiator are shown in figure 3. At the top are interferograms corresponding to the three temperature steps of the internal reference source. Notice that even the middle interferogram, corresponding to a temperature difference of 26mK, shows a clear interference pattern. These three interferograms together represent just less than one minute of data.

Power spectra are obtained from these interferograms by Fourier transform. These are shown in part **b** of figure 3, plotted as intensity (V/cm^{-1}) versus frequency (cm^{-1}). One extracts the calibration of the instrument from

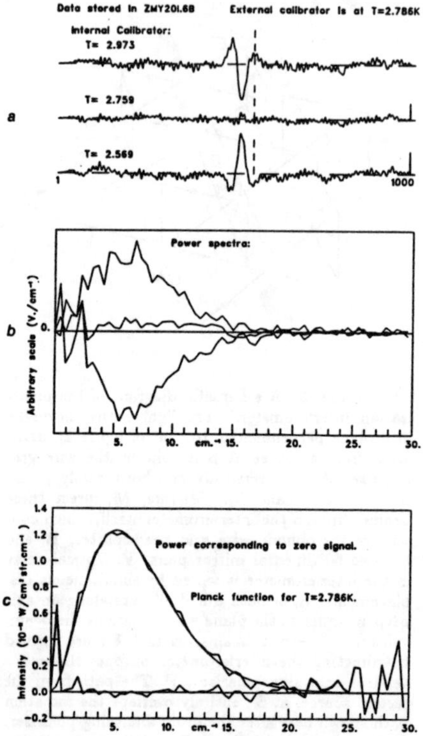

Figure 3. Laboratory data from the UBC spectrometer. The sky telescpope of the spectrometer was terminated with a blackbody simulator which was cooled with an independent liquid helium bath to a temperature near that of the CBR for this test. Interferograms obtained as the internal calibrator is cycled in temperature are shown in part **a** as a plot of detector voltage vs. array adress, which is proportional to optical delay (cm). Power spectra obtained from these interferograms by Fourier transform are shown in part **b**. The inferred brightness of the source filling the sky telescope is the power which would be required to null the instrument. This is plotted in part **c** along with the Planck function calculated for the measured physical temperature of the external load. These curves were not fit to each other. The difference between the two spectra is also shown. The data in this figure were collected in roughly one minute.

the differences between these power spectra by calculating the conversion factor between V/cm^{-1} and $W/cm^2 str.cm^{-1}$ using the assumed brightness of the Planckian spectrum emitted by the internal cavity. The measured brightness of the sky is the power, in $W/cm^2 str.cm^{-1}$, which would be required from the internal cavity in order to null the instrument. This is shown in part **c** of figure 3 along with the independently calculated Planckian spectrum corresponding to the reading of a thermometer mounted on the external calibrator. The difference between the measured and assumed spectra is also plotted in part **c**, giving a rough idea of our possible errors.

At the time of writing (Dec. 88) a number of small modifications to the instrument are being carried out to improve its performance. In particular a lack of null in one detector channel due to a radiation leak is being corrected and various thermometers are being recalibrated. Following this, final calibration runs will be made. It is expected to launch the instrument in the first half of 1989.

Development of the instrument has been funded by the Natural Sciences and Engineering Research Council of Canada, and by the University of British Columbia. The complex construction of this instrument, including electronics, was carried out almost entirely in the Physics Department workshops.

References:

Gush,H.P. "Rocket measurement of the cosmic background submillimeter spectrum."Phys.Rev. Lett.**47**,45(1981)

Gush,H.P. same title, in Gamov Cosmology, F.Melchiorri and R.Ruffini eds. Italian Physical Society, p260(1986).

Halpern,M., Benford,R., Meyer,S., Muehlner,D., and Weiss,R. "Measurement of the anisotropy of the Cosmic Background Radiation and diffuse galactic emission at mm and sub-mm wavelengths", Ap.J.**332**,596(1988).

Matsumoto,T., Hayakawa,S., Matsuo,H., Murakami,H., Sato,D., Lange,A.E., and Richards,P.L. "The submillimeter spectrum of the Cosmic Background Radiation" Ap.J.**329**,567(1988).

Smoot,G., Levin,S., Whitebsky,C., De Amici,G. and Rephaeli,Y. An analysis of recent measurements of the temperature of the Cosmic Microwave Background Radiation", Ap.J.**331**,653(1988).

Sunyaev,R.A. and Zel'dovich,Ya.B., "Microwave Background Radiation as a probe of the contemporary structure and history of the universe.", Ann. Rev. of Astron. and Astrophys., Burbidge, Layzer and Phillips eds. p.537(1980).

NEW MEASUREMENTS OF THE ANISOTROPY AND SPECTRUM OF THE MILLIMETER WAVE BACKGROUND

G.M. Bernstein, M.L. Fischer, and P.L. Richards
Department of Physics, University of California
Berkeley, California 94720, U.S.A.

J.B. Peterson
Department of Physics, Princeton University
Princeton, New Jersey 08540, U.S.A.

T. Timusk
Department of Physics, McMaster University
Hamilton, Ontario L8S 4M1, Canada

ABSTRACT. Recent measurements of the diffuse background at millimeter wavelengths indicate no departure from a Planck spectrum near the peak of the blackbody curve. Anisotropy measurements indicate no structure, at the 2% level, in the recently detected submillimeter excess.

We report here the results of an April 1987 balloon flight of an instrument designed to measure the spectrum of the cosmic background radiation from 1 mm to 3 mm. A description of the instrument can be found in Peterson, Richards, and Timusk [1]. Modifications were made to the apparatus and experimental procedure in order to identify and reduce systematic errors. Results from the latest flight indicate that two effects hamper the interpretation of the data. The first effect arises from refraction at the liquid helium surface, while the second is a small contamination of the beam by emission from warm optics. These systematic effects will be described in detail in a forthcoming publication; they are probably responsible for the non-Planckian spectrum measured by Woody and Richards [2]. Attempts to remove the systematic effects from our data yield upper limits to the CBR brightness temperature in 4 bands from 1 mm to 3 mm. There is no evidence for an excess of radiation near the 2.8 K blackbody peak.

Recent measurements have detected an excess of radiation in the submillimeter region [3]. If we assume that this excess has a spectrum extending into the millimeter region as fit by [4], then we may check its isotropy using data collected during our flight. Details of this analysis may be found in [5]. Our broad bandpass filter (0.5 to 4 mm) is used to obtain the upper limits on anisotropy shown in the figure. This level of isotropy strongly suggests that the submillimeter excess is not of local origin, but must rather originate at substantial redshift. Identification of the sources responsible for this excess is an exciting challenge for submillimeter astronomy.

This work was supported by NASA Grant #NSG-7205.

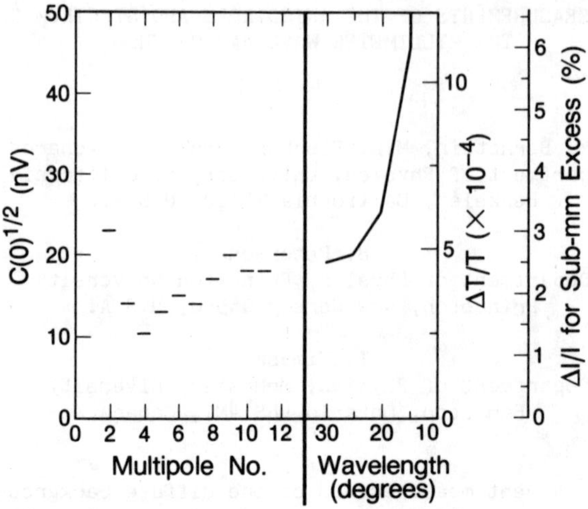

Figure 1. Anisotropy limits versus angular scale. Upper limits, at 95% CL, to the RMS sky fluctuation $\sqrt{C(0)}$ are shown for models with Gaussian fluctuations on various angular scales. Points on the left half are for models with fluctuations at multipole ℓ. The right side curve shows upper limits for models with fluctuations with angular wavelength λ. The right hand vertical axes give the calibration in terms of changes in the 2.74 K blackbody temperature, and in terms of variations in the submillimeter excess.

REFERENCES

[1] Peterson, J.B., Richards, P.L., and Timusk, T. (1985) 'The Spectrum of the Cosmic Background Radiation at Millimeter Wavelengths', Phys. Rev. Letters, 55, 332.

[2] Woody, D.P. and Richards, P.L. (1981) 'Near-Millimeter Spectrum of the Microwave Background', Ap. J., 248, 18.

[3] Matsumoto, T., Hayakawa, S., Matsuo, H., Murakami, H., Sato, S., Lange, A.E., and Richards, P.L. (1988) 'The Submillimeter Spectrum of the Cosmic Background Radiation', Ap. J. 329, 567-571.

[4] Hayakawa, S., Matsumoto, T., Matsuo, H., Murakami, H., Sato, S., Lange, A.E., and Richards, P.L. (1987) 'Cosmological Implication of a New Measurement of the Submillimeter Background', Publ. Astron. Soc. Japan 39, 941-948.

[5] Bernstein, G.M., Fischer, M.L., Richards, P.L., Peterson, J.B., and Timusk, T. (1988) 'Anisotropy of the Diffuse Background at Millimeter Wavelengths', Ap. J. Letters (to appear 1 Feb. 1989).

COSMIC BACKGROUND RADIATION TEMPERATURE AT 2.64 mm, 1.32 mm, AND 0.6 mm

Philippe Crane
European Southern Observatory, Munich

Marc L. Kutner
Rensselaer Polytechnic Institute, Troy

Dennis J. Hegyi
University of Michigan, Ann Arbor

J. C. Blades
Space Telescope Science Institute, Baltimore

E. Palazzi, and N. Mandolesi
Istituto TE.S.R.E., Bologna

ABSTRACT

Values for the Cosmic Background Radiation temperature at 2.64 mm, 1.32 mm, and at 0.556 mm have been determined using interstellar absorption lines. At 2.64 mm, the value for T_{CBR} is 2.796 (+0.19; -0.41) K, which includes all reasonable sources of error. At 1.32 mm, a value of 2.85 ± 0.10 K has been found in the direction of the star HD154368. The error is dominated by uncertainties in the equivalent widths, but includes a substantial uncertainty due to the saturation correction. Observations of interstellar CH toward ζ Oph have yielded a one standard deviation upper limit to T_{CBR} at 0.556 mm of 3.60 K.

1. INTRODUCTION

Interstellar absorption lines provide an interesting alternative technique for studying the Cosmic Background Radiation spectrum at wavelengths that are difficult or impossible using direct ground based techniques. The results from the interstellar absorption lines are subject to relatively few systematic errors. The major source of error is the statistical error in the optical absorption line strengths. This is especially true at the shorter wavelengths.

The results presented here were obtained over several years at the ESO 1.4 m CAT telescope using the Coudé Echelle spectrograph in a variety of combinations of cameras and detectors. Observing and analysis details can be found in the papers of Crane et al.[1),2)].

2. T_{CBR} AT 2.64 mm

The value of T_{CBR} at 2.64 mm was determined from the CN absorption lines at 3875 Å toward ζ Oph. In addition to a precise measurement of the R(0), R(1), and P(1) equivalent widths, a careful analysis of the saturation correction procedure has been made. Also a direct search for CN emission at 2.64 mm from the interstellar cloud toward ζ Oph was negative. This puts strong constraints on <u>any</u> source of local excitation which might excite the CN rotational levels above the CBR temperature.

The resulting value of T_{CBR} = 2.796 (+0.019; -0.041) K is the measured CN excitation temperature and includes uncertainties of 18.4 mK due to the equivalent widths, 5 mK due to a scale uncertainty, -18 mK for the saturation correction, and -31 mK for possible local excitation effects.

3. T_{CBR} AT 1.32 mm

The value of T_{CBR} at 1.32 mm has been determined from the CN absorption lines toward the star HD154368. The CN column density toward HD154368 is about 10 times greater than toward ζ Oph and therefore the R(2) and P(2) lines which are sensitive to T_{CBR} at 1.32 mm are 10 times stronger. The star is, however, about 10 times fainter than ζ Oph. Since the determination of very weak absorption lines is limited by detector performance and possibly by weak atmospheric lines, HD154368 is the best choice for measuring T_{CBR} at 1.32 mm.

The data also provide a measurement of the CBR at 2.64 mm. Since the excitation temperature is about 5 K, the saturation correction is large, and its uncertainty dominates the errors, the data do not provide an interesting value for T_{CBR} at 2.64 mm.

Toward HD154368, we have determined the equivalent widths of the R(1), P(1), R(2), and P(2) lines to be 15.12(0.14), 9.74(0.14), 0.81(0.11), and 0.55(0.11) mÅ respectively. Using an intrinsic linewidth of 14.3 mÅ (FWHM) determined from high resolution spectra for the saturation correction, yields a value of T_{CBR} = 2.85 ± 0.010 K. We have also searched for CN emission from this cloud at 2.64 mm and found nothing to an antenna temperature limit of $T_R^* < 0.012$ K. This provides a strong limit on the possible local excitation of CN at 1.32 mm.

4. T_{CBR} AT 0.556 mm

The CH molecule ground state rotational levels provide the possibility of studying the CBR spectrum at 0.556 mm. The $R_1(1)$ line at 4303.947 Å can be compared to the stronger $R_2(1)$ line at 4300.313 Å to yield a measure of the CBR temperature. For ζ Oph, the expected strength of the $R_1(1)$ line is on 0.005 mÅ if T_{CBR} is 2.8 K.

We have searched for this line and have established an upper limit of 0.030 mÅ which translates into the limit $T_{CBR} < 3.60$ K.

5. DISCUSSION

In view of the increasing interest in the short wavelength spectrum of the CBR, the results reported here provide an independent method of determining the CBR spectrum. The interstellar absorption line technique is a relatively simple and inexpensive alternative to the other techniques.

The present results are limited mainly by the detectors. Weak absorption lines must be detected against the stellar continuum. Non-uniformities in the detector response or low level semi-periodic noise has not been totally eliminated. Also, weak telluric features may represent a more fundamental limitation to this technique.

REFERENCES

1) Crane, P., Hegyi, D.J., Mandolesi, N., and Danks, A.C., Ap.J., 309, 822 (1986).
2) Crane, P., Kutner, M.L., Hegyi, D.J., and Mandolesi, N., 1989, Ap.J., submitted (1989).

FORMALDEHYDE MEASUREMENTS AT 2 mm

A. KOGUT, G.F. SMOOT
UNIVERSITY OF CALIFORNIA, BERKELEY

C.L. BENNETT, S.J. PETUCHOWSKI
NASA GODDARD SPACE FLIGHT CENTER

ABSTRACT: We have used observations of interstellar ortho-formaldehyde (H_2CO) in absorption against a bright continuum source to determine the temperature of the Cosmic Microwave Background Radiation (CMBR) at 2.1 mm wavelength. Using a LVG model to solve the equations of radiative transport in the clouds, we determined the temperature of the CMBR to be 3.2 ± 0.9 K, where the uncertainty is dominated by collisional effects from hydrogen molecules in the clouds.

Most experiments to measure the CMBR temperature are limited to sampling the local radiation field in the vicinity of Earth. Interstellar spectroscopy provides additional evidence for the non-local cause of the CMBR. We have extended the use of interstellar spectroscopy as a probe of the CMBR by observing interstellar formaldehyde.

Formaldehyde is an asymmetric rotor whose rotational energy levels show a more complex structure than a linear molecule (Fig 1). The doublet transitions at λ=6 cm and λ=2 cm, seen in absorption against a bright source of continuum radiation (an HII region), may be used to infer the level populations and hence the temperature of the local radiation field at 2.1 mm wavelength. The largest correction for local effects is the collisional excitation from molecular hydrogen (H_2), the main component of the molecular clouds in which H_2CO is observed.

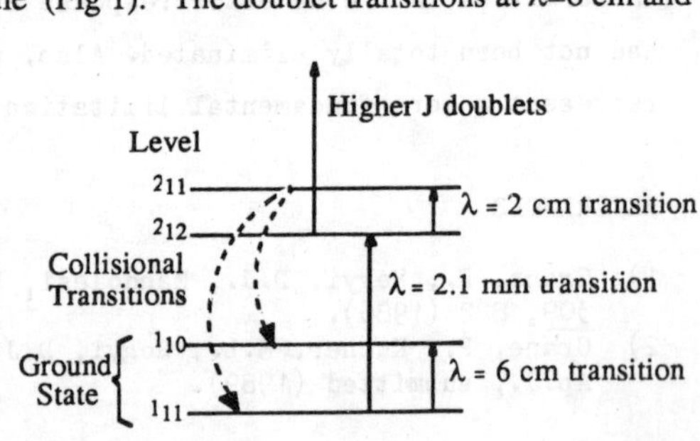

Fig 1. Rotational energy levels of H_2CO

We have observed the 6 cm and 2 cm transitions toward the HII region W51A (G49.5-0.4) using the VLA in C and D configuration. The high S/N of our observations allowed us to deconvolve the complicated morphology of the foreground clouds, and isolate individual condensations ("clumps") for analysis. We used the observed opacities, line widths, and clump sizes to solve the problem of radiative transport for each clump, using a spherically symmetric LVG model. The local radiation field is assumed to be the sum of a blackbody at temperature T_{CMBR} and a greybody from warm dust emission (scaled as λ^{-2}). The model includes the collisional excitation rates of Green *et al.* (1978) and an assumed abundance ratio $X(H_2CO/H_2)$ taken from surveys of molecular clouds.

Preliminary results from eight clumps yield a mean CMBR temperature of 3.2±0.3 K and a mean H_2 density of 2.0±0.3 cm^{-3}, where the statistical uncertainties are dominated by noise in the 2 cm opacity and in the determination of the clump radius. Of more concern are the systematic uncertainties associated with the modelling process, particularly the collisional excitation by hydrogen. We estimate these uncertainties by varying the associated parameters from their nominal values, re-calculating the CMBR temperature, and comparing the result to the mean T_{CMBR}. The results are shown in Table 1. Adding the statistical and systematic uncertainties in quadrature yields the CMBR result $T_{CMBR} = 3.2 \pm 0.9$ K.

Table 1: Estimated Systematic Effects

Effect	Nominal Value	Uncertainty	Error in TCMBR
H_2-H_2CO Collision Rates	1.5 * He-H_2CO rates of Green et al. 1)	±20%	0.2 K
H_2 Kinetic Temperature	11 K	±2K	0.6 K
$\frac{H_2CO}{H_2}$ Abundance	$2\times10^{-9}\left[\frac{n(H_2)}{8\times10^{4}\ cm^{-3}}\right]^{-0.5}$	Magnitude: $\pm0.5 \times 10^{-9}$	0.4 K
		Exponent: ±0.1	0.4 K
Dust Emission	Greybody at 15K	±50%	<0.05 K

Although the uncertainties are large, the result is in agreement with other CMBR measurements and excludes the possibility of a near-zero temperature. As understanding of the collisional effects and typical formaldehyde abundance grows, the systematics will be accordingly reduced, although the uncertainties will never approach the 1% level of CN. However, formaldehyde's ubiquity and ease of observation offers the exciting possibility of tracing T_{CMBR} as a function of redshift z in distant sources. The (1+z) scaling of CMBR temperature causes the signal to rise faster than the systematic uncertainties, which are dominated by molecular interactions. A search for a suitable source at modest redshift $z \approx 0.5$ to 1 is the next logical step for this research.

REFERENCES

1) Green, S., Garrison, B.J., Lester, W.A., and Miller, W.H., Ap. J. Suppl. Ser. 37, 321 (1978).

LOW-FREQUENCY MEASUREMENTS OF THE COSMIC MICROWAVE BACKGROUND RADIATION

Steve Levin, Marc Bensadoun, Marco Bersanelli, Giovanni De Amici, Alan Kogut, George Smoot, and Chris Witebsky

Space Sciences Laboratory and Lawrence Berkeley Laboratory

University of California at Berkeley

ABSTRACT

As part of an ongoing project to measure the spectrum of the Cosmic Microwave Background Radiation (CMBR), we have recently measured the antenna temperature of the CMBR at 1.5, 3.8, and 7.5 GHz. We briefly describe the measurement technique, and give preliminary results, as well as a summary of other recent measurements.

INTRODUCTION

Building on work done in the early 1980's as part of an international collaboration, we are engaged in a project to measure the low-frequency spectrum of the CMBR. By precisely determining the low-frequency spectrum, we hope to provide a better baseline for comparison with high-frequency measurements, and to distinguish among various models of the spectrum, including some of the models which account for the mm-wave distortion indicated by the results of Matsumoto *et al.* [3]

CONCEPT OF EXPERIMENT

We measure the CMBR temperature at each of several frequencies by comparing the radiometric power received from the sky with the power received from a liquid-helium-cooled calibration target and accounting for the contribution from other sky signals. After choosing the time and place of observation to minimize signals from all sources, the atmosphere and the galaxy remain as the major backgrounds.

The atmospheric signal is measured by observing the signal as a function of atmospheric path-length. We perform many tests to verify the atmospheric measurement and to make small corrections. In particular, we use large sheet-metal mirrors as shields to reject ground emission, and measure the remaining contribution by alternately adding and removing additional mirrors.

The galactic signal is modelled by a map extrapolated from measurements at lower frequencies (*e.g.* ref. [2]), and verified by comparing the model with the measured differences between observations at 15° E and 15° W of zenith.

In previous years, uncertainty in the antenna temperature of the cold calibration target was a significant source of error at low frequencies. In 1988 we used a new cold calibration target [1], with better low-frequency performance and a lower heat leak.

RESULTS

Preliminary results of our most recent (September 1988) measurements at 1.5, 3.8, and 7.5 GHz are shown in the table below, along with data from other experiments. Where appropriate, we have combined data from successive measurements into a single number, taking care to combine the statistical and systematic errors appropriately. Neglecting Matsumoto *et al.* [3] and treating the systematic errors at different frequencies as independent, the data fit a Planckian spectrum with temperature 2.74±0.02 K with $\chi^2 = 22$ for 13 degrees of freedom. The low-frequency ($\nu \leq 10$ GHz) measurements alone average to 2.63±0.04 K with $\chi^2/\mathrm{DOF} = 5.4/6$.

Reference	**Wavelength [cm]**	**ν [GHz]**	**T_{CBR} [K]**
Sironi, G., *et al.* 1987, in *13th Texas Symp. Rel. Astrophys.*, Ed. M.P. Ulmer	50.0	0.6	2.98 ± 0.55
Bensadoun (Smoot *et al.* 1989) [5]	20.5	1.46	2.38 ± 0.20
Sironi, G., Bonelli, G. 1986, *Ap. J.*, **311**, 418.	12.0	2.5	2.79 ± 0.15
De Amici (Smoot *et al.* 1989) [5]	7.9	3.8	2.63 ± 0.11
Mandolesi, N., *et al.* 1986, *Ap. J.*, **310**, 561	6.3	4.75	2.70 ± 0.07
Kogut (Smoot *et al.* 1989) [5]	4.0	7.5	2.51 ± 0.11
Kogut (Smoot *et al.* 1989) [5]	3.0	10.0	2.61 ± 0.06
Johnson, D.G, and Wilkinson, D.T. 1986, *Ap. J. Lett.*, **313**, L1.	1.2	24.8	2.783± 0.025
De Amici (Smoot *et al.* 1985) [4]	0.909	33.0	2.81 ± 0.12
Bersanelli (Smoot *et al.* 1989) [5]	0.333	90.0	2.60 ± 0.10
Meyer, D., Jura, M. 1985, *Ap. J.*, **297**, 119.	0.264	113.6	2.70 ± 0.04
	0.132	227.3	2.76 ± 0.20
Crane *et al.* 1988, *Proc. Third ESO/CERN Symposium*, in press.	0.264	113.6	$2.796^{+0.019}_{0.041}$
	0.132	227.3	2.76 ± 0.20
Matsumoto *et al.* 1987 [3]	0.116	259	2.799± 0.018
	0.0709	423	2.955± 0.017
	0.0481	624	3.175± 0.027

REFERENCES

[1] Bensadoun, M., *et al.* 1989, *Rev. Sci. Instrum.*, submitted.
[2] Haslam, C.G.T., *et al.* 1982, *Astr. Astroph. Suppl. Ser.*, **47**, 1.
[3] Matsumoto, T., *et al.* 1988, *Ap. J.*, **329**, 567.
[4] Smoot, G., *et al.* 1985, *Ap. J. Lett.*, **291**, L23.
[5] Smoot, G.. *et al.* 1989, *Ap. J. Lett.*, submitted.

SPECTRAL DISTORTIONS OF THE CBR DUE TO POWER LAW HEATING

by

James G. Bartlett and Joseph Silk

University of California, Berkeley

Spectral distortions of the cosmic background radiation (CBR) trace the thermal history of the universe[1]. Any energy source driving the electron-baryon plasma out of equilibrium with the photons will be compensated for by electron-photon interactions. If the energy deposition occurs late enough ($z < 10^6$ or so)[2] these interactions do not complete thermalization and leave a distortion today. We are investigating models where the heating of the plasma is produced by a source which varies with redshift as $(1+z)^p$. Three specific models with this property are superconducting cosmic strings[3], a decaying vacuum[4], and a rolling scalar field[5] coupled to photons. We are investigating all three, but only report here on the decaying vacuum model.

The plasma's thermal history is evolved according to

$$\dot{P} = \frac{1}{3}q - 5\frac{\dot{a}}{a}P - \frac{P}{t_{cool}}(1 - \frac{T_\gamma}{T_e}) \tag{1}$$

were q is the heat source provided by the model ($\propto (1+z)^p$), P is the electron pressure ($n_e kT_e$), and t_{cool} is the Compton cooling time. As an example, a decaying vacuum might result in a heat source given by,

$$q = 4xH\rho_{rad} \qquad x \equiv \frac{\rho_{vacuum}}{\rho_{vacuum} + \rho_{rad}} = const \tag{2}$$

The temperature history for $x = 0.00163$ is shown in fig. 1. Notice that there is no *global* recombination.

Once the thermal history of the plasma is obtained, the distortions of the CBR produced by Comptonization and bremsstrahlung can be calculated. They are described by the radiative transfer equation including expansion, bremsstrahlung emission and absorption, and Kompaneets' expression[6] for Comptonization. To make things more tractable, we simplify the Kompaneets term as follows: We assume that the Compton $y << 1$ and that $T_e >> T_\gamma$. The first assumption is demanded of the solution, while the second is valid for the example in fig. 1 for $z < 10^4$ or so. Under these conditions it is possible to write the transfer equation as

$$\frac{dI_\nu}{dt} = -3HI_\nu - c\alpha_\nu^{(ff)}I_\nu + cj_\nu^{(ff)} + \frac{kT_e}{m_e c^2}n_e\sigma_T cx\frac{\partial}{\partial x}x^4\frac{\partial}{\partial x}\frac{I_\nu}{x^3} \tag{3}$$

where $x = h\nu/kT_e$.

To actually solve this we break up the spectrum into three regions: In the high frequency range bremsstrahlung is ignored and the equation becomes Kompaneets' equation[6], with a solution given by the ZIS approximation[7]. In the mid-range frequencies both bremsstrahlung and Comptonization are important, but $\delta I_\nu/I_\nu$ is taken to be small, reducing the Comptonization term to $-2\dot{y}I_\nu$. At the lowest frequencies, where the last assumption fails, bremsstrahlung is dominating over Comptonization and we may use the same solution since the $-2\dot{y}I_\nu$ term is lost in the bremsstrahlung contribution. The two solutions (high ν and mid-low ν) are then matched at a point were the transition is the smoothest. Note that this is only the y-distortion. We have not yet finished work on the accompanying μ-distortion.

The result for the thermal history displayed in fig. 1. is shown in fig. 2. The data are taken from table 1 of Smoot *et al.*[8]. The Johnson and Wilkinson 1986[9] point is plotted with their conservative (larger) error bars.

REFERENCES

1. K. Chan and B. Jones, Ap. J., **195**, 1 (1975)
2. L. Danese and G. De Zotti, Astron. Astrophys., **107**, 39 (1982)
3. Ostriker and Thompson, Ap. J., **323**, L97 (1987)
4. K. Freese, *et al.*, Nuc. Phys., **B287**, 797 (1987)
5. P. J. E. Peebles and B. Ratra, Ap. J., **325**, L17 (1988)
6. A. S. Kompaneets, Sov. Phys. JETP, **4**, 730 (1957)
7. Zel'dovich, Illarionov, and Syunyaev, Sov. Phys. JETP, **35**, 643 (1972)
8. G. Smoot, *et al.*, Ap. J., **331**, 653 (1988)
9. Johnson and Wilkinson, Ap. J., **313**, L1 (1987)

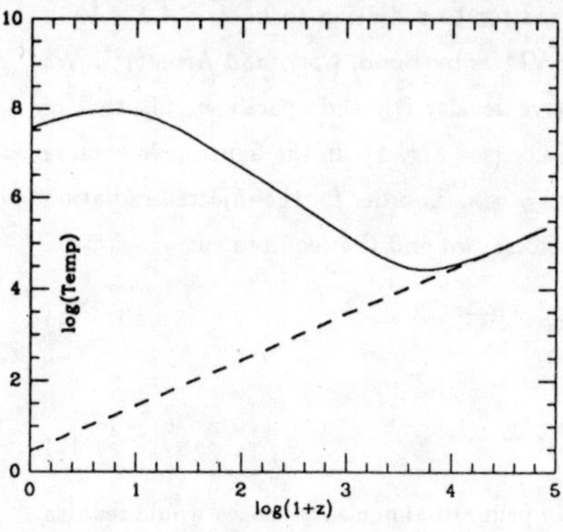

fig. 1
Solid line = electron temp.
Dashed line = unperturbed photon temp.
which is taken to be 2.78K today.
$\Omega_B = 0.1, \quad h_{100} = 0.50$

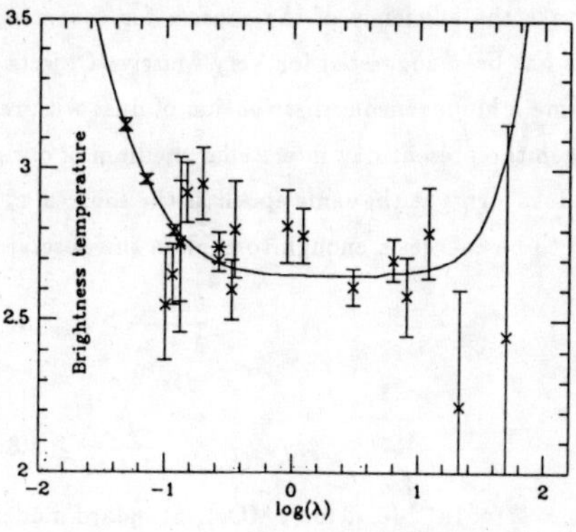

fig. 2

SPECTRAL DISTORTIONS OF THE COSMIC MICROWAVE BACKGROUND

Fred C. Adams[1], Katherine Freese[2], Janna Levin[2], and Jonathan C. McDowell[1]

[1]Harvard-Smithsonian Center for Astrophysics, Cambridge, MA 02138
[2]Massachusetts Institute of Technology, Cambridge, MA 02139

Presented by Katherine Freese

In 1987 Matsumoto *et al* [1)] reported a spectral distortion of the cosmic microwave background that corresponds to a 20% excess in energy density. The three boxes in Figs. 2 at wavelengths 1160, 710, and 480 μm are their data points; the fourth box at 260 μm is an upper limit. We show (Adams *et al*)[2)] that cosmic dust, absorbing and reradiating photons from an energy source such as Pop. III stars, provides a viable mechanism to explain the observed spectral distortion.[3)] We self-consistently numerically solve the equation of radiative transfer in an expanding universe and the equation of radiative balance for the two quantities: the dust temperature $T_D(z)$ and the spectrum of radiation I_ν at the present epoch. We assume a homogeneous, isotropic radiation field produced at z_* by energy sources of relative density Ω_*; we take the efficiency of the sources for converting matter to radiation to be $\epsilon = 4.1 \times 10^{-3}$ (this has been suggested for Very Massive Objects or VMOs by Bond, Carr, and Arnett)[4)]. We assume a homogeneous distribution of dust with relative density Ω_D and opacity κ_ν like that of dust in the present-day interstellar medium of our galaxy (see Fig. 1). In the figures, we assume the dust forms at the same epoch as the sources, $z_* = z_D = z$. In order for the injected radiation field to be energetic enough to explain the observed excess, we find the requirement

$$\frac{\epsilon\Omega_*}{1+z_*} \geq \Omega_{\rm excess} \simeq 2 \times 10^{-5} \tag{1}$$

or

$$\frac{\Omega_*}{1+z_*} \geq 4.8 \times 10^{-3} \tag{2}$$

for $\epsilon = 4.1 \times 10^{-3}$ (e.g. for VMOs). Standard models of primordial nucleosynthesis would require $\Omega_* \leq 0.20$ and hence $z_* \leq 40$.

For a given redshift of formation z, we find that with the single remaining parameter, Ω_D, we can fit the observed spectrum. In Figs. 2–4 we show the resulting sequence of models for the observed excess for redshifts in the range $20 \leq z \leq 50$. We plot the quantity $\Omega_\nu \equiv \frac{4\pi\nu I_\nu}{\rho_c c^3}$, the fraction of the energy density in photons, as a function of frequency ν. Note that the $10\mu m$ silicate feature in the dust opacity appears as an emission feature. This sequence of models can be understood by the requirement that the total optical depth be the same for each model ($\tau_\nu = O(1)$ at the excess):

$$\tau_\nu \propto \Omega_D (1 + z_D)^{n+3/2} \tag{3}$$

where n is the spectral index of the dust opacity ($\simeq 1$ at the excess). Quasar reddening and metallicity constraints require $\Omega_D < 10^{-4}$ and $z_D \gtrsim 25$. Thus dust distortion models with energy sources and dust formation in the redshift range $25 \leq z \leq 40$ remain consistent with existing cosmological constraints and can match the data of Matsumoto *et al* .[1]

REFERENCES

[1] Matsumoto,T., Hayakawa, S., Matsuo, H., Murakami, H., Sato, S., Lange, A.E., and Richards, P.L. 1988, *Ap. J.*, **329**, 567.

[2] Adams, F.C., Freese, K., Levin, J., and McDowell, J.C. 1989, MIT preprint CSR-AT-89-09.

[3] For previous work on this subject, see: Rowan-Robinson, M., Negroponte, J., and Silk, J. 1979, *Nature*, **281**, 635; Negroponte, J., Rowan-Robinson, M., and Silk, J. 1981, *Ap. J.*, **248**, 58; Wright, E.L. 1981, *Ap. J.*, **250**, 1; McDowell, J.C. 1986, *M. N. R. A. S.*, **223**, 763; Negroponte, J. 1986, *M. N. R. A. S.*, **222**, 19; Lacey, C.G., and Field, G.B. 1989, *Ap. J. Lett.*,in press; Draine, B.T., and Shapiro, P. 1989, in preparation.

[4] Bond, J.R., Carr, B.J., and Arnett, W.D. 1984, *Ap. J.*, **280**, 825.

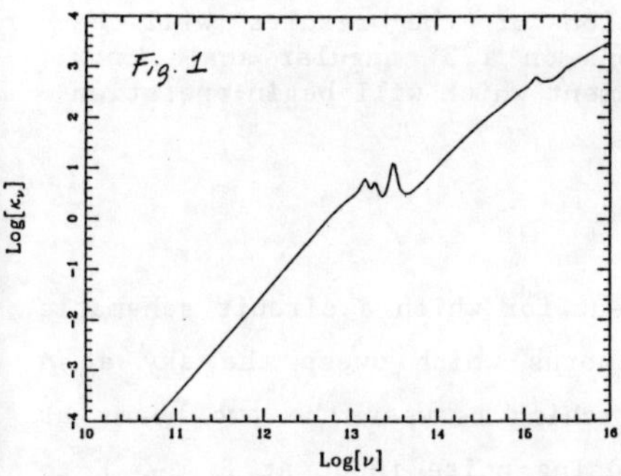

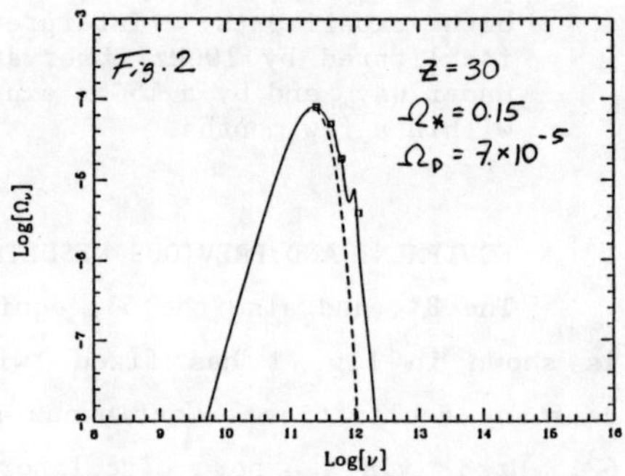

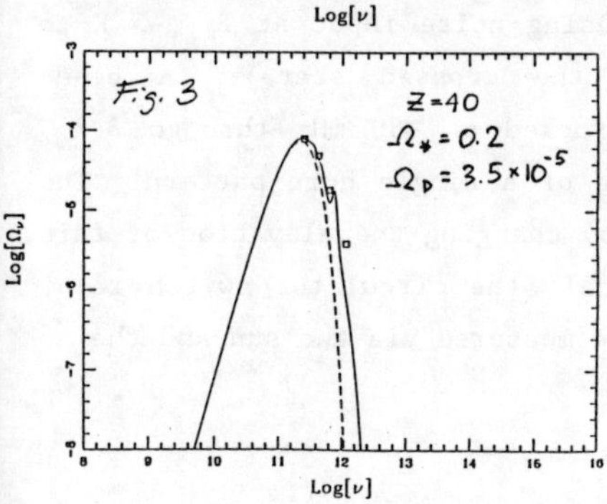

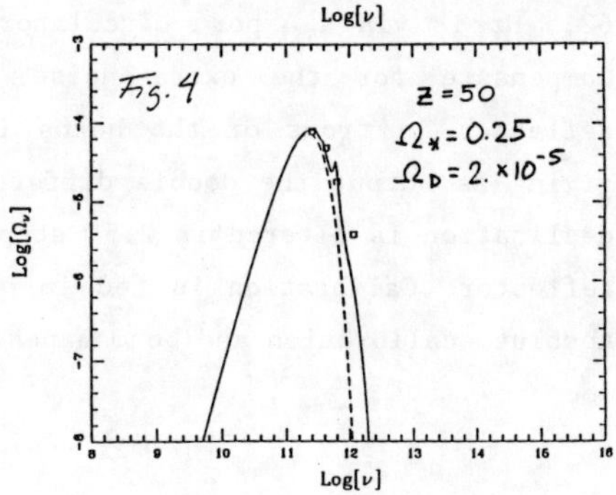

THE TENERIFE CMB FLUCTUATION EXPERIMENT: PROGRESS TO DATE

J E Beckman[1], R D Davies[2], A N Lasenby[3], R A Watson[2,1] and R Rebolo[1]

1. Instituto de Astrofisica de Canarias, Spain
2. Nuffield Radioastronomy Laboratories, Jodrell Bank, UK
3. Mullard Radioastronomy Dept., Cavendish Laboratory, Cambridge, UK

ABSTRACT

Two versions of the experiment, both working at 10.45 GHz, have been operational: the first with an 8.2° beamwidth and a throw of 8.3°, the second with a 5.6° beamwidth but an 8.3° throw, both being in triple-beam twin-horn configuration. Results from the former, giving an apparently significant detection of anisotropy on the 8° scale at declination 40° have been supplemented by new data from the 5° experiment at a similar sensitivity a $\Delta T/T$ level of $4x10^{-5}$. The presence of a feature at R.A. 14h 50m which does not correspond to any previously reported source is noted in the $\delta = 40°$ data from both experiments. Interpretation of the results will be facilitated by 10GHz observations on a 3° angular scale, now under way, and by a 15GHz experiment which will begin operation within a few months.

1. EQUIPMENT AND PREVIOUS RESULTS

The 8° (and also the 5°) equipment for which a circuit schematic is shown in Fig. 1 has fixed twin horns which sweep the sky at a constant declination. Continuous switching between the two beams at 62.5 Hz is via a 4-port circulator (using noise input at T_{bal}~5 K to compensate for the extra noise in the crossed state). A plane reflector in front of the horns is rocked at ~80 mHz through 8.3°, giving as output the double difference of a single horn pattern. The declination is altered in 2.5° steps by changing the elevation of this reflector. Calibration is fed in at half the circulator switch rate. Absolute calibration and beamshapes are measured via the sun and the moon.

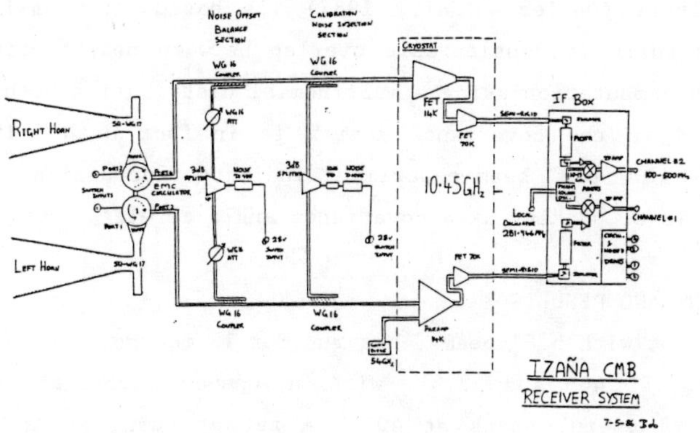

Fig. 1.: Schematic of the two-channel four-port circuit of the 10.45 GHz CMB receivers.

System temperature is 70-80 K, and bandwidth 0.5 GHz. A run to achieve limiting brightness temperature in the range $\Delta T/T<5\times10^{-5}$ lasts ~2 months. Daytime data is not badly affected by noise. We measured sidelobe responses from the sun; the worst case reached $\Delta T/T = 10^{-4}$, (in agreement with computed estimates); we reject data taken with the sun above the horizon. The moon cannot contribute more than $\Delta T/T = 5\times10^{-6}$ and in practice never enters the principal sidelobe. In Fig. 2 we show co-adding of samples taken on the same sky fields, compared with the predicted rms for gaussian noise. Agreement is good, indicating "well behaved" instrument noise and no other

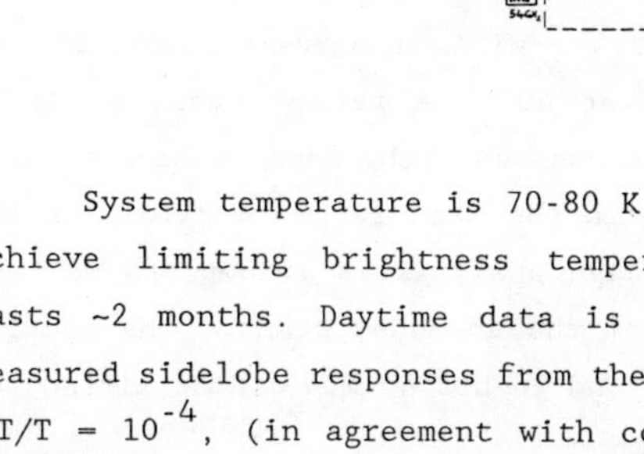

Fig. 2: Measurements of co-added output summed for the same sky point, against a theoretical fit for the Gaussian case.

unexpected contributions. For 24 hr cover in R.A. we space a run through the year. We have observed nine strips: at δ = -15° -5° 0° 5° 15° 25° 35° 40° and 45°, of which two, at 0° and 40° are deeper. All scans have a strong galactic plane crossing, and a weak one. Published analysis was of the section with most co-added scans at δ = 40°, from R.A. = 12 hrs to 16 hrs 40 min, (galactic latitude is >35° in this strip). That analysis (Davies et al., 1987) is based on Gaussian covariance, but as there is considerable overlap between neighbouring fields a likelihood computation using a multinomial distribution with a broad diagonal band in the covariance matrix is in fact used. The results for the $\delta = 40°$, 8.3° beam experiment indicate fluctuations at a $\Delta T/T \sim 3.5 \times 10^{-5}$ level peaking at a covariance angle of ~12°.

3. NEW EXPERIMENT AND RESULTS

A new experiment with 5.6° beams, has run for 18 months. We have scans at $\delta = 35°$ 37.5° 40° and 42.5°, with an uneven coverage; the maximum number is 88 useful scans at 40°. A set of scans at these declinations with their $\Delta T/T$ levels has been published (Watson et al., 1988). The 40° scan shows a $\Delta T/T$ in the same range, 4.2×10^{-5} as the previous 8° data at $\delta = 40°$. Although analysis is incomplete we have an interesting result to report from the co-added scans. The section of this sum (Fig. 3) between 12 hrs and 16 hrs 45 min can be fitted by

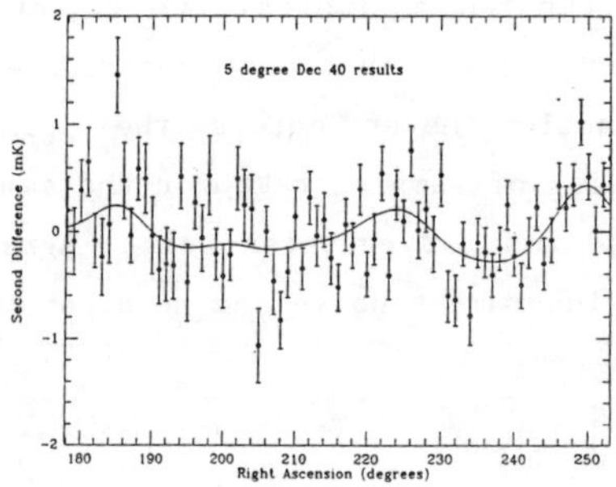

Fig. 3: Average of 88 scans between = 12 hrs & 17hrs at $\delta = 40°$ for the 5.6° experiment. Error bars on individual points are shown. Best fit shows two "point-source" peaks plus an extra peak at $\alpha = 14h50m$.

a curve with three "bumps", all at sub mk levels. Two of these, at R.A.'s of 12 hrs 20 min and 16 hrs 30 min, correspond to known point sources. The third, at 14 hrs 50 min, does not correspond to any known source, but does correspond to a feature appearing in the $\delta = 40°$ scan of the 8° beam experiment.

What could this bump be? A point source causing this amplitude must be in the range 5-10 Jy. The bump seen at $\alpha = 16$ hrs 30 min corresponds to a blend of two objects: 1633+38 (6 Jy) and 3C345 (10 Jy). No point source above 1 Jy is known near $\delta = 40°$, $\alpha = 14$ hrs 50 min. Extrapolations of the effect of grouped sub-Jy sources, from lower frequency observations by Condon (1984) are 10 times lower than the feature. A more probable origin is extended galactic emission. We extrapolated from the 408 MHz survey using a spectral index of -2.8 (cf. Lawson et al., 1987, who shows this to be a conservative index for b >10°; our b is > 30° throughout this scan). In Fig. 4, the 5° scan, reconstructed to take out the profiles of a single horn, is compared with point source predictions and the 408 MHz survey extrapolation. The feature at R.A.= 14 hrs 40 min is not, apparently, matched by either point sources or the galaxy. To dismiss the presence of local sources, especially the extended galactic plane emission, as causes of the bump, is premature, given the uncertainty in the spectral index. Two of the peaks in Fig. 4 correspond to diffuse galactic

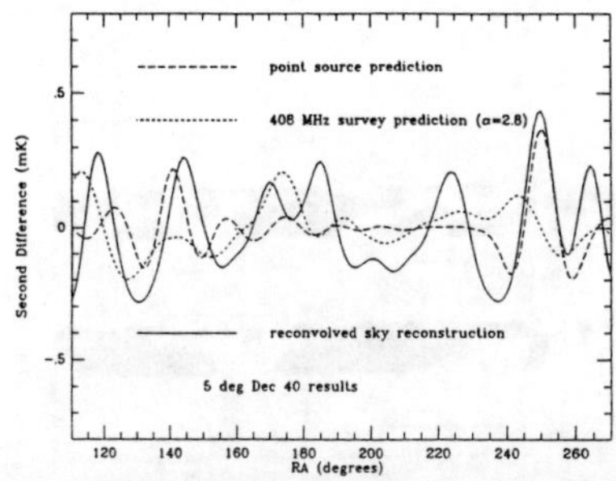

Fig. 4: The observed δ =40° scan for the 5.6° beam experiment compared with fits by a frequency - extrapolated galactic model, and beam-diluted point sources.

emission, reinforcing the view that this cannot easily account for the 14 hr 40 min bump, we await higher frequency CMB data at = 40° before reaching a conclusion. True background features should have frequency independent amplitudes. There is a further possibility: the imprint of large-scale structure on the CMB. The bump coincides with the southern part of the Bootes void (Kirshner et al., 1981). A rapid estimate of the effect of a void on the background temperature is $\Delta T/T \sim (\Delta v/c)^3$ where Δv is the radial velocity range of the Hubble flow from near to far surface of the void. For $\Delta v \sim 7\mathrm{x}10^3$ km s^{-1}, $\Delta T/T \sim 2\mathrm{x}10^{-5}$. The steep dependence on $\Delta v/c$ means that a slightly deeper patch of void could yield a well-contrasted bump. The calculation is for a truly empty void, and this effect seems less probable than either local effects or "true" inhomogeneities at the surface of last scattering.

Fig. 5 shows two-dimensional reconstructions of the sky around δ= 40° and δ = 0°, on both 8° and 5° angular scales. Details, limitations and systematic and random uncertainties inherent in these maps, as in our single scans, will be available in Davies et al (1989).

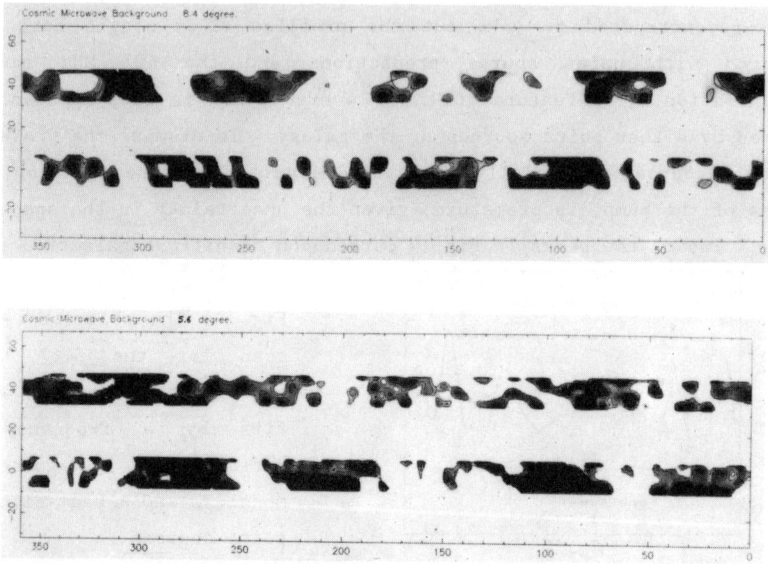

Fig. 5: Two dimensional reconstructions of the sky around $\delta = 0°$ and $\delta = 40°$ for the 8.3° and 5.6° experiments. Sensitivity is not yet uniform on these maps.

4. COSMOLOGICAL IMPLICATIONS

If anisotropies in the CMB are present, with $\Delta T/T \sim 3\times10^{-5}$, on degree scales, the most striking conclusion would be that standard cold dark matter dominated models would be excluded. Although primordial nucleosynthesis of ^{4}He and ^{7}Li appear to set an upper limit of $\Omega_b < 0.1$ to the possible baryon density, models with neutron-proton segregation at the quark-hadron phase transition may permit $\Omega_b = 1$, i.e an inflationary baryon-dominated universe. Baryon dominated universes with adiabatic primordial fluctuations seem to be ruled out by these results, together with the OVRO upper limit of $\Delta T/T < 1.9\times10^{-5}$ on a coherence scale of 2.5 arcmin (Readhead et al., 1989);

baryonic isocurvature models cannot be excluded. Indeed a detection with the Tenerife experiment would be consistent with such models for a highly restricted range of power spectral indices around n = -1.

If we also claim that the bump at α = 14 hrs 50 min, δ = 40° is telling us about the spatial distribution of fluctuations, this appears to rule out a gaussian spectrum. If the bump were identifiable with a local feature, we would be left with a lower level of general fluctuations, in the range $2\text{-}3\times10^{-5}$. This brief discussion is clearly speculative, giving a feel for the "reach" of the measurements and what could be established as more data becomes available.

5. FUTURE CONFIGURATIONS

We have recently set up a 10.76 GHz 3° beam experiment which has now started to collect data. In a few months (early 1989) a 15 GHz beamswitched system with 4° beams will be operational at the Tenerife site, with the priority aim of looking at the region of the δ = 40° bump. Later in 1989 a 30 GHz system with 3° beams will be tested at Tenerife, before transfer to the low water vapour site of the British Antartic Survey near the South Pole.

6. REFERENCES

Condon, J.J., Ap. J. 287, 461 (1984)

Davies, R.D., Lasenby, A.N. Watson, R.A., Daintree, E.J., Hopkins, J., Beckman, J.E., Sanchez Almeida, J., Rebolo, R., Nature 326, 467 (1987)

Davies, R.D., Watson, R.A., Hopkins, J.A., Lasenby, A.N., Beckman, J.E., Rebolo, R., In preparation (1989)

Kirshner, R.P., Oemler, a., Schechter, P.L., and Shechtman, S.A., Ap. J. 248, L57 (1981)

Lawson, K.D., Mayer, C.J., Osborne, J.L., and Parkinson, M.L., M.N.R.A.S., 225, 307 (1987)

Readhead, A.C.S., Lawrence, C.R., Meyers, S.T., Sargent, W.L.W., Hordebach, H.G., and Moffett, A.T., Ap.J. (In Press) (1989)

Watson, R.D., Rebolo, R., Beckman, J.E., Davies, R.D., Lasenby, A.N., Large Scale Structure and motions in the Universe, Eds. Giuriccin, Mezzetti, Mardirossian, Ramella. Trieste (In Press)

THE ARGO 1988 EXPERIMENT

P. de Bernardis, P. Calisse, A. Iacoangeli, R. Maoli,
S. Masi, F. Melchiorri, M. Perciballi
Dipartimento di Fisica, Universita' di Roma, Italy
A. Boscaleri, V. Natale, G. Valmori
I.R.O.E. - C.N.R., Firenze, Italy
A. Pitas
M.I.T. - Cambridge, Mass.

ABSTRACT

This is a progress report on the far infrared balloon Telescope ARGO 1988. It is a 1.2 m cassegrain telescope coupled to 3 millimetric and submillimetric detectors, which has been optimized for the search of diffuse radiation anisotropies.

Our group is working on a project devoted to the search of extragalactic backgrounds (EGB) anisotropies[1,2,3,4,5]. A conspicuous EGB has been discovered recently in a rocket experiment in the submm region [6], with an energy density of the order of 10% of the CBR. Several theoretical scenarios producing a submm background have been suggested: comptonization of the CBR photons[7]; decaying particles in the early universe[8], hot regions before the recombination[9]; emission from intergalactic dust in the pregalactic (pop. III) stars scenario[10].

Anisotropy measurements, at several different wavelengths, should be very useful in selecting different models: for example in the pop. III scenario, the submm background anisotropies are produced by fluctuations of both the optical depth and the temperature of the dust. The expected anisotropy increases with the observation frequency[11] and decreases with the angular separation[12]. The expected level of these anisotropies is quite high, and at high galactic latitudes should be larger than the fluctuations of patchy interstellar dust emission.

An experiment devoted to this search must have a large throughput ($A\Omega$) in order to enhance the sensitivity to diffuse radiation; on the other hand the beamsize should not be too large, so as to compare sky regions not too far away from each other. This requires a large collecting area, i.e. a telescope. The atmospheric stability is quite poor in the submm windows: for this reason the anisotropy of the EGB must be investigated from extremely dry sites or from above the atmosphere.

Following these ideas we have developed a balloon-borne FIR telescope so as to study both CBR and EGB anisotropies. The instrument is shown in Fig.1. The Cassegrain telescope has an aluminum (1.2 m diameter) primary mirror (f/0.5) and a wobbling secondary mirror. The total f/# is f/3. The telescope is fixed at an elevation of 45°. Three f/4.5 photometers have been used at the focus of the telescope, with a large entrance window (45 mm): in this way the field of view is 1°.0. The f/# mismatching has been introduced in order to reduce the diffraction from the secondary mirror[13]. The secondary wobbling allows for a 1°.8 synusoidal beam-switching on the sky at a frequency of 17 Hz. A 0.3^{o}K ^{3}He bolometer was used for the 2.2 mm band[14]; a 1.0^{o}K ^{4}He bolometer for the 0.8 mm band and a 1.6^{o}K ^{4}He bolometer for the low pass band (0.3 mm cut-on). The three detectors are used respectively for CBR anisotropies, submm background anisotropies and dust emission monitoring. Thermal gradients on the primary mirror should result as a source of variable offset[15] somehow difficult to model. A special large baffle was used, with a mylar superinsulation and heat pipes, in order to thermalize the telescope also in the presence of sun, moon and earth variable enlightments.

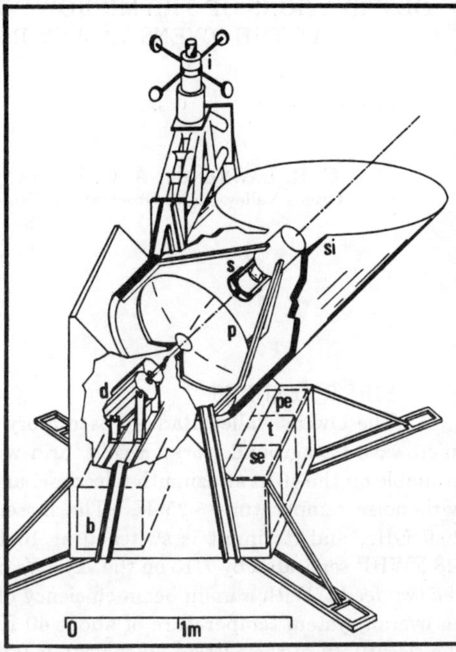

An inertial azimuth pointing system , capable of 15' accurancy has been developed for sky pointing and slow scanning . An engineering balloon flight was launched from Aire sur l'Adour (France) on Sept. 20 , 1988. The aim of the flight was to test the modulator , the baffles system and the pointing system. Due to an unexpected trajectory, the flight lasted for 2 hours only. However we tested successfully the above mentioned systems . A transmediterranean flight , 20 hours long , is planned for summer 1989 .

This project is entirely supported by Piano Spaziale Nazionale - Agenzia Spaziale Italiana.

Fig.1: a sketch of the ARGO 1988 experiment. p = primary mirror; s = secondary mirror assembly; si = super insulation baffle; d = detector assembly; b = battery pack; i = inertial motor for azimuth pointing; se = signal electronics; t= telemetry; pe = pointing electronics.

References

1) Melchiorri F., Melchiorri B., Ceccarelli C., Pietranera L., 1981, *Ap.J.Letters*, **250**, L1.
2) Ceccarelli C., Dall'Oglio G., de Bernardis P., Masi S., Melchiorri B., Melchiorri F., Moreno G., Pietranera L., 1983, *Ap.J.Letters*, **275**, L39.
3) de Bernardis P., Masi S., Melchiorri F., Moreno G., 1984, *Ap.J.*, **278**, 150.
4) de Bernardis P., Masi S., Malagoli A., Melchiorri F., 1985, *Ap.J.*, **288**, 29.
5) Masi S., Dall'Oglio G., de Bernardis P., de Santis E., Epifani M., Giovannozzi E., Guarini G., Melchiorri F., Boscaleri A., Natale V., Guidi I., *Search for EGB: a ballon borne 4-band FIR differential photometer with large throughput* 1988, *submitted to Astron. and Astrophys.*
6) Matsumoto T. , Hayakawa S. , Matsuo H. , Murakami H. , Sato S. , Lange A. , Richards P.L. , 1988 , *Ap.J.* , **329** , 567 .
7) Smoot G., Levin S.M., Witebski C., De Amici G. , Rephaeli Y. 1988, *Ap.J.*, **331**, 653.
8) Kawasaki M. and Sato S., 1988, *Publ. Astron. Soc. of Japan*, **39**, 837.
9) Daly R.A., 1988, *Ap.J.Letters*, **333**, L1.
10) Hayakawa S., Matsumoto T., Matsuo H., Murakami H., Sato S., Lange A., Richards P.L., 1988, *Publ. Astron. Soc. of Japan*, **39**, 941.
11) Panek M., Rudak B., in *Large Scale Structures and Motions in the Universe*, Giuricin et al. eds, Trieste, 1988.
12) Hogan C.J., Bond J.R., 1987, in *The post-recombination Universe*, Kaiser N., Lasenby A.N., Cambridge U.K., p.141 (Kluwer, Dordrecht)
13) de Petris M., Gervasi M., Liberati F., 1989, *New far infrared and millimetric telescopes devoted to differential measurements with large chopping angle in the sky submitted to Appl. Optics*
14) Mason C., Ceccarelli C., Masi S., Dall'Oglio G., Ferri G., Radford S.J.E., 1986, *IR.Phys.*, **26**, 273.
15) de Bernardis P., de Petris M., Epifani M., Gervasi M., Guarini G., Masi S., 1988, *IR.Phys.*, **28**, 243.

OBSERVATIONS OF THE MICROWAVE BACKGROUND RADIATION AT THE OWENS VALLEY RADIO OBSERVATORY

C. R. LAWRENCE, A. C. S. READHEAD, AND S. T. MYERS
Owens Valley Radio Observatory, California Institute of Technology

1. FIRST RESULTS

The Owens Valley Radio Observatory (OVRO) has several advantages for microwave background work: a cold, arid winter climate; lots of observing time available on the 40 m altazimuth telescope; and an excellent K-band maser receiver with noise temperature ~25 K. The maser is tuned to a center frequency of 20.0 GHz, and its input is switched at 10 Hz between two symmetric feeds of 1$'\!.$8 FWHP separated by 7$'\!.$15 on the sky. We measure the power difference between the two feeds. With a main beam efficiency of 0.47, a bandwidth of 400 MHz, and an overall system temperature of about 40 K (including atmosphere and MWB), the sensitivity of the system on the sky is 9 mK in one second.

We observed eight fields ($\alpha = 1^{\rm h}, 3^{\rm h}, \ldots, 15^{\rm h}$, $\delta = 89°$) with a standard double switching scheme[1], measuring the difference between the temperature of a 1$'\!.$8 patch of sky and the average of two similar patches 7$'\!.$15 away on either side. In 24 hrs all fields were observed for 2 hrs at upper culmination, and fields 5–11 were observed for another 2 hrs at lower culmination. This upper/lower culmination scheme was designed as a check for systematic errors. None has been found. The telescope moves only in azimuth when the beams are swapped on the sky every 40 s, so the two reference positions turn into little 30° arcs during a 2 hr period.

Since 1985 170 days of background observations have been scheduled, of which 102 days were lost to weather and 44 days to equipment problems. Allowing for a 30% overhead for calibration and telescope slews, this leaves 17 days of integration time. There is a weak source near one field. If 15 days of data for the other seven fields are combined, we find $\overline{\Delta T} = -11 \pm 10\,\mu$K, about 25% above the thermal noise limit. The excess noise comes from coherent atmospheric fluctuations on time scales of a few minutes to about two hours.

A complete description of these observations and their results has been submitted for publication[2]. Using a Bayesian analysis with a uniform prior density for $\theta_{\rm sky}$ (the dispersion of assumed Gaussian sky fluctuations as measured by our triple beam), we find a 95% probability that $\theta_{\rm sky} < 58\,\mu$K. If the sky consisted of uncorrelated, uniform 2$'$ patches, this triple-beam limit would give "$\delta T/T$"< 1.7×10^{-5}. For the more realistic generic case of a Gaussian autocorrelation function for the sky, $C(\phi) = C_0 \exp\{-\phi^2/2\phi_c^2\}$, we can calculate the limit on $C_0^{1/2}$

as a function of the *coherence angle* ϕ_c. This limit reaches its lowest value of 1.7×10^{-5} (in units of $\delta T/T$) at $\phi_c \approx 2'\!.5$. For smaller ϕ_c a single beam averages over several fluctuations; for larger ϕ_c the beam separation is too small to span even a single fluctuation. In either case, the limits on C_0 are less stringent.

2. FUTURE OBSERVATIONS

As mentioned above, no systematic differences between observations at upper and lower culmination have been seen, and observations of all twelve fields at $\delta = 89°$ will continue. Adjustments will be made to the pointed positions to avoid sources revealed by VLA maps at 5 GHz. Since atmospheric fluctuations are uncorrelated over periods longer than an hour or two, we expect that sensitivity will continue to increase as $t^{-1/2}$, and that we can reach levels at least a factor of two lower than present limits.

For Gaussian fluctuations, instrumental sensitivity proportional to $t^{-1/2}$, and a Bayesian statistical analysis, the limits placed on $\theta_{\rm sky}$ are lowest when a given total integration time is divided among 10–20 fields. For fluctuations with a power law tail with index greater than -2, a larger number of fields is better (the number of large fluctuations goes up faster than the sensitivity of the individual fields goes down). To test for such a power-law tail we are also observing 96 fields at $\delta \approx 88°$, with lower sensitivity per field. Results from this program should be available late in 1989.

In 1989 we will begin observations with a new 5.5 m telescope dedicated to MWB observations. A 32 GHz maser receiver mounted at the Cassegrain focus will have two beams of 8′ FWHP separated by 24′ on the sky. We expect the total system temperature to be about 40 K, and anticipate that long term atmospheric fluctuations will be lower at 32 GHz than at 20 GHz. Overall sensitivity on the sky will be about 5 mK in one second, and with a dedicated system more observing time will be available each year. We expect this system to reach the 15 μK level, where source confusion or integrated Sunyaev-Zel'dovich dips from clusters at high redshifts will likely be detected. Most models predict larger fluctuations on 24′ scales than on 7′ scales, so observations with this system should provide stringent constraints on models.

REFERENCES

1. Lake, George and Partridge, R. B. 1980, *Ap. J.*, **237**, 378.
2. Readhead, A. C. S., Lawrence, C. R., Myers, S. T., Sargent, W. L. W., Hardebeck, H. E., and Moffet, A. T. 1988, submitted to *Ap. J.*.
3. Franceschini, A., Toffolatti, L., Danese, L., and De Zotti, G. 1988, preprint.

SEARCHES AT SHORT WAVELENGTHS FOR SMALL SCALE ANISOTROPIES IN THE 3 K AND SUBMILLIMETER BACKGROUNDS

R. B. Partridge
Haverford College
Haverford, PA 19041

Reviewed here are some published and unpublished microwave measurements which set limits on $\Delta T/T$ fluctuations in the 3 K cosmic background (CBR), and may help to constrain models for the submillimeter flux recently reported by Matsumoto et al (1988).

VLA measurements at $\lambda = 6$ cm on angular scales $12'' \leq \theta \leq 60''$ have been made by Martin and Partridge (1988) and Fomalont et al (1988). After instrumental noise has been subtracted (see Knoke et al, 1984; Fomalont et al, 1988), some sky variance remains. Both faint discrete sources and $\Delta T/T$ fluctuations contribute to σ^2_{sky}. The two groups used different approaches to correct for the contribution by discrete sources (see Partridge, 1988). We extrapolated source counts made at 20 cm, and assumed that such sources could make only positive contributions to the measured flux in the VLA maps (Martin and Partridge, 1988). We found that sources could explain only $\sim 1/2\ \sigma^2_{sky}$, and tentatively ascribed the rest to fluctuations in the CBR (see table). Fomalont et al allowed for the contribution of the side lobes of faint sources as well; these permit negative as well as positive contributions to a synthesis map. Allowing contributions of both signs "soaks up" more of σ^2_{sky}; hence they report only upper limits on $\Delta T/T$.

Recent work by Danese reported here (see Franceschini et al, 1988) shows that sources do make contributions of both signs (so that our estimate of $\Delta T/T$ is too high), but also that a better fit to Fomalont et al's VLA data is obtained with a non-zero value of ΔT. The truth probably lies somewhere between the two sets of results.

Reference	λ, cm	Pol.	Angular Scale	Limits on on ΔT/T, $\times 10^4$	Limits on Flucs. in Sub mm $\begin{pmatrix}\alpha = 1\\ \alpha = 2\end{pmatrix}$
Martin & Partridge (1988)	6	I	18"-80"	1.7±0.5	
"	"	I	36"-160"	1.3±0.2	
Fomalont et al (1989)	"	I	12"	< 8.5	
"	"	I	18"	< 1.2	
"	"	I	30"	< 0.8	
"	"	I	60"	< 0.6	
Partridge et al (1988)	"	Q	60"-160"	< 0.4	
"		U	"	< 0.5	
"		V	"	< 0.6	
Hogan & Partridge (1989)	2	I	5".3-48"	< 6.3	< $\begin{pmatrix}.2\\ \sim 6\end{pmatrix}$
"	2	I	10"-48"	< 3.2	
"	2	I	18"-50"	< 1.6	
Sheth & Partridge (1989)	0.37	I	~ 70"	< 3.5	< $\begin{pmatrix}.01\\ .05\end{pmatrix}$

Nowakowski, Martin and I (1988) have used our 6 cm data to set limits on linearly polarized fluctuations in the CBR (see table).

One way to check the effect of faint sources is to work at shorter wavelength, since $T \propto \lambda^{2.8}$ for typical synchrotron sources. Hence Hogan and I (1989) used the VLA at 2 cm to reobserve a field studied at 6 cm. When the data were smoothed to 18" scale, they agree well with the 6 cm results (see table). Unfortunately, the 2 cm receivers were so noisy we were unable to make a crisp test of the effect of sources. So we are joining forces with Fomalont et al to make a deep map at λ = 4 cm, where the receivers are less noisy. We should reach $\Delta T/T \sim 6.0\text{-}1.5\times 10^{-5}$ on scales of 10"-40", since the receivers are less noisy.

The maximum-resolution (5".3) 2 cm observations can be used to constrain one model of the submillimeter flux--emission from dust in high-redshift galaxies (Bond, Carr and Hogan, in preparation). In this model, shot noise fluctuations $\Delta\tau/\tau \propto \theta^{-1}$ would be expected.

The limits set by our 2 cm results depend on the emissivity of the dust, which we take to be $\varepsilon \propto \lambda^{-\alpha}$.

3 mm limits. As part of an unconnected program, R. Sheth and I used the NRAO 12-meter to observe several ~ 70" patches of nominally blank sky at λ = 3.7 mm. The observations improve previous 3 mm results (Boynton and Partridge, 1973) by ~ 5, and limit $\Delta\tau/\tau$ to 1% or 5% for α = 1 or 2. A far tighter constraint on the Bond et al model is set by the higher frequency work of Kreysa reported here.

Limits on the large-scale isotropy of the submillimeter flux may also be set by scaling limits on the dipole and quadrupole moments of the CBR (Lahav and Partridge, 1988). We used the 3.3 mm results of Lubin et al 1983) to show that the submillimeter flux has a dipole moment < 4% or 20% for α = 1 or 2. The quadrupole component is smaller, < .5% or < 2%. Tighter limits on the anisotropy of the submillimeter excess may be set by using higher frequency measurements of the CBR (see Halpern et al, 1988). These limits rule out a Galactic origin for the submillimeter excess.

REFERENCES

Bond, J. R., Carr, B. J., and Hogan, C. 1988, in preparation.

Boynton, P. E., and Partridge, R. B. 1973, Ap. J., 181, 243.

Fomalont, E. B. et al 1988, A. J., 96, 1187.

Franceschini, A., Toffolatti, L., Danese, L., and DeZotti, G. 1989, in press at Astron. and Astrophys.

Halpern, M., Benford, R., Meyer, S., Muehlner, D., and Weiss, R. 1988, Ap. J., 332, 596.

Hogan, C., and Partridge, R. B. 1989, in preparation for Nature or Ap. J. Lett.

Knoke, J. E., Partridge, R. B., Ratner, M. I., and Shapiro, I. I. 1984, Ap. J., 284, 479.

Lahav, O., and Partridge, R. B. 1988. M.N.R.A.S., 325, 1p.

Martin, H. M., and Partridge, R. B. 1988, Ap. J., 324, 794.

Matsumoto, T. et al 1988, Ap. J., 329, 567.

Partridge, R. B. 1988, in Third ESO/CERN Symposium, Bologna, Italy; Reidel Publ. Co.

Partridge, R. B., Nowakowski, J., and Martin, H. M. 1988, Nature, 331, 146.

SMALL SCALE ISOTROPY OF THE COSMIC MICROWAVE BACKGROUND AT 230 GHz

E. Kreysa, R. Chini
Max-Planck-Institut für Radioastronomie
Auf dem Hügel 69
5300 Bonn 1

The new generation of large mm- and submm-telescopes can probe the small scale isotropy of the Cosmic Microwave Background (CMB) at high frequencies. At 230 GHz (1.3 mm wavelength) the IRAM 30 m telescope[1)] has a beam diameter of 11" (FWHM). At cm-wavelengths, this beamsize is achieved only by interferometers. The most sensitive continuum receivers at submm- and short mm-wavelengths are ^{3}He-cooled bolometers. With good atmospheric conditions the MPIfR-bolometer[2)], operating at 1.3 mm on the 30-m telescope can reach a flux density of 50 mJy in one second of integration time.

Recently we observed all those radio-quiet quasars that had been detected in at least three IRAS-bands by Neugebauer et al.[3)]. There are 25 sources in the complete sample. With a focal plane chopper, operating at 8.5 Hz and 30" amplitude, an ON-ON observing mode was used. Atmospheric opacity was monitored by load-switching and sky dips. Neptune und Uranus were our primary calibrators. Our aim was to achieve a statistical error of about 1 mJy for each source. The final results in Fig. 1 show that only 5 sources were detected.

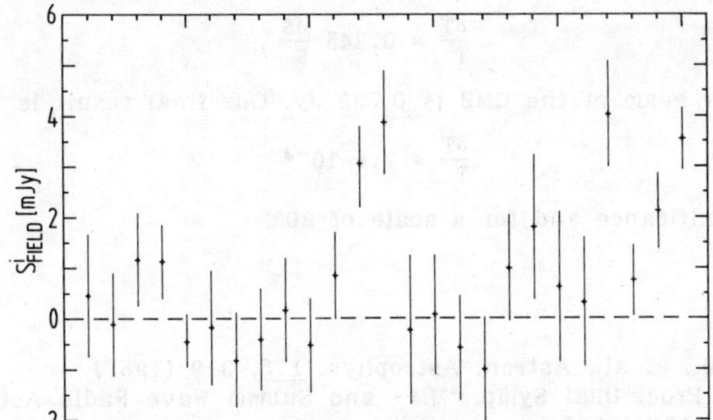

Figure 1. Final averages for all 25 fields observed

The relevance of this result for radio-quiet quasars is discussed in another paper[4]. Here we use the null result for most of the fields centered on the quasars in order to deduce an upper limit for the anisotropy of the CMB. The detected flux density at 1.3 mm of four sources is consistent with an extrapolation of the nonthermal radio spectrum. We exclude these sources from our sample, although this procedure is not quite satisfactory. The data for the remaining 21 fields are consistent with normally distributed errors as shown in Fig. 2.

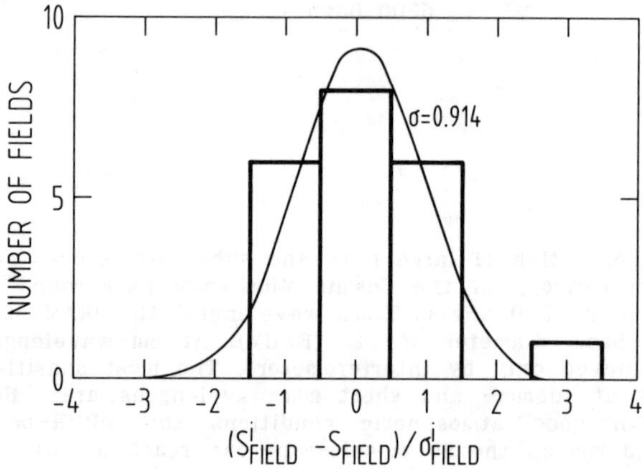

Figure 2. Scatter of results for 21 fields.

With the statistic used by Boynton and Partridge[5] we derive an upper limit for the fluctuation of the sky of 0.94 mJy at 95 % confidence level. ON–ON measurements test the statistical sky fluctuations with an efficiency of $(1.5)^{1/2}$ at the scale of the beam separation[6]. Therefore δS_{rms} = 0.77 mJy. For a blackbody of 2.75 K, observed at 230 GHz, the flux scale can be converted to a temperature scale

$$\frac{\delta T}{T} = 0.245\ \frac{\delta S}{S}$$

The flux S per beam of the CMB is 0.733 Jy. Our final result is therefore

$$\frac{\delta T}{T} \leq 2.6\ 10^{-4}$$

at a 95 % significance and for a scale of 30".

REFERENCES

1. Baars, J.W.M., et al., Astron. Astrophys. 175, 319 (1987)
2. Kreysa, E., Proc. URSI Symp. 'MM– and Submm Wave Radio Astronomy', Granada, p. 153 (1985)
3. Neugebauer, G., et al., Ap. J. 308, 815 (1986)
4. Chini, R., et al., Astron. Astrophys. (1989), submitted
5. Boynton, P.E. and Partridge, R.B., Ap. J. 181, 243 (1973)
6. Uson, M.J. and Wilkinson, D.T., Ap. J. 283, 471 (1984)

COSMIC MICROWAVE BACKGROUND SESSIONS SUMMARY

George F. Smoot

Lawrence Berkeley Laboratory and Space Sciences Laboratory
University of California at Berkeley

ABSTRACT

The Particle Astrophysics Workshop sessions included discussions of measurements of the Cosmic Microwave Background Radiation (CMBR) spectrum and angular distribution. There was keen interest in the reported high frequency spectral distortion and its interpretation, and in improved measurements of the CMB anisotropy, as well as, discussion on the proper statistical techniques and experiment design.

INTRODUCTION

The CMB workshop was an exciting experience. We were all impressed by the number of participants, the level of activity, and the sophistication of the efforts. Since the discovery of the CMB in 1965 we have seen a large growth in the discoveries/measurements, technology, people, effort and funding in the field. The future looks even more positive for the results expected to be forthcoming.

SPECTRUM

The main topic of discussion was the sub-mm wavelength excess reported by the Berkeley-Nagoya collaboration (Matsumoto *et al.* [1] 1988). Figure 1 summarizes the spectrum measurements reported to date. The energy contained in this excess is roughly 10 to 20% of the energy in the CMB. This is a very large distortion; attempts to model or explain it have run into serious difficulties from the need to employ resources that are large even for the universe. Nevertheless there have been a number of different attempts. The most popular of these are:

(1) Early dust - Several groups, including Freese *et al.* at this conference, have found that they need to process $\Omega_{baryon} \cong 0.2$ of material through massive stars with maximum energy efficiency and then eject the processed matter as dust which absorbs the starlight and reradiates the energy to produce the sub-mm excess. One needs to fine tune the parameters of the model carefully to get the proper distortion. These stars need to form at a very early epoch, on the order of a redshift z ~ 30 to 40.

(2) Compton Distortion - Again we need $\Omega_{baryon} \cong 0.2$ which has to be heated to a high temperature by an ionizing energy source - e.g. a hot IGM. The

problems with this are that the energy requirements are quite large and one can produce a high isotropic cosmic X-ray background and, if not very careful, too much far-infrared flux.

(3) Vacuum decay - Jim Bartlett presented results of calculations of the distortions that would result from a decay of the vacuum at a rate that corresponds to about 4% of the energy of the CMB per expansion time. The natural scale of the photon wavelength is the horizon size and thus below the plasma cutoff frequency. The energy goes into heating the plasma and the universe never goes through recombination. This produces a distortion that fits the Berkeley-Nagoya excess at short wavelengths and also produces an excess at long wavelengths which falls just below current observational limits.

(4) Superconducting Cosmic Strings - Chris Thompson showed that superconducting cosmic strings produce distortions very much like those of vacuum decay, characteristic of a continuous energy -release mechanism.

(5) Particle decay - While particle decay can produce this and other types of spectral distortions - i.e. adding excess photons or excess energy - Joe Silk has pointed out that normal stellar evolution and parameters rule out many reasonable classes of potential particle decays as the source of the sub-mm excess.

Small-angular-scale anisotropy measurements by Kreysa and Chini at 230 GHz indicate that the number of sources needed to provide the reported sub-mm excess is greater than 10^4 times the number of bright galaxies. Measurements by Bernstein *et al.*[2] and by Halpern *et al.* indicate that on large angular scales any sub-mm anisotropy is less than 10^{-2}.

These examples show the lengths to which modelers have had to go to explain the reported sub-mm excess and why any discovery of spectral distortions in the CMB will be very important and exciting in terms of our view of the universe.

New molecular and atomic line work was reported by Crane *et al.* and Kogut *et al.* These experiments sample the CMB at distant points in the galaxy, usually by observing a cold cloud in absorption against a bright source and taking into account the local excitation (often negligible compared to the CMB). Crane *et al.* made improved measurements and analysis of the CN molecule to derive CMB temperatures toward ζ Oph at 2.64 mm wavelength and toward HD154368 at 1.32 mm. In addition they observied CH toward ζ Oph for an improved limit at 0.056 mm wavelength of $T_{CMB} < 3.6$ K (68% confidence level). Shortly after the conference Kaiser and Wright presented new data on CN line widths that they indicated would raise the Crane *et al.* 2.64 mm results to about 2.85 K, which is in close agreement with the preprint by Meyer, Roth, and Hawkins. Kogut *et al.* reported a measurement of the CMB temperature using the formaldehyde (H_2CO) molecule in cold clouds of W51. The result obtained, $T_{CMB} = 3.2 \pm 0.9$ K was not very accurate compared to the CN or direct observations. Much of the error has to do with the unknown parameters of the clouds and will eventually be reduced as the W51 complex is studied but will never approach the 1% errors of the 2.64 mm CN results. There remains the interesting possibility of using formaldehyde as thermometer to trace the (1+z) redshift dependence of T_{CMB} since the systematic errors are unlikely to rise as quickly with redshift and formaldehyde is very common. Meyer *et al.* [3] have been attempting to use atomic carbon to trace the T_{CMB} redshift evolution and found a 90% c.l. upper limit of $T_{CMB} < 16$ K at 0.61 mm wavelength for (1+z) = 2.776 where one would expect $T_{CMB} \sim 7.5$ K.

Steve Levin *et al.* reported new and updated low-frequency spectrum results made at White Mountain using coherent microwave receivers. A summary of recent CMBR spectrum measurements is given in the table below and in Figure 1 (a and b).

A Summary of Recent CMBR Spectrum Measurements

Reference	Wavelength [cm]	ν [GHz]	T_{CBR} [K]
Sironi, G., *et al.* 1987, in *13th Texas Symp. Rel. Astrophys.*, Ed. M.P. Ulmer	50.0	0.6	2.98 ± 0.55
Bensadoun (Smoot *et al.* 1989) [5]	20.5	1.46	2.38 ± 0.20
Sironi, G., Bonelli, G. 1986, *Ap. J.*, **311**, 418.	12.0	2.5	2.79 ± 0.15
De Amici (Smoot *et al.* 1989) [5]	7.9	3.8	2.63 ± 0.11
Mandolesi, N., *et al.* 1986, *Ap. J.*, **310**, 561	6.3	4.75	2.70 ± 0.07
Kogut (Smoot *et al.* 1989) [5]	4.0	7.5	2.51 ± 0.11
Kogut (Smoot *et al.* 1989) [5]	3.0	10.0	2.61 ± 0.06
Johnson, D.G, and Wilkinson, D.T. 1986, *Ap. J. Lett.*, **313**, L1.	1.2	24.8	2.783± 0.025
De Amici (Smoot *et al.* 1985) [4]	0.909	33.0	2.81 ± 0.12
Bersanelli (Smoot *et al.* 1989) [5]	0.333	90.0	2.60 ± 0.10
Kogut, Smoot, Bennett, & Petukowski 1989	0.21	150	3.2 ± 0.9
Meyer, D., Jura, M. 1985, *Ap. J.*, **297**, 119.	0.264	113.6	2.70 ± 0.04
	0.132	227.3	2.76 ± 0.20
Meyer, D., Roth, K., & Hawkins, I. 1989,	0.132	227.3	2.83 ± 0.07
Crane *et al.* 1988, *Proc. Third ESO/CERN*	0.264	113.6	$2.796^{+0.019}_{-0.041}$
Symposium, in press.	0.132	227.3	2.85 ± 0.10
and this workshop	0.0559	537	<3.6 K (1σ)
Matsumoto *et al.* 1988 [1]	0.116	259	2.799± 0.018
	0.0709	423	2.955± 0.017
	0.0481	624	3.175± 0.027

Neglecting Matsumoto *et al.* [1] and treating the systematic errors at different frequencies as independent, the data fit a Planckian spectrum with temperature 2.74±0.02 K with χ^2 = 22 for 13 degrees of freedom. The low-frequency ($\nu \leq 10$ GHz) measurements alone average to 2.63 ± 0.04 K with χ^2/DOF = 5.4/6.

In the workshop panel session and discussions we all agreed that confirming or disproving the reported distortion is the highest priority in the spectrum field. Fortunately, several experiments will soon provide new measurements: Lange reported that the Berkeley-Nagoya collaboration will fly a second-generation rocket-borne experiment in August. Gush and Halpern are preparing their rocket-borne experiment for a flight soon. The Cosmic Background Explorer satellite is scheduled for launch this summer (1989). Each of these experiments is expected to have very small error bars. It is also possible that Crane *et al.* will be able to get improved limits or a measurement at 0.56 mm using CH.

Our second conclusion is that by precisely determining the low-frequency spectrum, one can provide a better baseline for comparison with high-frequency measurements, and can distinguish among various models of the spectrum, including some of the models which account for the mm-wave distortion indicated by the results of Matsumoto *et al.* [1].

We also had a recomendation by panel member Wilkinson that in the future spectrum results be presented with two error estimates: (1) the best estimate - with statistical and systematic errors added in quadrature where appropriate and (2) a conservative error estimate where one assumes a worst case correlation of systematic errors and adds them linearly.

Figure 1. Cosmic Microwave Background Flux versus Frequency

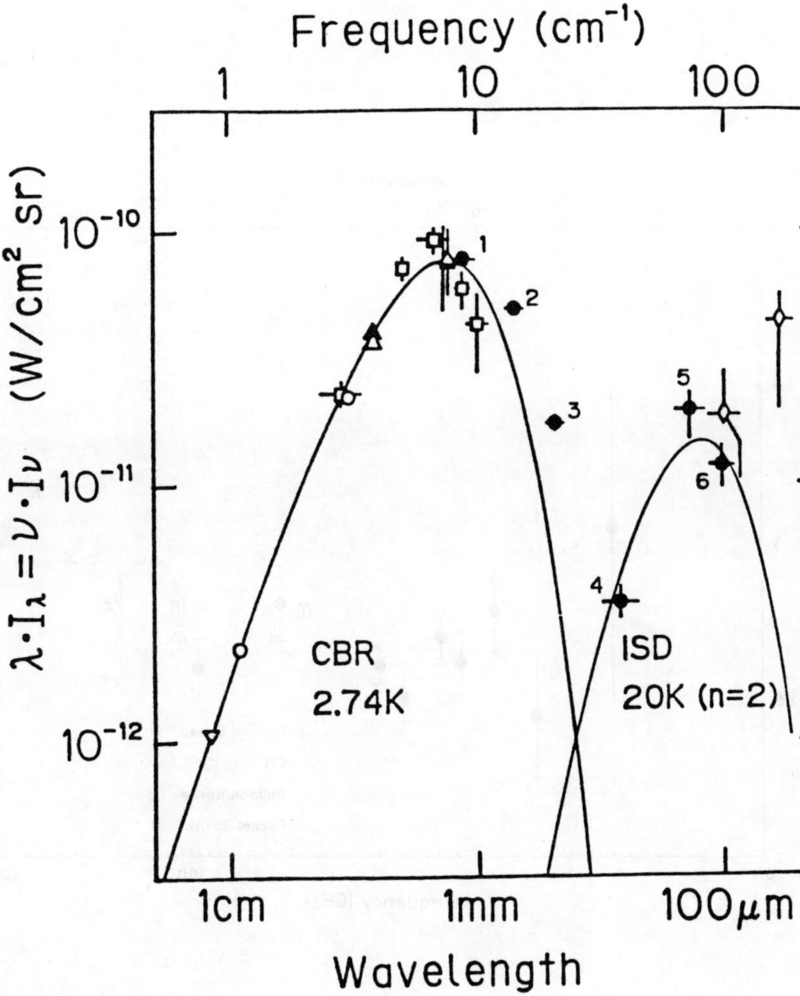

Figure 2. Cosmic Microwave Background Temperature versus Frequency

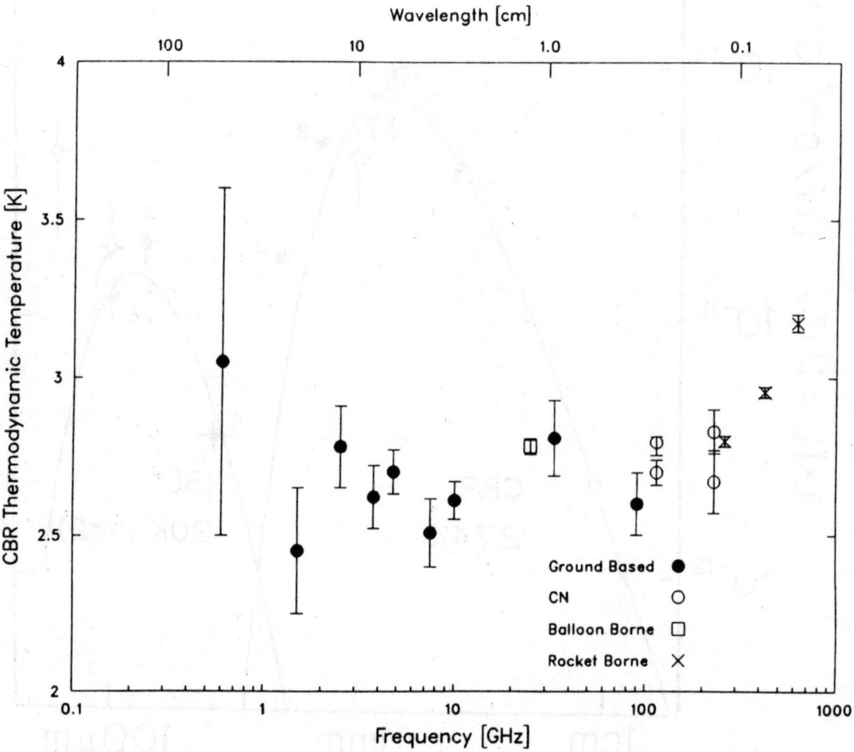

ANISOTROPY

The anisotropy of the CMB generally maps the surface of last scattering. The surface of last scattering is roughly $\Delta z = 80$ thick, corresponding to a subtended angle of about $\Delta\ \theta = 8'\ \Omega^{-1/2}$. The horizon size subtends an angle of about 2°. To date all measurements of the anisotropy of the CMB have yielded only upper limits to very low levels ($\Delta T/T < 10^{-4}$) with the exception of the first order anisotropy ($\Delta T/T \approx 10^{-3}$), which is thought to be due to the motion of the solar system relative to the cosmological comoving frame (distant matter and radiation), and the Sunyaev-Zel'dovich effect (scattering of CMB photons by hot intergalactic gas in specific galactic clusters). This isotropy caused problems for the isothermal and adiabatic theories of galaxy formation in the context of the standard (1970's) Big Bang model; the standard model of the early universe nows includes an inflationary phase to remove the small horizon size, requires that $\Omega = 1.0$, and allows for cold dark matter or other sources to produce $\Omega = 1$ and the observed structure of the universe. Inflation takes care of the horizon, causality, and flatness problems and also predicts that there will be extremely little vorticity/angular momentum in the universe.

Many efforts are now underway to measure the CMB anisotropy on nearly all angular scales and with increasing accuracy. As a result of the improvement and sophistication of the experiments, more effort has to go into the understanding of the experiment, the statistical interpretation of the results and the astrophysics of interfering sources. At the 10^{-4} level the experimentor must choose the operating wavelength and sky coverage to minimize and model the galactic interference. Danese summarized the interference of discrete sources on small and medium angular scale anisotropy measurements and showed a contour plot of expected interference level versus wavelength and angular scale. Not surprisingly, this plot was very similar to the plot of galactic background versus wavelength. It showed that there is a very limited wavelength range in which to make measurements at the 10^{-5} level and that at the 10^{-6} level there is confusion at any frequency of observation. Nevertheless, experimenters are pressing forward with new experiments and detectors that will achieve this level of sensitivity. Our group concluded that there will be detections soon and that proper interpretation will be paramount in avoiding confusion. In addition to careful design; future experiments must obtain spectral information by multifrequency observations.

As the experiments get harder, due both to the instrumentation and the required astrophysical modeling, interpreting the results more carefully becomes worthwhile. We spent some time in discussions of technique of estimators, the maximum liklihood and the power of various statistical tests - Lawrence includes a summary of some of this in his section as does Partridge. Likewise the experiment and its interpretation need to be matched to the type of model being tested. The statistical distribution of signal or power on the sky depends very much on the model type. For example primordial inflation will produce Gaussianly distributed fluctuations on the sky very much like the older adiabatic and isothermal models, while strings, both normal and superconducting, Compton-cooling (Sunyaev-Zel'dovich effect), dust and some evolved inflation models can produce non-Gaussian distributions on the sky. Our group concluded that we need theorists to make up some sample maps such as the cosmic string maps presented by Al Stebbins and co-workers. In their case they went on to show how the strings might show up in the data of a couple of different experiments. This pointed out the need for close coordination between theorists and experimentalists in both the interpretation of experimental results and the design of future experiments. The experimenter needs to choose carefully between experiments with a few field samples, with rough optimum number of samples on the order of 10 fields, and making maps or very many samples. This will depend upon a trade-off between experimental concerns and the type of model to be tested or limits to be set.

Most experiments to date have published limits that apply to gaussianly distributed features. These results are summarized in figure 3.

Beckman reported results of the I.A.C.-Jodrell CMB fluctuation experiment at Tenerife on the Canary Islands which had published reports (Davies *et al.* 1987) of anisotropy on a scale of 8° measured at 3 cm wavelength. The group has new data on angular scale of 5° that give about the same effect. Working group members commented that this could be an early example of a positive anisotropy result being confusing because of a possible galactic contamination. The Jodrell-I.A.C. group has said they would have new measurements at 6-cm wavelength over the same region to help distinguish between a galactic and cosmological origin.

Timbie, Bernstein, Cottingham, Cheng, Lawrence, Partridge, Kreysa, and Starobinsky all gave reports of various efforts to measure isotropy from the very large to very small angular scale. In general new, lower, and improved upper limits or interpretations were presented. These people and Masi, DeBernardis, and Fischer described new experiments that were underway, under construction, or in design. At present there are a large number of impressive projects in the works: On the large angular scale we can expect the Cosmic Background Explorer (COBE), to be launched by NASA this summer, to make very good sky maps on a scale of >7°; it will be followed in a couple of years by the USSR's Relict 2 mission which will cover the same angular range but at 5 frequencies instead of 3. On angular scales of roughly 1° (0.2° to 10°) there are ground-based and balloon-borne experiments in progress to be followed by the Japanese ISAS satellite, the USSR's Aelita mission and possibly the Italian IBIS mission. On smaller angular scales ground-based radiotelescopes and dedicated observatories such as the Caltech-JPL-Owens Valley facility, Hat Creek, and the VLA can be expected to yield new and improved results in the next few years.

We concluded that all angular scales are important: at 1 arcsecond one can compare dust versus other sub-mm background models; one arcminute is the scale for observing the Sunyaev-Zel'dovich effect; 10 arcminutes is on the scale of the recombination/last scattering surface thickness and of galaxy clusters; and anything greater than about 5° is primeval and reflects the large-scale structure of the universe.

Other areas of observational cosmology are closely tied to CMB anisotropy. For example, large-scale velocity flows (streaming) would imply power in the CMB $\Delta T/T$, while the current upper limits on $\Delta T/T$ put an upper limit on this potential streaming. Likewise large-scale structure predicts $\Delta T/T$ power in the CMB and its upper limits restrict the large-scale structure. Radio source counts, dust, etc and CMB work will have to merge as the limits get better and more positive detections are made.

CONCLUDING REMARKS

There is a lot of new technology - e.g. improved bolometers - that is going to affect the field soon, but there are so many areas and specialties that coherent receivers and alternate technologies will continue to play an important role, as will a myriad of approaches such as ground-based, balloon-borne, rocket-borne, and satellite-based experiments. Many new groups are joining the field and new techniques being applied. In many areas the scale of the effort is increasing, while in others one or two people are doing work that contributes significantly. These are signs of a vital and healthy field exciting and important enough to attract this much effort and growth. A number of experiments in the next few years will make significant improvements in our knowledge of the cosmic microwave background radiation as well as producing some new positive results that will be grist for theorists' mills as well as motivation for new experiments and a new understanding of the evolution and structure of the universe.

REFERENCES

[1] Matsumoto, T., *et al.* 1988, *Ap. J.*, **329**, 567.

[2] Bernstein, G, *et al.* 1989, *Ap. J.*, **3xx**, xxx.

[3] Meyer, D., Black, J., Chaffee, F., Foltz, C. and York, D., 1987, *Ap. J. Lett.*, **308**, L37.

[4] Smoot, G., *et al.* 1985, *Ap. J. Lett.*, **291**, L23.

[5] Smoot, G., *et al.* 1989, *Ap. J. Lett.*, submitted.

Figure 3. Cosmic Microwave Background Anisotropy Limits

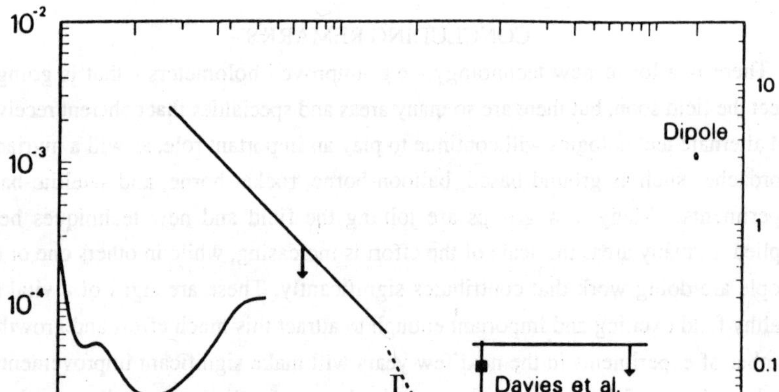

SUPERNOVAE

A SNOWFLAKE IN HELL

S. E. Woosley

Board of Studies in Astronomy and Astrophysics
University of California at Santa Cruz
Santa Cruz, CA 95064

ABSTRACT

A model for Type I supernovae is discussed that, while providing generally similar explosions, allows some diversity based upon the fractal dimension of the burning front and details of the initial conditions. A few recent developments in the theory of Type II supernovae are also reviewed.

I. INTRODUCTION

When Carl Pennypacker asked me to give this talk his instructions were not too specific. "Just talk about supernovae," he said, "but say something about fractal geometry and, oh yeah, the Hubble constant," and almost as an afterthought, since the meeting was on particle astrophysics, he suggested that I might say something about neutrinos.

Well that left quite a lot of room to work in, and while I'm still wondering how to relate neutrinos to the Hubble constant, I can relate the Hubble constant to Type Ia supernovae and the supernovae to fractal geometry, and will do so and, just for good measure, will mention neutrinos as well. But one thing was lacking - a really snazzy title. There have already been too many conference papers - things like "The Physics of Supernovae" or "Supernova 1987A: A Progress Report," and such. This called for a new name. So I came up with the one you see above. At least in the future those few hapless people reading my *Vitae* will wonder what it was about.

In this case Hell is provided to us by Bob Kirshner who pointed out in a marvelous, witty talk in Aspen last year, the kind I wish I could give, that, since the melting point of brimstone is something less than 1000 K, (actually I looked it up; it's only 386 K) and since supernovae develop temperatures in their cores around 10^{10} K (Type I) to 10^{11} K (Type II), a supernova is some ten to 100 million times hotter than Hell. That is, a supernova interior is probably as extreme a version of Hell as we are likely to find this side of the Big Bang. As for the snowflake, well that comes later.

II. TYPES OF SUPERNOVAE

The noted pioneer of supernova morphology, Fritz Zwicky, categorized supernovae into 5 classes, Types I thru V. Some of these classes only had a few members and others were quite irregular. Today we have made great progress towards grand unification and there are only two types of supernovae. Type II has hydrogen in its spectrum near peak light and Type I doesn't. Simple enough, but of course each of these types is now divided into at least three subclasses, so one wonders about the nature of progress. Some recent reviews of supernova types are Doggett and Branch (1985), Branch (1986), Woosley and Weaver (1986a), and Harkness and Wheeler (1989).

Type I can be Type Ia which is what everyone used to call Type I, those with the classical regular hump-tail light curve with its long regular exponential decline and spectra that vary only a little from event to event. They are probably still the most common Type I event, but then there is Type Ib which has a light curve very much like Ia, but a little fainter. The Ib's can play havoc with the Hubble constant if you don't have a spectrum and can't tell which Type I you're seeing. The Ib's notably lack the 6150 Å absorption feature at peak light, a feature that is generally attributed to Si II. They also have spectra in which neutral oxygen emission becomes a very prominent feature at late times (again unlike the Ia's). So far the Ib's are only found in spiral galaxies near regions of recent star formation. They are also the only Type I's that have been seen as strong radio emitters which may suggest a lot of circumstellar matter in their vicinity when they explode. Besides the Ia's and Ib's there may be further segregation of the Ib's according to the strength of the helium lines (Harkness and Wheeler 1989) and some Type I's don't seem to be either Ia or Ib but have other unusual characteristics such as a very brief light peak. These are sometimes called Ip's for "I-peculiars" (Branch 1986). Possible examples include some famous supernovae, SN 1885, SN 1954a, and SN 1957a.

The Type II's are at least as diverse. The simplest, most easily understood cases are the IIp's where "p" here stands for "plateau" and denotes the existence of a roughly two month long epoch of nearly constant luminosity of roughly 5×10^{42} erg s^{-1}. This plateau probably is the release of radiation deposited in the envelope of a massive red supergiant by the shock wave being released as the envelope expands and recombines. There is also, in theory at least, a brief spike, several hours in duration, of hard ultraviolet emission that accompanies shock break-out. So far the only observational evidence for this comes from Supernova 1987A where the declining tail of this emission was observed. At late times a long exponential tail from radioactive decay is probably present in most Type II's. It has certainly been seen very clearly in Supernova 1987A. There can also be Type II-L, where "L" stands for "linear" in which the plateau is either very brief or non-existent, and the light curve declines at an almost constant rate from the earliest times. These are probably massive stars that have lost almost all of their hydrogen envelopes. Finally there can be a large range of variations about these standard classes. Supernova 1987A, for example, which for reasons that we now understand, was born faint, grew brighter, and then declined exponentially and SN 1961v, which no one really understands (Branch and Greenstein 1971; Utrobin 1984). The variations are presumably caused by variations in a) the mass of the presupernova star; b) the amount of mass loss that star experiences before exploding; c) whether the star is a red or blue supergiant (SN 1987A was blue when it exploded, most others are expected to be red); d) the amount of radioactivity produced in the explosion; e) the amount and density of circumstellar material; e) the amount of dust through which we observe the supernova which is typically found in a region of young stars where dust is abundant; and, finally, f) perhaps variations in the explosion mechanism itself. It is therefore easy to understand why there may be considerable variation in Type II's. See also Doggett and Branch (1985) and Harkness and Wheeler (1989) for recent reviews.

Since the Type II's are found in regions of recent star formation, since theoretical models based upon massive star models give good agreement with observations, and since the progenitor star for Supernova 1987A had a main sequence mass of about 20 $M_{\odot}$, there is general agreement that Type II supernovae come from massive stars. Most people assume that the explosion mechanism must involve the known instability of the iron core of a highly evolved star to gravitational collapse. That certainly was the case for SN 1987A (we saw the neutrino burst) and probably is the case for most other Type II supernovae (though see §IV for another possible model for Type II supernovae).

The Type II mechanism must therefore typically involve the inefficient coupling of some small fraction (few tenths of a per cent) of the gravitational binding energy of the neutron star formed by the collapse (about 3×10^{53} erg) to the kinetic energy of ejection of all layers external to the iron core. Just how that coupling proceeds in detail is still an unsolved problem. As of April, 1989 no one is really sure how it works.

With Type Ia's we have a very different situation. Whereas most everyone agrees on the evolutionary scenario leading to Type II's but has difficulty with the explosion mechanism, people do not understand the exact evolutionary scenario that leads to a Type Ia though most everyone agrees that the mechanism is a degenerate thermonuclear runaway in an carbon-oxygen white dwarf. Evidence supporting this hypothesis has been reviewed by Woosley and Weaver (1986a) and includes a) the occurrence of Type Ia supernovae in regions, such as elliptical galaxies, where the rate of star formation is very low; b) the lack of hydrogen lines in their spectra; and c) the good agreement with observations of the spectra and light curves calculated using such a model. Since a solitary white dwarf will not by itself become unstable, the evolution must provide a carbon-oxygen white dwarf accreting mass from a binary companion at the proper rate. A helium white dwarf or a neon-magnesium white dwarf will not do. The helium white dwarf detonates and gives too brief a light curve and a spectrum deficient in intermediate mass elements. A neon-magnesium dwarf would collapse upon ignition to a neutron star.

But then we are left with the following conundrum. If all Type Ia's are the explosion of carbon-oxygen white dwarfs that approach the Chandrasekhar mass, where is there room for variation in the observed properties of resultant supernova? Until recently that was not a problem because everyone presumed, on the basis of sparse data, that all Type Ia supernovae *were* identical. Now that our data base is just beginning to encompass a few moderately well studied Type Ia's, it seems that is not true (Branch 1987; Filippenko 1989). Though similar, the spectra of Type Ia's at peak light do show variation and there may also be some variation in their rate of decline and peak luminosities. Only to the extent that we understand these variations can we rely on Type Ia supernovae as standard candles.

III. FLAME PROPAGATION IN A CARBON-OXYGEN WHITE DWARF

The observable consequences of a degenerate thermonuclear runaway in a carbon-oxygen white dwarf are most sensitive to the speed at which a nuclear combustion front propagates through the star. The burning does not proceed as a detonation wave but as a subsonic *deflagration* front, which we may simply refer to as a "flame." A deflagration may be distinguished from a detonation by several properties: 1) the deflagration is subsonic; 2) pressure is constant across the flame rather than increasing as it does in a detonation; and 3) the density decreases behind a deflagration wave while it increases behind a detonation wave. Thus, in a star, any outwardly moving deflagration is Rayleigh-Taylor unstable. Because the flame is subsonic the white dwarf expands and quenches the runaway before the entire white dwarf is consumed. This is one strength of the model, that there exist regions that experience intermediate burning temperatures in the explosion (3 to 5×10^9 K) as the flame dies out. These regions produce silicon and calcium which are quite prominent in the spectrum at peak light.

Given that the white dwarf expands faster than the flame moves, the amount of carbon and oxygen that are converted into ^{56}Ni is set by the competition between expansion (at the sound speed) and burning. The number of grams that burns to ^{56}Ni (plus the little bit that burns to silicon and sulfur) determines the kinetic energy of the explosion (just the total nuclear energy release minus the initial binding energy of the white dwarf, about 5×10^{50} erg). This sets not only the expansion rate of the supernova but also the luminosity of the supernova at peak, which is directly proportional to the mass of ^{56}Ni produced in the explosion (Arnett, Branch, and Wheeler 1985). This is a powerful constraint on the Hubble constant since sufficient ^{56}Ni must be produced to unbind the star at speeds of several thousand km s^{-1}, at least 0.4 $M_\odot$, implying a lower bound on L_{peak} of about 9×10^{42} erg s^{-1} and H_o less than about 80 km s^{-1}.

The spectrum and nucleosynthesis of the supernova are also determined by the flame speed. A supernova that burns most of its matter at high density will experience a great deal of electron capture and make a lot of ^{54}Fe and ^{58}Ni. In order not to have too much capture it is helpful if the flame is initially born slow and then accelerates. Then the white dwarf is "inflated" before the flame front goes through most of its mass. Later we shall discuss a prescription that gives such a behavior. The ionization states that appear in the late time spectrum, the strengths of the intermediate mass element lines in the peak light spectrum, and the widths of all the lines are all very sensitive to the flame speed. A more rapidly expanding supernova will reach lower density earlier and since cooling goes as the square of the density while heating by radioactive decay goes as the first power of the density, supernovae that expand too fast will be too hot and display emission lines of highly ionized iron that are not seen (Woosley, Axelrod, and Weaver 1984). Supernovae with too fast a flame speed, *e.g.* detonation or even a nearly sonic deflagration, are disallowed upon this basis.

So essentially all observable quantities in a Type Ia supernova are determined by the rate at which a nuclear flame moves through a degenerate carbon-oxygen white dwarf. Two supernovae having different effective flame speeds will have different observable properties. What is the actual value of this critical parameter?

From microzoned numerical calculations as well as analytic expressions, Weaver and I (1986b, 1989) have derived quite accurately the speed at which a flame sheet moves through a high density mixture of carbon and oxygen *provided* that the flame surface is smooth and unwrinkled. That speed is

$$v_{cond} \approx 50\,(\frac{X_{12}}{0.5})^{0.8}(\frac{\rho}{2\times 10^9\ \mathrm{g\ cm^{-3}}})\ \mathrm{km\ s^{-1}}. \qquad (1)$$

The flame sheet is very thin, a few times 10^{-4} cm thick. Put the above speed into a model for an exploding white dwarf, however, and you get nothing that resembles a Type I supernova. The flame is far too subsonic. The white dwarf pulses and a portion becomes unbound but the kinetic energy is nothing like that of a supernova.

Figure 1. A Koch snowflake

So we come at last to the snowflake, the Koch snowflake to be specific (Fig. 1). The flame in the supernova we have been describing is not smooth; it is very convoluted and deformed. There are several reasons for this but the most straightforward is the fact that burning in a subsonic flame lowers the density as it raises the temperature. In a gravitational field this leads to Rayleigh-Taylor instability. Since unburned fuel crosses the thin flame sheet with a microscopic normal speed that is everywhere given by eq. (1), the amount of matter burned will increase linearly with the surface area of the wrinkled flame and the effective flux of matter across the flame rather than being $4\pi r^2 \rho v_{cond}$ will instead be $A\rho v_{cond}$ where A is the generalized area that may be much greater than $4\pi r^2$ even when the average radius of the burned out region is nearly r (just as a radiator has much more area than just its planar length times width). In short, a Type Ia supernova is entirely a consequence of the Rayleigh-Taylor instability. By increasing the area, we are increasing the rate at which matter in the star burns. If the area is increased enough, the effective flame speed can be increased to an appreciable fraction of the sound speed.

We are interested in calculating how rapidly matter crosses some highly deformed surface of area A. We know the density, we know that the flame is very thin, and we know the normal velocity everywhere. If we just knew the area of the flame, we would know how fast matter was being burned. It is important to note the existence of two length scales within the supernova. One is the size of the (deformed, quasi-simply connected) burned out core, the volume of the white dwarf that has already burned to ^{56}Ni. We call the typical radius of this, r_b, which is also λ_{max}, the biggest deformation that can be formed, the largest scale for the Rayleigh-Taylor instability. Less apparent is a minimum length scale for instability,

$$\lambda_{min} = \frac{4\pi v_{cond}^2}{g_{eff}}, \tag{2}$$

with $g_{eff} = g(r)(\Delta\rho/\rho) = (4/3)\pi G r \rho(\Delta\rho/\rho) \sim 10^9 r_7$ cm s^{-2} with r_7 the distance to the center of the star in 100's of kilometers. A instability having smaller length scale will be burned by a flame moving at the conductive speed in a time shorter than its growth time. All scales of deformation between λ_{min} and λ_{max} can exist. Initially, during the first 20 km, λ_{min} is greater than λ_{max} and there is no Rayleigh-Taylor instability but this is such a small region as to be both unrealistic (the flame does not really ignite at a geometric point) and unimportant. While the larger instabilites grow slower they move a greater distance and thus are responsible for the overall growth of the burning region. The smaller instabilities "digest" the material so that the large lobes do not merely break off (become non-linear) and float away. We presume that between λ_{min} and λ_{max} there are no other natural length scales so deformations of all scales exist and are maintained as the flame propagates.

So now our problem is well defined. There is a surface whose area we wish to determine. The surface may be highly convoluted and not even simply connected. We also have a minimum tile size with which to cover that surface, λ_{min}^2. This is a problem in fractal geometry. Time out for an example from Mandelbrot's (1983) classic book on the subject!

First a one-dimensional example: start with an equilateral triangle. At the center of each face of that triangle place an equilateral triangle whose sides are one third the length of the previous triangle. Continue this until the length of the sides of the triangles is infinitesimal, or reaches some lower limit. The end result is something which looks like a snowflake (Fig. 1). The question now is "How does the perimeter of our figure increase as the length of the ruler decreases?" (If we were dealing with a three dimensional object, we would be asking how the surface area of our object was increasing). The fractal dimension gives us the answer to this question. The formula for this quantity is

$$D = \frac{-log(\text{increase of perimeter (or area) for each operation})}{log(\text{scale change})} + 1 \ (or\ 2) \quad (3)$$

where the "2" is used when dealing with areas and the "1" is used for perimeters. For our example, the perimeter increases by a factor of (4/3) with each operation, and the change in scale between consecutive triangles is (1/3). Using these numbers gives us a fractal dimension of 1.262 for our snowflake. Consider a more relevant but similar example where one uses pyramids instead of triangles (three rather than two dimensions). The scale change in *surface area* would be (3/2) for a length scale change of (1/2) (each face of the tetrahedron can be ruled into 4 equilateral triangles of side L/2 having vertices at the midpoints of the original triangular faces. The new figure covers up the center of these 4 triangles but exposes three new faces so the area increases by six-fourths). This object has fractal dimension D = 2.585. If the original length of a side was L and the smallest subdivision has side l, then the area is the area of the original tetrahedron ($3^{1/2}L^2$) times a factor $(L/l)^{D-2}$.

Returning to our supernova problem, the above analysis suggests an area of the burned out region equal to that of the undeformed surface ($4\pi r_b^2$) times a factor $(\lambda_{max}/\lambda_{min})^{D-2}$ with D between 2 and 3 but probably not far from 2.6 (the above example with pyramids had all length scales present with no large change in scale in each generation). When people study turbulence, the fractal dimension that they get is 2.6 - 2.7 (Mandelbrot 1983), so our example is not unrealistic.

This implies, were we to map our deformed flame onto a spherical surface having an equivalent area, than the sphere would grow with a speed $v_{eff} = v_{cond}(\lambda_{max}/\lambda_{min})^{(D-2)}$. Substituting $\lambda_{max} = r_b$, the radius of the sphere, and our previous expression for v_{cond} gives

$$v_{eff} = \text{Const}\ X_{12}^{(1-2n)} \rho^{(0.8-0.6n)} r_b^{2n} \quad (4)$$

where $n = D - 2$. For a density of 2×10^9 g cm^{-3}, a carbon mass fraction 0.5, and a fractal dimension between 2.5 and 2.7 this gives a reasonable range of effective velocities, 250 r_7 to 480 r_7 km s^{-1}, with r_7 the radius in 100's of kilometers. Equation (4) remains valid so long as the density and fractional decrease in density across the flame remain constant. Note that by the nature of the mapping, the effective speed can be supersonic even though microscopically the flame is subsonic throughout the star. Fractal dimensions near 2.5 have the interesting property that the radius of the burned out mass increases exponentially with time. This suggests extreme sensitivity to the initial conditions.

Nucleosynthesis and light curve calculations have been carried out by Woosley (1989) using various ratios of flame speed to the speed of sound. A variation of a factor of two in the flame speed can make up to a factor of two difference in peak luminosity (Models DF1 and DF2 of that work, for example, have flame speeds of 1/2 and 1/4 the sound speed, $C_s \sim 9000$ km s^{-1}, produce 0.89 $M_\odot$ and 0.51 $M_\odot$ of ^{56}Ni respectively and have peak luminosities 2.8×10^{43} erg s^{-1} and 1.5×10^{43} erg s^{-1} respectively). Even smaller variations might be expected to produce significant spectroscopic differences.

The above equation suggests that a factor of two variation in effective flame speed could result from variation in D between 2.5 to 2.7. Our ignorance of turbulence is such that variations of this magnitude could easily exist. Perhaps more important however, is the range in initial conditions that might exist in a real white dwarf that is running away. Obviously there is some dependence on composition. A larger carbon abundance gives a greater laminar conductive flame speed, but this makes λ_{min} larger which may reduce the area of the deformed flame, so in the end the composition dependence is not straightforward.

The greatest uncertainty though involves our unrealistic assumption that the runaway ignites at an infinitesimal point at the center of the star. It does not. The runaway is actually preceded by a stage of convective carbon burning, which for lack of anything better, is described using mixing length theory. As the runaway develops, an increasing energy flow must be carried by convective blobs which must circulate faster and faster. There comes a point however, when the time scale for accelerating the runaway becomes shorter than the time it takes for a convective blob to circulate (the Brundt-Väisälä frequency). This happens when the central temperature reaches about 6×10^8 K and the time scale is 10 - 100 seconds. The next stage is critical. By the time the temperature has risen to 1.0×10^9 K the time scale has dropped to 0.01 sec and it takes a "convective blob" longer than this to rise a pressure scale height at the speed of sound. The runaway becomes localized. Between 0.6 and 1.0 billion degrees, the situation is very complex. Blobs may rise and continue to burn as they rise. Thus the runaway may spread throughout macroscopic regions of the star, like sticks of dynamite borne by balloons, before the runaway becomes localized. Obviously this can lead to diversity in the explosion characteristics based upon very reasonable small deviations in the initial conditions. It may also explain the "mixing" artificially invoked by Branch *et al.* (1985) in order to get a good match to the observed spectrum at peak light. Unburned material may exist at comparable speeds as burned material because of a complex angular dependence to the location of the flame front, or it may simply be entrained in an explosion that initiates at many points (see also Müller and Arnett 1986).

And if this is not enough one must also consider the effect of the URCA process. Trace nuclei convected through a shell of critical density during the carbon burning that precedes and initiates the runaway may capture electrons inside that critical density and beta-decay outside. An important example is the URCA cycle $^{21}\mathrm{Ne}(e^-,\nu)^{21}\mathrm{F}(e^-\bar{\nu})^{21}\mathrm{Ne}$ which emits a neutrino-pair for each round-trip across the critical density shell. Iben (1978ab) has shown that cooling caused by this process may briefly delay the runaway and lead to off-center ignition when it finally occurs. The abundance of ^{21,22}Ne, ^{23}Na, ^{25}Mg and other important URCA nuclei will depend upon the initial metallicity of the star that made the white dwarf.

Given all this, it is not surprising that some diversity exists in Type Ia supernova, even if they are all carbon deflagrations in white dwarfs near the Chandrasekhar mass.

IV. TYPE Ib AND TYPE I 1/2 SUPERNOVAE

This is a section in which we learn that, contrary to traditional paradigms, core collapse in a massive stellar core can sometimes give a supernova called Type I and carbon deflagration, in certain conditions, can be made to power a supernova of Type II. Consider first the Ib's.

Except for the fact that, on the average, their light curves are roughly 1.5 to 2 magnitudes fainter in the blue, SN Ib are very similar photometrically to SN Ia. Chief differences are spectroscopic - Type Ib lacks the 6347 and 6371 Å blend of Si II lines near peak light and displays, at times later than 100 days or so, a very strong emission line (6300 Å) of [O I]. Thus far Type Ib supernovae have all been discovered in spiral galaxies, always in or near H II regions. They also seem to have unusual properties in their infrared spectra and two Type Ib's are the only Type I's ever to be observed as strong radio sources. There is some diversity in the light curves of SN Ib's; SN 1985F, for example, seems to have declined much more slowly after peak than 1964L (Ensman and Woosley 1988), but the lack of bolometric data for both makes this an unreliable premise.

Many models have been proposed to explain Type Ib supernovae (see Woosley and Weaver 1986a for a review). The favored model of the day is that Type Ib's are explosions of massive stars that, owing to mass exchange in a binary system or moss loss by stellar wind, have lost their hydrogen envelope. This would account for the lack of hydrogen lines, the association with star forming regions, the oxygen emission spectrum at late times, and possibly the strong radio emission associated with the supernova blast wave interacting with its own pre-explosive ejecta. Because the amount of ^{56}Ni synthesized in the explosion of a massive star is much smaller (SN 1987A, a supernova with a 6 $M_\odot$ helium core, ejected only 0.075 $M_\odot$ of ^{56}Ni, about 6 times less than Model DF2), it is expected that such explosions would have fainter light curves than Type Ia. Why the Si II feature should be suppressed in the spectrum at peak light remains unclear; the explosion of a massive star should make as much silicon as a carbon deflagration. But this aspect of the spectrum has not yet been accurately modelled. Filippenko (1988) has also found, in the unusual case of SN 1987K, what might be regarded as the "missing link" - a supernova that initially had a spectrum like that of a Type II, presumably the explosion of a single massive star, but at late times lacked hydrogen lines and looked like a Ib.

The greatest concern regarding the massive helium core model for SN Ib's, as has been recently emphasized by Ensman and Woosley (1988), is that, for any credible explosion energy, the ejecta expand too slowly to give the observed rapid decline in luminosity after peak observed for all Type Ib's except perhaps 1985F. Though bolometric data for Ib's is unfortunately quite scarce, Ensman's study appears to rule out helium cores of greater than 6 $M_\odot$ as Type Ib progenitors. Further, since helium cores lighter than about 4 $M_\odot$ produce very little ^{56}Ni, it seems that the progenitors of Ib's would have to lie in the narrow mass range 4 to 6 $M_\odot$. Even so, the decline after peak is too slow unless rather severe clumping is allowed in the ejecta in order to facilitate γ-ray escape.

For this reason, as well as the possible existence in nature of more than one kind of Type Ib progenitor, it remains important to consider alternative models. Two models based upon white dwarfs, either detonating or deflagrating, have been discussed by Branch and Nomoto (1986) and Woosley (1986, 1989).

We now turn to examine briefly the kind of supernova that might result should an asymptotic giant branch star prove incapable of expelling its entire hydrogen envelope before its carbon core grows to 1.38 $M_\odot$. Because its spectrum would contain hydrogen, such an explosion would certainly be categorized as Type II. The explosion mechanism, however, would not be core collapse but carbon deflagration just as we have discussed for Type Ia's. The light curve as well as spectrum would be greatly modified by the massive but tenuous hydrogen envelope that tamps the explosion. Iben and Renzini (1983) have called such a supernova - Type I by mechanism, Type II by observation - a Type I 1/2 supernova.

Recently Woosley and Iben (1988) have evaluated the probable light curve for such a model. The light curve has also been given by Woosley (1989). As a result of a nuclear explosion the star and core are totally disrupted with a final kinetic energy of 1.0×10^{51} erg. Core velocities are greatly tamped by the hydrodynamical interaction with the envelope and end up moving slower than 2000 km s^{-1} throughout. The hydrogen envelope on the other hand has higher velocities ranging to above 10,000 km s^{-1} in the outer layers. The resultant light curve, powered at plateau by the release of internal energy in th hydrogen envelope and on the tail by the decay of ^{56}Co, was virtually indistinguishable from that of an ordinary Type II-p supernova, SN 1969l, except that the ratio of the bolometric luminosity on the tail to that on the plateau was about three times larger than observations. This reflects the greater amount of ^{56}Ni created in the explosion.

Based purely upon the observational data, there is no reason to exclude a large fraction of Type II supernovae from being powered by this sort of mechanism. One key difference, however, is that there would be no neutron star or black hole remnant and no neutrino burst.

V. NEUTRINO BACKGROUND AND NEUTRINO NUCLEOSYNTHESIS

We now know, even though we very strongly suspected before that the collapse of an iron core in a Type II supernova produces several times 10^{53} ergs of neutrinos over a period of several seconds and that the temperature, at least of the electron antineutrino component is near 5 MeV. One interesting aspect of these neutrinos is that the total of all supernovae that have occurred since the Big Bang provides a background of electron antineutrinos that could in principle be detected. Calculations by Woosley, Wilson, and Mayle (1986) suggest a flux near 250 cm^{-2} s^{-1}. Unfortunately terrestrial nuclear reactors have a comparable or greater flux (Lagage 1985) and, perhaps even more unfortunately, almost all these neutrinos are to be associated with the era of galaxy formation so that a red-shift factor of perhaps 3 must be introduced into the energy. The large background from terrestrial β-decay then makes the supernova neutrino background undetectable by any experiments that might be constructed in the near future. There is still hope for seeing the effects of the neutrino bursts from past supernovae, however, simply by buying a tube of toothpaste or a can of *Boraxo*.

Calculations by Wilson and Mayle at Livermore, and others show that the μ- and τ-neutrinos have a temperature about a factor of two greater than the electron neutrinos (they escape more freely from deeper layers). These neutrinos have sufficient energy and flux in the heavy element shells overlying the collapsing core to induce nuclear transmutation by inelastic neutral current scattering. A representative flux of μ- and τ-neutrinos and their antiparticles from the core is $10^{38}\tau^{-1}r_9^{-2}\exp(-t/\tau)$ cm^{-2} s^{-1} where τ is the characteristic e-folding time scale for the neutrino luminosity of the core, about 3 sec. A typical inelastic neutrino cross section for ^{56}Fe if the μ and τ neutrino temperature is $\sim$8 MeV is 7×10^{-41} cm^{-2}. This scales approximately as atomic mass for other nuclei. Thus for a shell at $r_9 = 1$ (10,000 km) the rate at which a neutron or a proton is ejected from a given heavy nucleus is approximately 0.01 sec^{-1}. For typical radii ranging from several thousand kilometers for the silicon shell to around 100,000 kilometers for the helium shell, fractions of from few percent down to a few thousandths of a percent of the resident nuclei will be converted into daughters having one less nucleon. The back reaction of the neutrons and protons with the products must be followed. Specific nuclei due almost entirely to the ν-process are ^{7}Li, ^{11}B, ^{19}F, ^{138}La, and ^{180}Ta. Significant amounts of ^{10}B, ^{15}N, ^{22}Na, ^{26}Al, ^{31}P, ^{35}Cl, ^{39,40}K, ^{51}V, ^{45}Sc, ^{49}Ti, ^{50}V, ^{55}Mn, ^{59}Co, and ^{63}Cu are also produced (Woosley *et al.* 1989), so much so that, within the uncertainties of the model, they might also owe their origin predominantly to the ν-process. Despite our success in producing many intermediate mass isotopes, we found that the neutrino-induced r-process in the helium shell recently suggested by Epstein, Colgate, and Haxton (1988) is quite small in any of the realistic scenarios we explored.

An illustrative case is ^{19}F (Woosley and Haxton 1988). With an effective temperature of about 8 MeV, the inelastic scattering reactions ^{20}Ne(ν, ν'p)^{19}F and ^{20}Ne(ν, ν'n)^{19}Ne eject nucleons from abundant ^{20}Ne nuclei. Because of the large parent to daughter ratio in the solar abundances (3100), sufficient fluorine is produced that, even accounting for the amount that is destroyed as the shock passes through the neon shell, the remainder is adequate to account for its cosmic abundance.

VI. CONCLUSIONS

Starting from very similar initial thermodynamic conditions, which themselves depend upon the metallicity, carbon abundance, and the rate of mass accretion, carbon deflagration models can provide a range of outcomes. An element of chaos is introduced into the results by an uncertain but likely variation of the fractal dimension of the propagating flame front and the near exponential sensitivity of the flame speed to complex initial conditions set up during the transition from convective energy transport to a fully developed deflagration wave. However, in order to fit the shapes of typical Type Ia light curves seen in the past, expansion must occur rapidly enough but not so rapid as to consume all of the intermediate mass elements. Thus the ^{56}Ni mass needs to be in the range 0.5 to 1.0 $M_\odot$ and the peak luminosity needs to be in the range $2.5 \pm 1 \times 10^{43}$ erg s^{-1}, implying a Hubble constant 60 ± 20 km s^{-1} Mpc^{-1} (Woosley 1989).

Some other interesting aspects of current models for Type I and II supernovae have been reviewed. What the field really desperately needs now is more *bolometric* light curves. Supernova 1987A has shown us just how much we can learn if we have these. Type Ib is a prime example. There does not exist a single bolometric light curve for Type Ib showing the transition from peak to tail in sufficient detail that one can really make a ruling on the success or failure of the massive stellar core model.

It is always better to catch a supernova earlier than later. Obviously it is difficult to tell what the peak luminosity of the supernova was if the event is already past the peak, but measuring the full width at half-maximum requires an even earlier discovery. The shape of the light curve peak is very useful in narrowing the range of peak luminosities in the models and thus in pinning down the Hubble constant.

Bolometric observations of Type II's other than SN 1987A are also lacking and, again, early detection giving the full duration of the plateau will ultimately aid in quantifying the explosion energy and the mass of the hydrogen envelope that was in place when the star exploded. Bolometric observations of the tails of several Type II's of (at least approximately) known distance will tell us the spread of ^{56}Ni that is produced in various explosions, helping to calibrate our theories of stellar nucleosynthesis, constrain the explosion model, and perhaps ultimately giving a diagnostic of the mass of the star that exploded (generally speaking a brighter ratio of tail to plateau luminosity is expected for more massive stars).

As interesting as the light curves are, more information is contained in the spectrum. Spectra are harder to model (and to measure!) than the light curve but many clever people are now working on the problem, having cut their teeth on SN 1987A. It seems very likely that spectral histories will ultimately prove the tightest constraints on the Type Ia model and considerably narrow the allowed range for the Hubble constant. Already just seeing lines of intermediate mass elements in the spectrum at peak light has told us a great deal (ruling out helium and carbon detonation models, for example).

The spectrum, when correctly analyzed, also gives unique and valuable information on the explosion energy for, nucleosynthesis in, and compositional mixing of all Types of supernovae, yet high quality data for all supernovae save 1987A are sadly lacking.

Finally one needs a great deal of theory to make sense of all the supernova observations, to build on the lessons learned from SN 1987A, and ultimately to come a full understanding of the entire zoo of these magnificent creations. Let's get to work!

This work has been supported by the National Science Foundation (AST 88-13649) and by NASA (NAGW 1273).

REFERENCES

Arnett, W. D., Branch, D., and Wheeler, J. C. 1985, *Nature*, **314**, 337.

Branch, D. 1986, *Ap. J. Lettr.*, **300**, L51.

Branch, D. 1987, *Ap. J. Lettr.*, **316**, L81.

Branch, D. and Greenstein, J. 1971, *Ap. J.*, **167**, 89.

Branch, D., and Nomoto, K. 1986, *Astron. and Ap.*, **164**, L13.

Branch, D., Doggett, J. B., Nomoto, K., and Thielemann, F. -K. 1985, *Ap. J.*, **294**, 619.

Doggett, J. B., and Branch, D. 1985, *Astron. J.*, **90**, 2303.

Ensman, L. M., and Woosley, S. E. 1988, *Ap. J.*, **333**, 754.

Epstein, R. I., Colgate, S. A., and Haxton, W. C. 1988, *Phys. Rev. Lettr.*, **61**, 2038.

Filippenko, A. V. 1989, submitted to *Pub. Astron. Soc. Pac.*

Filippenko, A. V. 1988, *Astron. J.*, in press.

Harkness, R. P. and Wheeler, J. C. 1989, in *Supernovae*, ed. A. Petschek, (Berlin: Springer Verlag), in press.

Iben, I. Jr. 1978a *Ap. J.*, **219**, 213.

Iben, I. Jr. 1978b, *Ap. J.*, **226**, 996.

Iben, I., and Renzini, A. 1983, *Ann. Rev. Astron. and Ap.*, **21**, 271.

Lagage, P. O. 1985, *Nature*, **316**, 420.

Mandelbrot, B. B. 1983, *The Fractal Geometry of Nature*, (New York: W. H. Freeman and Co.)

Müller, E., and Arnett, W. D. 1986, *Ap. J.*, **307**, 619.

Utrobin V. P. 1984, *Astron. and Ap.*, **98**, 115.

Woosley, S. E. 1986, in *Nucleosynthesis and Chemical Evolution*, 16th Advanced Course, Swiss Society of Astrophysics and Astronomy, ed. B. Hauck, A. Maeder, and G. Meynet, (Geneva Observatory), p. 1.

Woosley, S. E. 1989, in *Supernovae*, ed. A. Petschek, (Berlin: Springer Verlag), in press.

Woosley, S. E., Axelrod, T. S., and Weaver, T. A. 1984, in *Stellar Nucleosynthesis*, ed. C. Chiosi and A. Renzini, (Dordrecht: D. Reidel), p. 263.

Woosley, S. E., and Weaver, T. A. 1986a, *Ann. Rev. Astron. and Ap.*, **24**, 205.

Woosley, S. E. and Weaver, T. A. 1986b, in *Radiation Hydrodynamics in Stars and Compact Objects*, Proc. IAU Symposium 89, ed. D. Mihalas and K. -H. Winkler, Berlin: Springer Verlag), p. 91.

Woosley, S. E., and Weaver, T. A. 1989, *Ap. J.*, in preparation.

Woosley, S. E., Wilson, J. R., and Mayle, R. M. 1986, *Ap. J.*, **302**, 19.

Woosley, S. E., and Iben, I. Jr. 1988, *Bull. Am. Astron. Soc.*, **19**, 1036.

Woosley, S. E., and Haxton, W. C., 1988, *Nature*, **334**, 45.

Woosley, S. E., Hartmann, D., Hoffman, R. D., and Haxton, W. C. 1989, in preparation for *Ap. J.*.

RECENT SPECTROSCOPIC OBSERVATIONS OF SUPERNOVAE

Alexei V. Filippenko
Department of Astronomy
University of California
Berkeley, CA 94720 U.S.A.

Historically, supernovae (SNe) have been divided into two major spectral classes. The fundamental criterion is the presence (Type II) or absence (Type I) of hydrogen at *any* time during the spectral development of the SN, but in practice the classification has usually been done when the object is near maximum brightness. Significant differences in the average light curves of the two spectral types are known to exist.

During the past few years SNe I have been segregated into two subclasses, Ia and Ib, with SNe Ib lacking the deep absorption trough near 6150 Å exhibited by SNe Ia during the first month past maximum.[1] Moreover, the late-time spectra of SNe Ia and Ib are completely different, the latter being dominated by emission lines of intermediate-mass elements (e.g., O, Ca, Na, Mg) rather than by Fe and Co. SNe Ib tend to be fainter (by 1 − 2 mag) and redder than SNe Ia near maximum brightness; they also have distinct infrared light curves, and some are radio sources. Moreover, unlike SNe Ia, they appear almost exclusively in late-type spiral galaxies, close to (or superposed on) H II regions. Several spectroscopically distinct varieties of SNe Ib exist.[2] In light of these differences, it is clearly very important to exclude SNe Ib when attempting to use SNe I as standard candles.

The spectra of different SNe Ia were long thought to be very homogeneous, when compared at the same phase. This is consistent with the hypothesis that the deflagration of a C-O white dwarf with a mass equal to the Chandrasekhar limit is responsible for the sudden outburst of light. It also provides support for the use of SNe Ia as standard candles. During the past two years, however, several investigations have led to the demise of SN Ia homogeneity. The most convincing evidence is that the measured wavelength of the 6150 Å trough can differ widely among objects observed at the same phase.[3] Poorly determined times of maximum cannot account for all of the observed differences, which in some cases amount to over 30%. This result, in fact, is not unexpected; theoretical calculations show that the velocity of the deflagration wave need not be the same in all SNe Ia.[4] It is puzzling, therefore, that the measured absolute magnitudes of SNe Ia exhibit very small scatter in the recent analysis by Leibundgut.[5]

Although Zwicky[6] identified one or more examples of SNe Types III, IV, and V, these are now generally regarded as peculiar SNe II,[7] which are known to exhibit great heterogeneity in their spectroscopic and photometric properties. An object of particular interest is SN 1987F in the spiral galaxy NGC 4615.[8] The spectrum of this SN II resembled those of type 1 Seyfert nuclei and QSOs, especially at late times (9 − 12 months past maximum). Forbidden lines, normally quite strong at this phase, were very weak, and Fe II emission was prominent. A crude estimate showed that the mass of the envelope is no less than 5 $M_\odot$, and might be $\gtrsim 30\ M_\odot$, consistent with the slow decline of the object's visual brightness. In some respects SN 1987F slightly resembled SN 1961V, Zwicky's "Type V" SN,[6] but there were many important differences as well. On physical grounds, it is not yet clear whether we should continue to focus so much attention on the mere presence or absence of hydrogen; after all, there may even exist SNe in which a white dwarf surrounded by a hydrogen envelope undergoes carbon deflagration ["Type I 1/2 SNe"[9]]. A more physically illuminating way to observationally classify SNe is desirable.

SN 1987K in the Virgo spiral NGC 4651 demonstrates, better than any other single object, that our classification scheme is, at the very least, incomplete.[10] Figure 1 shows that its spectrum changed with time from that of a SN II to that of a SN Ib! At late times no trace of broad Hα was present in SN 1987K. Moreover, its forbidden emission lines were considerably broader than those in the "plateau" Type II SN 1986I (Fig. 1*c*), as expected in the absence of a massive hydrogen envelope that normally decelerates the inner layers. These observations suggest that at least *some* SNe Ib are closely related to SNe II, the mass of the hydrogen envelope being the dominant variable. If the progenitor of SN 1987K was a single star, perhaps it was not quite massive enough to lose all of its hydrogen due to winds, yet sufficiently massive to lose most of the envelope; this could include the range $20 - 40\ M_{\odot}$.[11] On the other hand, if the progenitor was part of a binary system, significant mass transfer may have occurred even if its initial mass was much smaller. In either case it appears that core collapse was the most likely explosion mechanism, just as in normal SNe II. Further studies of the connection between SNe II and Ib are in progress.

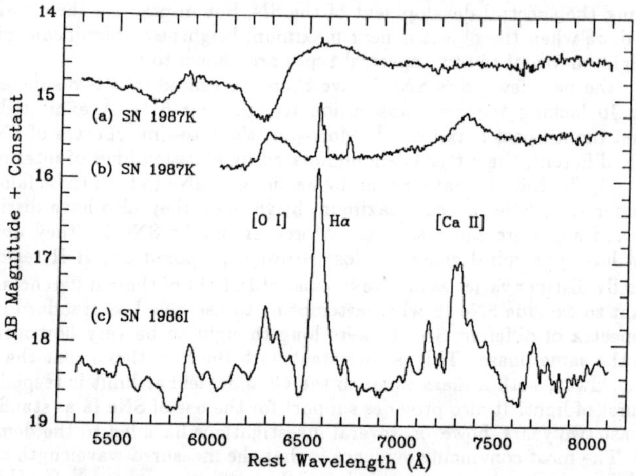

Figure 1: Spectra of SN 1987K roughly one week (*a*) and seven months (*b*) past maximum brightness, and of the Type II SN 1986I about nine months past maximum (*c*). Both SNe were discovered by the Berkeley Automated Supernova Search Team.[12] The narrow emission lines prominent in spectrum (*b*) are produced by the superposed H II regions.

My research has been supported by the California Space Institute, most recently through grant CS–41–88, and by NSF Core Block grant AST 86–14510 to Lick Observatory.

1) Porter, A. C., and Filippenko, A. V. *Astron. J.*, **93**, 1372 (1987).
2) Wheeler, J. C., Harkness, R. P., Barker, E. S., Cochran, A. L., and Wills, D. *Astrophys. J. (Letters)*, **313**, L69 (1987).
3) Branch, D., Drucker, W., and Jeffery, D. J. *Astrophys. J. (Letters)*, **330**, L117 (1988).
4) Woosley, S. E. In *Supernovae*, ed. A. G. Petschek (Berlin: Springer-Verlag), in press (1989).
5) Leibundgut, B. *Ph.D. thesis*, Universität Basel (1988).
6) Zwicky, F. In *Stars and Stellar Systems*, Vol. 8, ed. L. H. Aller and D. B. McLaughlin (Chicago: Univ. Chicago Press), p. 367 (1965).
7) Oke, J. B., and Searle, L. *Annu. Rev. Astron. Astrophys.*, **12**, 315 (1974).
8) Filippenko, A. V. *Astron. J.*, in press (1989).
9) Iben, I. Jr., and Renzini, A. *Annu. Rev. Astron. Astrophys.*, **21**, 271 (1983).
10) Filippenko, A. V. *Astron. J.*, **96**, 1941 (1988).
11) Woosley, S. E., and Weaver, T. A. *Annu. Rev. Astron. Astrophys.*, **24**, 205 (1986).
12) Perlmutter, S., *et al.* In *Instrumentation for Ground-Based Optical Astronomy*, ed. L. B. Robinson (Berlin: Springer-Verlag), p. 67 (1987).

Model Atmospheres for Supernovae and the Distance to SN 1987A

Ronald G. Eastman

Harvard-Smithsonian Center for Astrophysics

Abstract

The early spectra of SN 1987A have been analyzed using an atmosphere code which solves simultaneously the non-LTE equations of statistical equilibrium and the time independent equation of radiation transport. We examined the evolution during the first 10 days of an atmosphere where the density is a power law of the radius. The model atmospheres are used to derive the distance to SN87A as 50 ± 6 kpc.

To study the atmospheres of supernovae, we have implemented an algorithm for simultaneously solving the combined equations of statistical equilibrium and comoving frame radiation transport in an expanding atmosphere. The code takes advantage of the large velocity gradients in supernovae to simplify the coupled statistical equilibrium and radiation transport equations. In the present set of calculations, the temperature distribution was obtained by solving the transfer equation and thermal balance constraint under the assumption of LTE. This temperature distribution was then used in the next stage of the calculations, where the assumption of LTE was relaxed for hydrogen and helium. This method produces lab frame luminosities which vary from more exact treatments by less that 5%.

The transfer equation solved assumes homologous expansion, spherical geometry, and is correct to order $\beta \equiv v/c$. Also, it is assumed that

$$\frac{DI_\nu}{Dt} \equiv \frac{\partial I_\nu}{\partial t} + v\frac{\partial I_\nu}{\partial r} \approx 0, \tag{1}$$

where I_ν is the radiation specific intensity at frequency ν.

We have computed model atmospheres for conditions appropriate to SN87A at times less than 10 days. The outer layers of Woosley's model 10H (Woosley, 1988) can be represented by a simple power law density distribution of the form

$$\rho(v) = \rho_0\left(\frac{v}{v_0}\right)^{-\gamma}\left(\frac{t}{t_0}\right)^{-3} \tag{2}$$

with $\rho_0 = 2.9 \times 10^{-11}\ \mathrm{g\,cm^{-3}}, \gamma \approx 9, v_0 = 10^9 \mathrm{cm\ s^{-1}}$ and $t_0 = 1$ day. The luminosity in the observer frame is taken from observations of SN87A by Menzies *et al.* (1987) and Blanco *et al.* (1987).

The main sources of continuum opacity at early times are bound-free photoionization of hydrogen and helium, and bremsstrahlung opacity. The main scattering opacity in the optical and infrared is electron scattering. However, in the ultraviolet, line scattering dominates the opacity. The calculations include contributions to the scattering opacity from 63,000 bound-bound transitions in heavy elements.

Figure 1 compares the observations with the synthetic spectrum for the assumed density distribution at $t = 10$ days (top and middle spectra). The overall agreement is not too bad, although several features are in discord with the observations, perhaps as a result of departures from LTE for the elements that give rise to those features. One problem with this model at $t = 10$ days is that it doesn't have enough flux in the B-band, because the effective temperature is slightly too low. The situation is improved by decreasing the density everywhere by a factor of 4. The synthetic spectrum which results from from doing this is also shown in figure 1 (bottom spectrum).

To get the distance to SN87A we have used our models to derive correction factors to the Expanding Photosphere method of estimating distances to type II supernovae (Kirshner and Kwan, 1974). The basic idea is that the flux emerging from the photosphere is Planckian, though possibly diluted because of multiple scatterings (Hershkowitz, Linder and Wagoner, 1986) and other non-LTE effects. The observed flux can be written as

$$f_\nu^{obs} \approx \theta^2 \xi \pi B_\nu(T_c) \tag{3}$$

where $\theta \approx v_{phot}t/D$ with t being the time after core collapse, v_{phot} is the photospheric velocity, D is the distance from supernova to observer, and ξ is a dilution factor derived from the models. The value of v_{phot} is determined by comparison of synthetic spectra with observations. The present analysis uses $UBVR_cI_c$ photometry. Using magnitudes, the approximation becomes

$$m_\lambda \approx -2.5\log\xi\theta^2 - 2.5\log\pi B_\lambda(T_c) + C_\lambda. \tag{4}$$

Applying the correction factors derived from models to data on SN87A gives the results shown in figure 2, which displays the distance measured at several different times (dots). The dashed line shows the mean distance determined in this way. The optical continuum has a color temperature which after the first 2 days appears somewhat hotter than the color temperature fitted to $UBVR_cI_c$ photometry. It turns out that the $V - I_c$ color temperature gives a good fit to the optical continuum. If we assume that the flux coming from the photosphere is a dilute blackbody at the $V - I_c$ color temperature, then we get a new set of correction factors from the models and another distance determination for SN87A, which is also shown in figure 2 (triangles). The mean distance to SN87A determined by both methods is 50 kpc, with a standard error of 5 and 6 kpc, respectively. This is in reasonable accord with the recent results of Walker and Mack (1988) who observed RR Lyraes in NGC 1786 and obtained an LMC distance of 48.8 kpc, and of Feast and Walker (1987), who measured the distance from LMC Cepheids as 49.4 kpc. The ability to find type II supernovae at large redshift and accurately determine their intrinsic luminosities make them ideally suited for cosmological distance studies.

My collaborator on this work is Robert Kirshner. I would like to thank Michael Kurtz for his invaluable logistic support, and Robert Kurucz for sharing his atomic data with us. This work was supported in part by the Smithsonian Institution through a Smithsonian Predoctoral Fellowship, by the National Science Foundation through grant NSF AST-8516537, and by NASA through grants NAG5-642 and NAG5-841.

References

Blanco, V. M., *et al.* 1987 *Ap.J.* **320**, 589.

Menzies, J. W., *et al.* 1987, *MNRAS* **227**, 59P-49P.

Branch, D. and Patchett, B. 1973, *MNRAS* **161**, 71.

Eastman, R. G. and Kirshner, R. P., 1988 (in preparation).
Hershkowitz S., Linder E., and Wagoner R. 1986, *Ap.J.* **301**, 220.
Kirshner, R. P. and Kwan, J. 1974, *Ap.J.* **193**, 27.
Saio, H., Kato, M. and Nomoto, K. 1988, *Nature* **334**, 508.
Woosley, S. E. 1988, *Ap.J.* **330**, 218.
Walker, A. R. 1987, *MNRAS* **225**, 627.
Walker, A. R. and Mack, P. 1988, preprint.

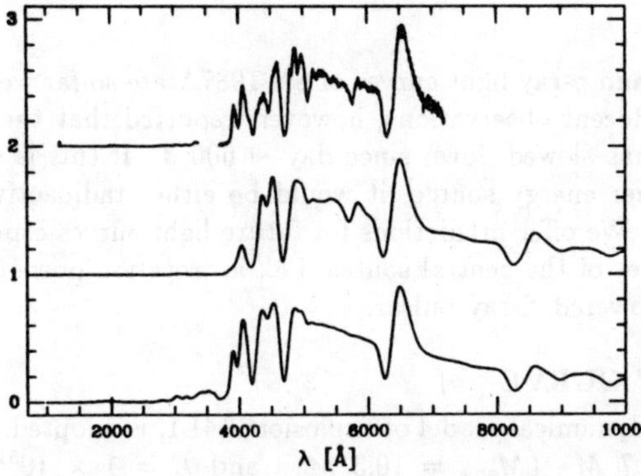

Figure 1 - Observed spectrum of SN87A at $t = 10$ days (top) curve, and synthetic spectra of two model atmospheres (middle and lower curves), which differ only in that the later has 1/4 the gas density of the former.

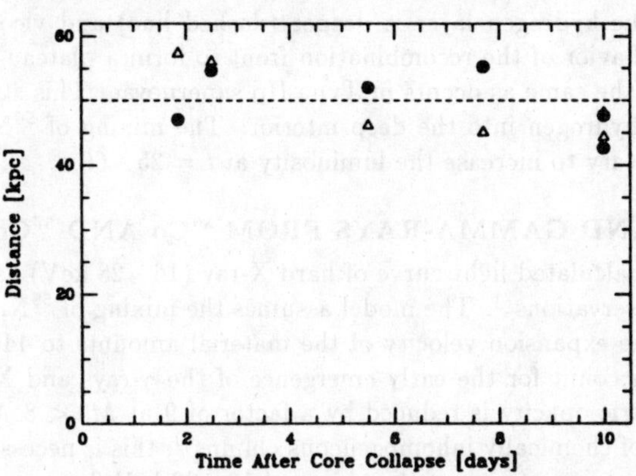

Figure 2 - Distance to SN87A determined via modified Expanding Photosphere Method from observations and models of the supernova at several different times.

MODELS FOR GAMMA-RAYS, X-RAYS AND OPTICAL LIGHT FROM SN 1987A

Ken'ichi Nomoto, Toshikazu Shigeyama, and Shiomi Kumagai
Department of Earth Science and Astronomy, University of Tokyo

1. ABSTRACT

The optical, X-ray and γ-ray light curves of SN 1987A are so far well modeled by the decay of ^{56}Co. Recent observations, however, reported that the decline of the optical light curve has slowed down since day $\sim$ 600 d. If this is due to the contribution from another energy source, it would be either radioactive ^{57}Co or the buried neutron star. We offer predictions for future light curves especially due to ^{57}Co and various types of the central source, i.e., an rotation-powered isolated pulsar or an accretion-powered X-ray pulsar.

2. OPTICAL LIGHT CURVE

The updated hydrodynamical model of explosion, 14E1, is adopted, where the ejecta mass is $M = 14.7\ M_\odot$ ($M_{\rm env} = 10.3\ M_\odot$) and $E = 1 \times 10^{51}$ erg[5),6)]. Explosive nucleosynthesis for the 6 $M_\odot$ helium core produces radioactive species of ^{56}Ni (0.075 $M_\odot$), ^{57}Ni (0.0043$M_\odot$), and ^{44}Ti ($1.2 \times 10^{-4}\ M_\odot$), if the baryon mass of the neutron star is 1.60 $M_\odot$[2)].

As seen in Figure 1, the calculated optical light curve for 14E1 is successful in reproducing the plateau-like peak. Such a plateau is formed by the hydrogen recombination front; its radius is almost stationary at $\sim 8 \times 10^{14}$ cm for 60 d - 100 d, if hydrogen is mixed down to $M_r \sim 1\ M_\odot$ (solid curve in Fig. 1). The plateau lasts longer if the hydrogen layer is deeper (dashed line) and vice versa (dotted line). Such a behavior of the recombination front to form a plateau of the light curve is essentially the same as occurs in Type IIp supernovae. This strongly indicates the mixing of hydrogen into the deep interior. The mixing of ^{56}Ni into outer layers is also necessary to increase the luminosity at t = 25 - 60 d.

3. HARD X-RAYS AND GAMMA-RAYS FROM ^{56}Co AND ^{57}Co

Figure 2 shows the calculated light curve of hard X-ray (16 - 28 keV) as compared with the Ginga observations[7)]. The model assumes the mixing of ^{56}Ni up to $M_r = 13.5\ M_\odot$ where the expansion velocity of the material amounts to 4400 km s^{-1}; this is required to account for the early emergence of the γ-rays and X-rays. Moreover, the photoelectric opacity is reduced by a factor of 9 at $M_r < 8\ M_\odot$ by assuming the formation of chemically inhomogeneous clumps[3)]; this is necessary to account for the approximate constancy of the observed 16 - 28 keV flux.

The X-ray flux due to the degradation of the line γ-rays (122 keV and 136 keV) from the ^{57}Co decay (dash-dotted curve in Fig. 2) exceeds that from the ^{56}Co decay (dashed) at $t > 700$ d and slows down the decline of the X-ray light curve. Its bolometric luminosity is proportional to the mass of ^{57}Co, which is $L_{\rm opt} \sim 5 \times$

10^{37} erg s^{-1} at $t \sim 600$ d for the present model. This might be contributing to the recent change in the slope of the optical light curve[1)].

4. RADIATION FROM THE BURIED NEUTRON STAR

We consider two cases for radiation from the newly born neutron star. These are an isolated pulsar with a surrounding synchrotron nebula and an X-ray pulsar with some ejected material falling back. [See ref. 3) for the case with black body radiation.] We assume that the pulsar nebula has the Crab Nebula-like spectrum of $E^{-2.2}$ and the X-ray-pulsar has the SMC X-1-like spectrum. Figure 3 shows the emergent spectra of these pulsar components for $L_{X\gamma} = 3.5 \times 10^{37}$ erg s^{-1} together with those of ^{56}Co and ^{57}Co, where $L_{X\gamma}$ denotes the input luminosity from the energy range of 1 keV $< E < 10^5$ keV. Figure 4 shows the 16 - 28 keV light curves powered by the X-ray pulsar (dash-dotted curves for $L_{X\gamma} = 1 \times 10^{38}$ and 3.5×10^{37} erg s^{-1}) and an isolated pulsar (dashed curve for $L_{X\gamma} = 1 \times 10^{38}$ erg s^{-1}). These curves, which are shifted in proportion to $L_{X\gamma}$, predict that Ginga will detect the emergent X-ray flux if $L_{X\gamma} > \sim 2 \times 10^{37}$ (5×10^{37}) erg s^{-1} for the X-ray (isolated) pulsar.

Suppose that the reported change in the optical luminosity[1)] is due to an additional $L_{\rm opt}$ from the neutron star. If it is an accretion-powered X-ray pulsar and $L_{\rm opt} > 2 \times 10^{37}$ erg s^{-1}, the X-rays will be observed by Ginga since $L_{X\gamma} \sim L_{\rm opt}$. For a rotation-powered pulsar, $L_{X\gamma}$could be lower than $L_{\rm opt}$ depending on how the spin-down power is thermalized by the collision between the pulsar wind and the dense ejecta and converted into radiation. Future X-ray light curves will provide quite a useful material to diagnose which physical process is taking place around the newly born neutron star.

If the X-rays from the buried pulsar will become observable with Ginga, a search for pulsations in the X-ray flux is crucially important. As the optical depth of the supernova decreases with expansion, fraction of the unscattered photons over the total photons in the 16 - 28 keV range increases as seen in Fig. 4 for the isolated pulsar (dashed) and X-ray pulsar (dash-dotted). If the fraction exceeds $\sim$ 10 - 40 percent depending on the flux level, Ginga can detect modulations for a 95 percent confidence level[4)]. This would be possible at $t > 1000$ d even for a weak flux.

5. REFERENCES

1) Hamuy, M., Suntzeff, N.B., Martin, G., and Gonzalez, R. 1988, *IAU Cir.* 4680.
2) Hashimoto, M., Nomoto, K., and Shigeyama, T. 1989, *Astr. Ap.*, in press.
3) Kumagai, S., Shigeyama, T., Nomoto, K., Itoh, M., Nishimura, J., and Tsuruta, S. 1989, *Ap. J.*, in press.
4) Mitsuda, K., Dotani, T., Itoh, M., Tsuchiya, T., and Shimizu, S. 1988, in *Physics of Neutron Stars and Black Holes*, ed. Y. Tanaka, p. 491.
5) Nomoto, K. *et al.* 1988, in *Physics of Neutron Stars and Black Holes*, ed. Y. Tanaka (Tokyo: Universal Acad. Press), p. 441.
6) Shigeyama, T. 1989, Ph.D. thesis, University of Tokyo.
7) Tanaka, Y. 1988, in *Physics of Neutron Stars and Black Holes*, ed. Y. Tanaka, (Tokyo: Universal Acad. Press), p. 431.

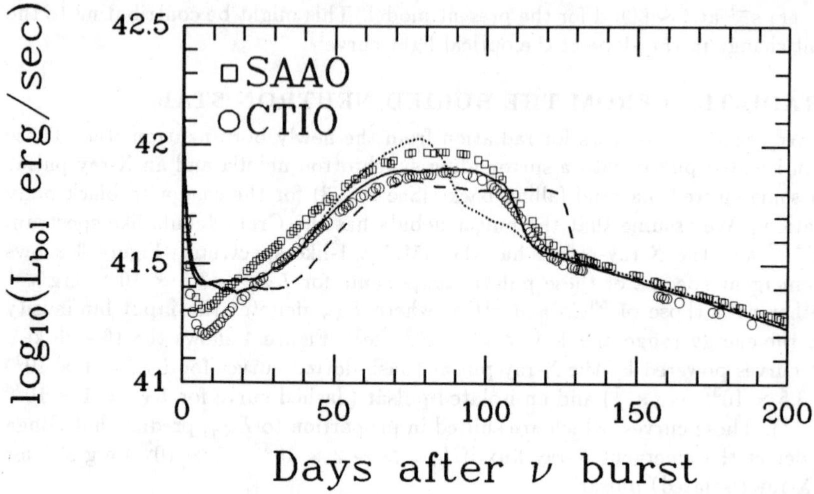

Figure 1: Bolometric light curve for 14E1. The solid, dashed, and dotted curves assume mixing of hydrogen down to the center, $M_r = 1\ M_\odot$, and $3\ M_\odot$, respectively.

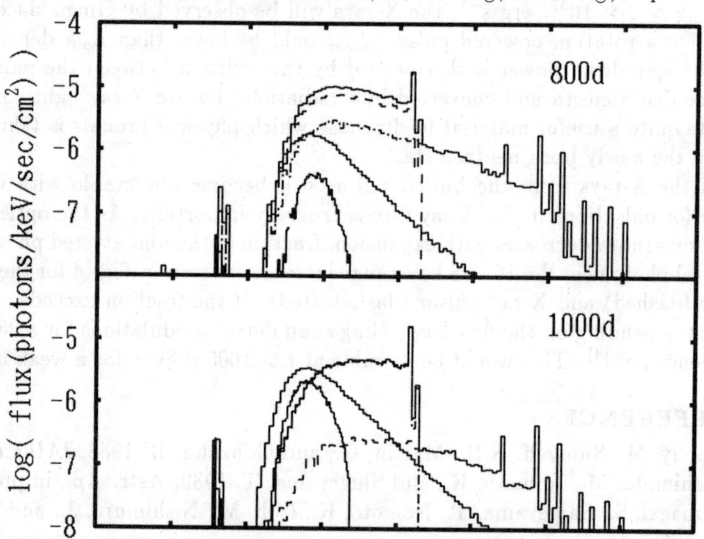

Figure 3: Emergent X-ray spectra for t = 800 d - 1000 d from the pulsar nebula and the X-ray pulsar as compared with the cobalt components. The input luminosities of these sources are $L_{X\gamma} = 3.5 \times 10^{37}$ erg s^{-1} at 1 keV $< E < 10^5$ keV.

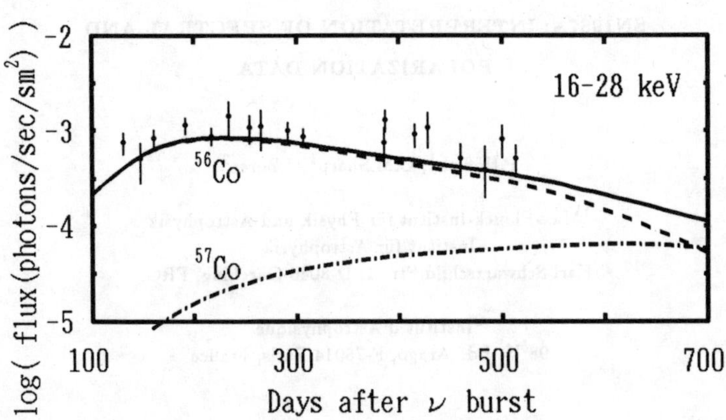

Figure 2: Calculated X-ray (16 - 28 keV) flux originated from the decays of ^{56}Co (dashed), ^{57}Co (dash-dotted), and their sum (solid). The reduction of the photoelectric opacity by a factor of 9 is assumed for $M_r < 8\ M_\odot$.

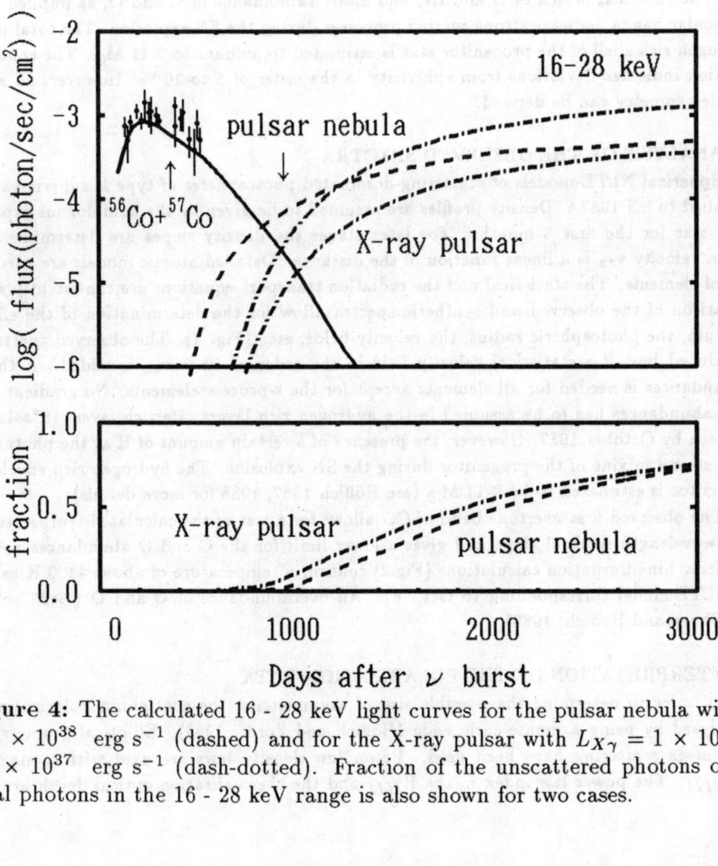

Figure 4: The calculated 16 - 28 keV light curves for the pulsar nebula with $L_{X\gamma} = 1 \times 10^{38}$ erg s^{-1} (dashed) and for the X-ray pulsar with $L_{X\gamma} = 1 \times 10^{38}$ and 3.5×10^{37} erg s^{-1} (dash-dotted). Fraction of the unscattered photons over the total photons in the 16 - 28 keV range is also shown for two cases.

SN1987A: INTERPRETATION OF SPECTRAL AND POLARIZATION DATA

P.Höflich[1], C.M.Sharp[1], J.Zorec[2]

[1]Max-Planck-Institut für Physik und Astrophysik
Institut für Astrophysik
Karl Schwarzschild Str. 1, D-8046 Garching, FRG

[2]Institut d'Astrophysique
98bis, Bd. Arago, F-75014 Paris, France

ABSTRACT: In order to determine the chemical and density profiles of SN1987A during the first few months, spherical NLTE- model calculations have been carried out. In general, a third of solar abundances have to be assumed. However, a strong enrichment of the s-process elements is needed. The chemical profile of H and He, and an overabundance of C and O, as implied by the CO-molecular bands, indicate strong mixing processes during the SN-explosion. The total mass of the hydrogen rich shell of the progenitor star is estimated to amount to 9-11 $M_{\odot}$. The continuum polarization indicates deviations from sphericity in the order of 5 to 10 %. However, no specific large scale geometry can be derived.

1. ANALYSIS OF THE OBSERVED SPECTRA

Spherical NLTE-models of scattering-dominated photospheres of type II supernovae have been applied to SN 1987A. Density profiles are assumed to be given by the homologous expansion of a B3I star for the first 3 months. For later stages the density slopes are determined. The expansion velocity v_{ex} is a linear function of the distance. Detailed atomic models are used for a number of elements. The statistical and the radiation transport equations are consistently solved. A comparison of the observed and synthetic spectra allow for the determination of the effective temperature, the photospheric radius, the velocity fields, etc. (Fig. 1). The observed spectra can be reproduced best if a statistical velocity field in the order of 10 % v_{ex} is added. A third of solar abundances is needed for all elements accept for the s-process elements. No gradient of the chemical abundances has to be assumed in the hydrogen rich layers. He rich layers (5*solar) are already seen by October 1987. However, the presence of a certain amount of H at the photosphere indicates strong mixing of the progenitor during the SN-explosion. The hydrogen rich envelope of the progenitor is estimated to be 9-11 $M_{\odot}$ (see Höflich 1987, 1988 for more details).

The observed first overtone band of CO allows for a test of the calculated temperature by the long wavelength spectral slope and gives a lower limit for the C and O abundances near the photosphere. Line formation calculations (Fig.2) confirm a temperature of about 4300 K as given by the NLTE-model corresponding to Oct. 87. An overabundance of C and O ($\geq$25*solar) is needed (Sharp and Höflich, 1988).

2. INTERPRETATION OF THE POLARIZATION DATA

In order to determine the possible size of asymmetries, the continuum polarization has been analysed by using a monte-carlo code (Höflich and Zorec, 1988). Elliptical geometry and pure Thomson scattering have been used. Power law density laws are used with a maximum radius R_{HII}. The power law index $\tilde{n}$, the R_{HII} and the thermalization optical depth are given

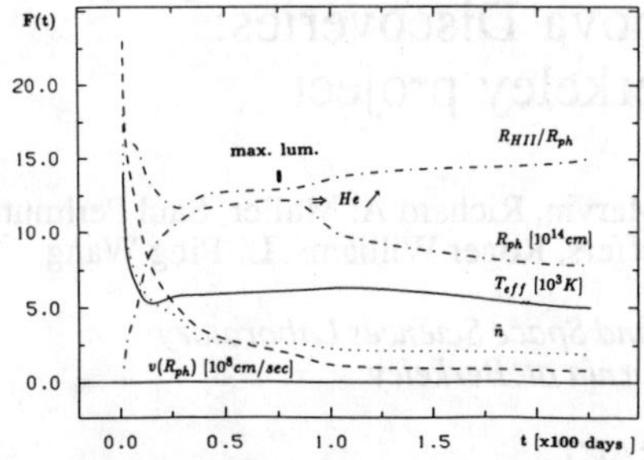

Figure 1: The distance R_{ph} at which the optical depth is 1 for true absorption at 5000 Å, T_{eff}, the velocity field v at R_{ph}, the corresponding power law index $\bar{n}$ of the density slope and the distance R_{HII} up to which H is mainly ionized are given as a function of time (in days).

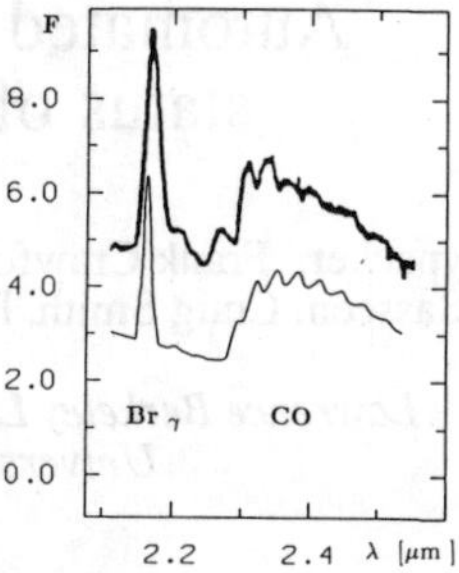

Figure 2: IR-Spectrum as observed by ESO between 1st and 6th of October 1987 (upper curve) in comparison with the reddened, synthetic spectrum ($E_{B-V} = 0.15^m$).

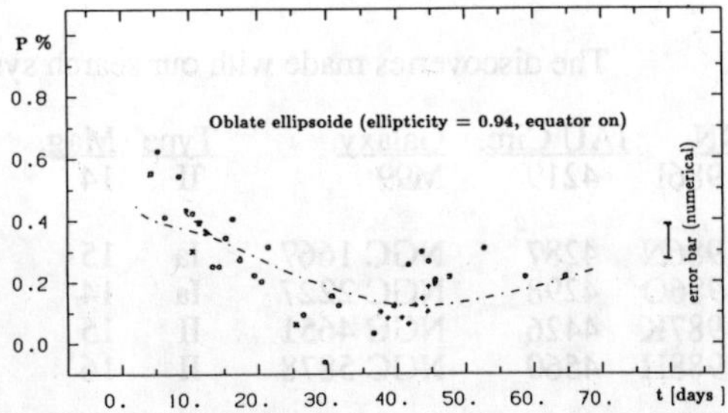

Figure 3: Comparison of the observed (Mendez et al., 1988) and calculated polarization in the V-band, using the density and ionization profiles of Fig.1. An interstellar polarization correction of 0.41 % is used (Schmidt, 1976; region VIII)

by the atmospheric models (Fig. 1; and Table 1 of Höflich, 1988). The size and time dependence of the polarization can be understood by an oblate ellipsoid (Fig. 3). Here, the increase of the calculated polarization after day 45 is due to an increase of R_{HII}/R_{ph} of the NLTE-models. However, a certain part of the increase of the observed polarization is due to line scattering. Note that the same size of agreement can be achieved by assuming a prolate ellipsoid (ellipticity 0.9). Furthermore, there is no reason to assume any rotational geometry. Therefore, the polarization data allows only an estimate of global deviations from sphericity to be in the order of 5 to 10 % . This value is consistent with the assumed statistical velocity, but it should be regarded as a first approximation only, because of its strong dependency on the interstellar correction, the thermalization optical depth and the real geometry.

3. REFERENCES

Höflich,P., Proc.Astron.Soc.Austr., Canberra, in press (1988)

Höflich,P., Proc. of the ESO workshop on SN1987A, Garching, July 1987, 447 (1987)

Höflich,P.; Zorec,J., Proc. of the Peckers colloquium, Paris, ed. P. Delache and S. Laloe, *Editions Frontieres*, in press (1988)

Mendez et al., Ap.J., 334, 295 (1988)

Schmidt,Th., Astr.Ap.Suppl., 24, 357 (1976)

Sharp,C.M.; Höflich,P., Proceedings of the IAU Conference, Baltimore, in press (1988)

Automated Supernova Discoveries: status of the Berkeley project

Carl Pennypacker, Frank Crawford, Heidi Marvin, Richard A. Muller, Saul Perlmutter
Tim Sasseen, Craig Smith, Richard Treffers, Roger Williams, Li-Ping Wang

Lawrence Berkeley Laboratory and Space Sciences Laboratory
University of California at Berkeley

ABSTRACT

The Berkeley Automated Supernova Search project has discovered six supernovae. In 1988 approximately 600 galaxy-years of observations were accumulated, and no supernovae within our galaxy list were missed. Planned improvements for the next year and a half are described, as well as a new deep search for supernovæ to $z \approx 0.4$.

DISCOVERIES TO DATE

The discoveries made with our search system[1,2] are listed in the table below.

SN	IAU Circ.	Galaxy	Type	Mag.	
1986I	4219	M99	II	14	Image 9 days earlier had no SN Spectral studies in ref. 6.
1986N	4287	NGC 1667	Ia	15	
1986O	4298	NGC 2227	Ia	14	Spectral Studies by ESO
1987K	4426	NGC 4651	II	15	Type II → Type Ib
1988H	4560	NGC 5878	II	16	First discovered with the completely automated system.
1988L	4590	NGC 5480	Ia,b,II?	16.5	Completely automated discovery

Four of the supernovae were found with a semi-automated version, and two so far with the completely automated system. Our present limiting magnitude for discovery is about $m_V \approx 17$ with a 60-second exposure at our Leuschner observatory site. Recently we have taken over 500 useful images in a single night.

As part of our coordinated follow-up program, we have organized a follow-up network of observers, including astronomers at Lick, Anglo-Australian Observatory, Palomar, McDonald, WIRO, Mt. Stromlo, Royal Grenwich Observatory, and Mt . Hopkins Observatory. Two of our supernovæ, 1986I and 1987K, were intensively studied spectroscopically. We are collaborating with theoretical groups to analyze the spectra. These groups include those of Wagoner, Hoflich, Eastman, and Wheeler.

PLANNED IMPROVEMENTS

We are currently assembling a new detector system to operate in a dark site with good seeing. This system will consist of a new 30" Ritchey-Chretien telescope, a new Texas Instrument CCD, a μVAX 3200 computer, and an 80386-based PC telescope control computer. With these improvements, our sensitivity will improve to 20th magnitude for a 20 second exposure. We will search galaxies in Virgo every clear night, and search additional galaxies on a several night cycle.

The improved sensitivity will allow us to see a supernova in Virgo at about 0.5% of maximum light, although this limit is sensitive rise time of the light curve. For supernova beyond Virgo, we will search galaxies to a red-shift of 10,000 km/sec. A supernova at the limit of our sample would have a $m_v \approx 17$ at maximum light; we should find these supernova at about 10% of maximum light. An automated photometry system will make broad-band photometric measurements of these supernovæ. One of our near-term goals of the search is to study Type Ia supernova light curves, and help determine if they can be used as standard candles. Finding them at this magnitude and then taking photometry will allow us to follow them up and over the peak, and to greatly increase our data base of Type Ia events. For a supernova rate of one per 100 galaxy-years, we will find about 10 Type Ia events per year, and a similar number of Type II events. We plan to coordinate both spectroscopic measurements and spectral analysis. The uniformity of Type Ia events will be studied, and the Baade method will be used on Type II supernovæ with the goal of using them as primary distance indicators for measurement of Hubble's constant H_0.

DEEP SEARCH WITH ANGLO-AUSTRALIAN OBSERVATORY

We are adapting our search software, and in collaboration with the Anglo-Australian Observatory, building a wide-field imaging/supernova discovery system for the Anglo-Australian Telescope (AAT). The system uses the 3.9 meter AAT, an f/1 focal reducer (designed by Harvey Richardson, Dominion Astrophysical Observatory), a Thomson CSF 1024 x 1024 pixel CCD (19μ x19μ pixels), and modified image processing software from the Berkeley Automated Search. Assuming the supernova rates of Cawson and Kibblewhite[4] , or Van Den Bergh, et al[5] , we should find a pre-maximum light distant supernova (at $z \approx 0.4$) every two nights or so. Photometric measurement will be made with other telescopes Elias[6] estimates Type Ia are uniform to about 15% ; with a dozen or so such supernovæ, we hope to make measurements of q_o to 15% by using them as standard candles. Measurement of nearby Ia supernovae will help to reduce systematics, to "calibrate" the candles, and thus improve the estimates of q_o.

SUMMARY

Over the next two years, we will bring to operation our new supernova detection system, and increase our search rate to several thousand galaxy-years per year. A separate search will look for deep supernovæ at $z \approx 0.4$. We will coordinate reduction and analysis efforts with other groups in supernova spectral analysis. One of our goals is determination of q_0 to an accuracy of 15% or better.

References

1. Perlmutter, S., F.S. Crawford, R.A. Muller, C.R. Pennypacker, T.P. Sasseen, C.K. Smith, R. Treffers, and R. Williams, "The Status of Berkeley's Real-Time Supernova Search", in *Instrumentaion for Ground Based Astronomy,*, Ed L.B. Robinson, Springer-Verlag, New York, (1988).
2 Perlmutter,S., et al. This volume.
3. Cawson, M., Ph.D. Thesis, Cambridge University, (1983).
4. Van Den Bergh, S., McClure, R.D., Evans,R., *Ap.J.*, **323**,44-33, (1987).
5. Elias, J., et al, *Astrophysical Journal,* 379, (1985).
6. Pennypacker, C., et al, *Astronomical Journal,* January, (1989).

A Strip Search for Supernovae: The CTI Survey

John T. McGraw
Steward Observatory, University of Arizona

ABSTRACT The CCD/Transit Instrument (CTI) is a fixed, meridian-pointing telescope which uses CCD detectors clocked in the "time-delay and integrate" mode to nightly produce an image of the same strip of the sky to V > 20. The CTI observes about 15 square degrees of the sky per night, or about 45 square degrees per year. The resulting image data are reduced and analyzed in a data base environment. The reduction procedures and the use of a unique data base environment for research in physical sciences are discussed.

A principal use of these data is the detection of supernovae (SNe). Because the CTI does not select target galaxies, the SN production rate derived from the data will not be biased in the same manner as earlier surveys. The SNe which will be detected will be faint, most of them at the limit of survey detectability. These distant SNe are excellent candidates for followup spectroscopy and imaging in an effort to use them as distance indicators. These distant SNe are clearly in the Hubble flow and are thus useful for measuring cosmological parameters.

THE TELESCOPE

The CTI is a 1.8 m, f/2.2 non-moving telescope utilizing CCD detectors to nightly produce an image of the same strip of sky. The optics consist of the primary, a parabaloid, and a Paul-Baker two-mirror corrector system which produces a one degree field-of-view. Two RCA CCDs (320 X 512 30-micron pixels) are placed in the focal plane end-to-end with their vertical shift directions aligned E-W. A filter assembly over each allows selection of optical bandpasses. One CCD observes in V every night while the other observes in one of B, R or I, depending upon the sky brightness. Thus, we obtain light curves in V for every detected object as well as mean magnitudes and colors. The CCD camera is controlled by a real-time program which automatically acquires flat-fields and biases, opens the roll-off building which houses the telescope, opens the mirror cover and acquires the data. The only action required of a person is to give permission to open - the program does not know if it is raining at the start of the night! A unique bi-metallic structure maintains the telescope in focus throughout the diurnal temperature variations. Details of the telescope and its operation are given in McGraw, Cawson and Keane (1986).

THE SOFTWARE SYSTEM

The vast amount of data produced by the CTI, up to 450 Mbytes per night, requires an efficient method of reduction and analysis. We have elected to work in a relational data base environment of our own construction. Pixel data are entered into the system from magnetic tape, the instrumental signature and cosmic rays are removed, and the data are analyzed with a unique algorithm which calculates parameters for "images" which are distinctly present above the background and for "segments" which are either parts of images or are less well-defined above the noise. For example, a galaxy would be an image while an H II region (or a supernova) in the galaxy would be a segment. Intermediate products of the CTI data pipeline are all data bases which can be queried by DB, the interactive, scientific file manager (Cawson, McGraw and Keane 1986).

The principal products of the pipeline are the Master and History lists which contain our noise-weighted "best" estimates of image parameters and the V light curve of each object, respectively. These can be queried to accomplish scientific programs. For example, the light curves of an ensemble of galaxies can be searched for fluctuations which might be due to SNe embedded within them.

THE CTI SUPERNOVA SEARCH

Our principal technique for discovering SNe differs from other searches in that we investigate the reduced light curve instead of examining pixel data directly. We first segregate galaxies from stars by utilizing surface brightness, size and color criteria. The light curves of these galaxies are then searched for high variance and for the signature of a SN - any one or both of a rapid rise followed by a slower decline. We do not use a "standard light curve" as template, but examine the selected light curves visually to determine plausibility and probability. This technique is necessary because the vast majority of the 40,000 galaxies in the CTI strip will be very distant ($z > 0.2$), faint ($V > 18$) and small in angular extent. The majority of SNe we detect will simply appear as brightness enhancements in these galaxies. This technique has required extreme care in producing our system of photmetric standards, which we have now refined to an rms uncertainty of 0.01 mag. at $V = 17.5$.

Thus far we have discovered two SNe, both of which were detected prior to maximum. Within the next few months we shall initiate a program to search the previous year's data to limit the amplitude of possible SNe in any of our target galaxies. These data will help define the SN production rate as the galaxies surveyed are not selected in any way. In future, early detection of SN events will allow us to notify other (pointing) telescopes, including the Hubble Space Telescope, of a target of opportunity. These distant events will be of particular use in calibrating the distance scale of the universe as they will be clearly in the Hubble flow, with velocities dominated by universal expansion and relatively unaffected by local infall.

In summary, CTI data will be used to:

1) measure the SN production rate as a function of galaxy color, morphology and redshift,
2) investigate SNe as distance indicators -
   a. SNe I as (calibrated) "standard candles" and
   b. SNe II (and I) expansion parallaxes,
3) announce pre-maximum detection of one or more "bright" ($V < 16$) SNe per year, and
4) derive galaxy counts in a homogeneous, continuous image.

REFERENCES

McGraw, J., Cawson, M. and Keane, M., Proc. SPIE 627, 60 (1986).
Cawson, M., McGraw, J. and Keane, M., Proc. SPIE 627, 79 (1986).

The Danish–Anglo–Australian Distant Supernova Search

W.J. Couch[1], L. Hansen[2], H.E. Jørgensen[2], H.U. Nørgaard–Nielsen[3], R.S. Ellis[4], A. Aragon[4]

[1] *Anglo-Australian Observatory*
[2] *Copenhagen University Observatory*
[3] *Danish Space Research Institute*
[4] *Physics Department, University of Durham*

1. INTRODUCTION

The potential use of Type Ia supernovae (SNIa) as standard candles for cosmological applications has long been recognised [1,2]. In 1986 we began a long term search for SNIa at large distances with the aim to exploit this potential and, ultimately, place useful constraints on q_0. We briefly outline the approach and status of our search; in particular, we take this opportunity to report our recent exciting discovery of a SNIa at a redshift of z=0.31. A fuller description of our programme and the events discovered to date can be found in articles published elsewhere [3,4,5].

2. SEARCH STRATEGY

Our search is being conducted by repeated deep optical CCD imaging of the cores of some 60 rich clusters of galaxies in the redshift interval $0.2 \leq z \leq 0.5$. Our choice to target our search on clusters offers the advantages that (i)a large number of galaxies (100-200) are monitored in any one exposure, and (ii) our search is concentrated on elliptical and lenticular galaxies which suffer negligible internal absorption.

The observations are made at La Silla, Chile, using an RCA CCD on the Danish 1.5m telescope. One hour V exposures are taken of each cluster once per month for the period they are accessible during August–April. Analysis of the data takes place immediately after the exposure: the new frame is compared with one taken at an earlier epoch by forming a *difference* frame once the two have been aligned and smoothed to the same seeing. The difference frame is then scanned by eye for residual stellar images. In the event of a detection the field is promptly re-imaged for confirmation; if positive, spectroscopy of the event is obtained immediately with the 3.9m AAT or 4.2m WHT to establish whether it is a supernova and to determine its type and redshift. When possible, these 2 telescopes are also used for photometric follow-up where their extra light grasp is crucial in following the light curve at late times (t_{obs} >30 days; see Fig. 2).

Despite the added difficulty that SN events are required to be detected above the often bright background of their host galaxy, we nevertheless expect to achieve a detection limit of V∼24 with our approach. This has in the course of our search been ably demonstrated by the detection of a genuine event at V∼23.6 (see §3). SNIa in our observed redshift range are, at maximum, expected to have V∼21 (z=0.2) and V∼23 (z=0.5) and take just over a month [allowing for (1+z) time dilation] to fade to a magnitude below peak. Consequently, with monthly monitoring and a V∼24 detection limit *all* SNIa events should be detected out to z=0.5.

The rate at which we expect to detect SNIa events can be predicted from the (uncertain) local rates [6,7,8]. These range from 0.2–0.8h^2 SNu (1 SNu = 1 SN per

$10^{10} L_{B_\odot}$ per century; $h = H_0/100$) for E/S0 galaxies; taking into account redshift effects (time dilation and K-corrections) and adopting a mean cluster luminosity within our CCD field of 5.0 x $10^{11} L_{B_\odot} h^{-2}$, our chances of a detection are between 0.013 and 0.053 per comparison.

3. CURRENT STATUS

We are now into our third observing season at La Silla and a total of 158 comparison frames have been secured. This has lead to the discovery of two SN events plus a number of faint blue variable objects most likely to be QSO's. Importantly, our SN detection rate and in particular that of SNIa (see below), is already significantly at odds with the highest of the local rates (Tammann [6]) which predicts we should have by now discovered 8 SNIa.

Our first SN event was detected in Feb 1988 in a faint (V=21.7) member of the cluster J1836.14RC at z=0.28. Being at V=23.6 at the time of detection and fading after that, it was too faint to be confirmed spectroscopically. We concluded on the basis of its photometric behaviour that it was most likely a Type II event near maximum although the possibility of it being a Type Ia detected a month beyond maximum could not be ruled out [4].

The second event, which was discovered in the z=0.31 cluster AC118 in August 1988, was substantially brighter. It had V=22.1 at detection, resulting in a 0.85mag overall brightening in the host galaxy (cf. 0.2 mag for the first event). A low dispersion spectrum of the event was kindly obtained for us by P.Hewett and S.Warren on the AAT and it is shown in Fig. 1. The identification of this event as a SNIa at z=0.31 is shown rather convincingly by the comparison in this figure of the spectrum with that of SN 1981B (Type Ia) 17 days after maximum [9] and shifted to z=0.31. Furthermore, a subsequent AAT spectrum taken of the host galaxy shows it to be at z=0.31.

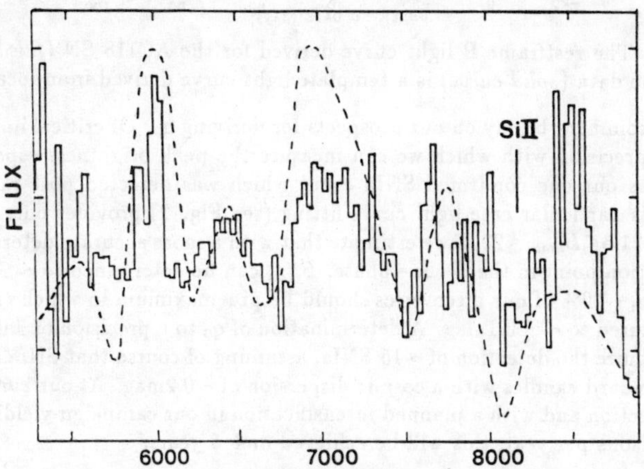

Figure 1 : A 4000sec FORS spectrum (*solid line*) obtained on the AAT of the SN event detected in the cluster AC118. The event has been identified as an SNIa at the redshift of the cluster, z=0.31, on the basis of the comparison with the spectrum of the nearby Type Ia SN 1981B shifted to this redshift (*dashed line*).

Quite extensive photometric monitoring of this latter event has also been pursued. A restframe B light curve constructed from observations made with both the Danish 1.5m and the WHT is shown in Fig. 2. The fit of a template light curve derived from nearby SNIa [10] is also shown. Clearly the most readily determined and useful quantity defined by the data is the slope of the initial decline ie.,the Pskovskii [11] β parameter. A linear least-squares fit to the data covering the first 10 days after detection gives β=6.8±0.1 mags/100 days which, when corrected for time dilation, places it close to the *middle* of the distribution of values observed for local SNIa. We emphasise in the context of the remarks made at this workshop by Leibundgut that there is certainly *no* evidence for the decline rate being higher than the local values.

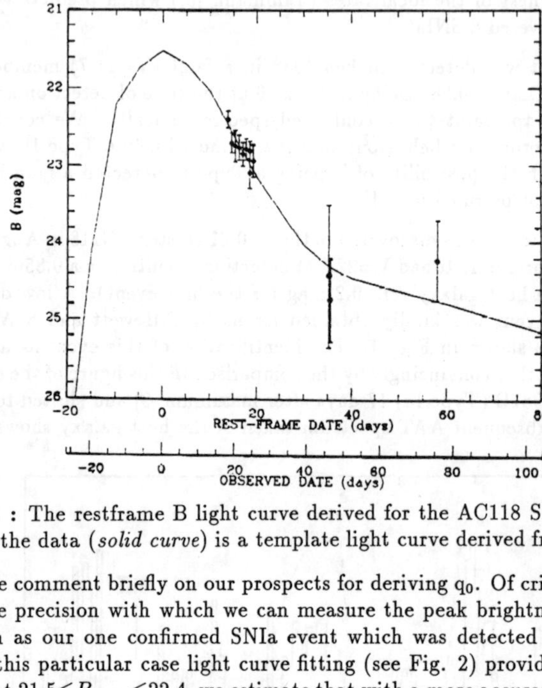

Figure 2 : The restframe B light curve derived for the AC118 SN (*filled circles*). Fitted to the data (*solid curve*) is a template light curve derived from local SNIa.

Finally, we comment briefly on our prospects for deriving q_0. Of critical importance here is the precision with which we can measure the peak brightness especially in cases such as our one confirmed SNIa event which was detected post–maximum. While in this particular case light curve fitting (see Fig. 2) provides only the constraint that $21.5 \leq B_{max} \leq 22.4$, we estimate that with a more accurate determination of the inflection point in the decline phase, B_{max} can be determined to $\sim \pm 0.3$mag. Furthermore, ~40% of our discoveries should be pre–maximum in which case B_{max} can be measured to $\sim \pm 0.1$mag. A determination of q_0 to a precision of ±0.1 would therefore require the detection of ~15 SNIa, assuming of course that SNIa are well–behaved standard candles with a cosmic dispersion of ~0.2mag. At our *current* rate of SNIa detection and with a planned intensification in our campaign yielding up to 200 comparisons per year, this will be achieved in 4–5 years.

REFERENCES

[1] Kowal,C.T.,*A.J.*,**73**,1021,(1968).
[2] Colgate,S.A.,*Ap.J.*,**232**,404,(1979).
[3] Hansen,L.,Nørgaard-Nielsen,H.U. and Jørgensen,H.E.,*ESO Messenger*,**47**,46,(1987).
[4] Hansen,L.,Jørgensen,H.E.,Nørgaard-Nielsen,H.U.,Ellis,R.S. and Couch,W.J.,*Astron. Astrophys. Lett.*,in press,(1989).
[5] Nørgaard-Nielsen,H.U.,Hansen,L.,Jørgensen,H.E.,Aragon Salamanca,A., Ellis,R.S. and Couch,W.J.,*Nature*,submitted,(1989).
[6] Tammann,G., in *Supernovae: A Survey of Current Research*,eds. Rees,M.J., Stoneham,R.J.,Reidel,Dordrecht,p371,(1982).
[7] van den Bergh,S.,McClure,R.D. and Evans,R.,*Ap.J.*,**323**, 44,(1987).
[8] Capellaro,E. and Turatto,M.,*Astron. Astrophys.*,**190**,10,(1988).
[9] Branch,D.,Lacy,C.H.,McCall,M.L.,Sutherland,P.G.,Uomoto,A.,Wheeler,J.C. and Wills,B.J.,*Ap.J.*,**270**,123,(1983).
[10] Cadonau,R.,Sandage,A. and Tammann,G., in *Supernovae as Distance Indicators*, ed. Bartel,N.,Springer,p151,(1985).
[11] Pskovskii,Y.P.,*Soviet Astron.*,**21**,675,(1977).

The First Year (Almost) of Real-time Automated Operation of the Berkeley Supernova Search

Saul Perlmutter, Frank Crawford, Heidi Marvin, Richard Muller, Carl Pennypacker, Tim Sasseen, Craig Smith, Li-Ping Wang
Lawrence Berkeley Laboratory and Space Sciences Laboratory
University of California at Berkeley

ABSTRACT

The Berkeley supernova search went through a reincarnation this year: a new real-time automated system went on line on 24 January 1988, with new software on a new computer running the mostly old CCD camera and telescope. These changes have been described elsewhere;[1] here we report on the results of the first 48 weeks of this system's operation. The new system was designed to keep records of every observation and analysis cycle performed at the telescope. The database thus generated makes it possible to characterize the coverage in time, number of galaxies, and sensitivity of our supernova search.

GALAXY COVERAGE. As of the beginning of December, 1988, the Berkeley search has collected almost 19,000 images of 1,320 galaxies. These galaxies are selected from the ZCAT redshift catalog[2] on the basis of distance (<7500 km/sec) and absolute magnitude (brighter than -18^{mag} for spirals, -20^{mag} for elipticals). The first few months of the year had weeks of clear weather, during which we managed more than 900 observations per week. Since than the weather has only allowed a third of this rate for a typical week. The real-time image processing successfully checked for supernovae in about 80% of these images during dark time and about 50% of the images near the night of the full moon. The processing success rate in the last few months climbed to about 90% on good weeks.

The figure shows the time-integrated coverage of our search as a function of the time that we have been searching. *Galaxy-years* is the sum over all galaxies of the time during which we would have detected a supernova in that galaxy. Each observation with no supernova thus adds the time since the last observation of that galaxy, with a maximum of 45 galaxy-days, the time for a supernova in most of our galaxies to fade below our detection threshold.

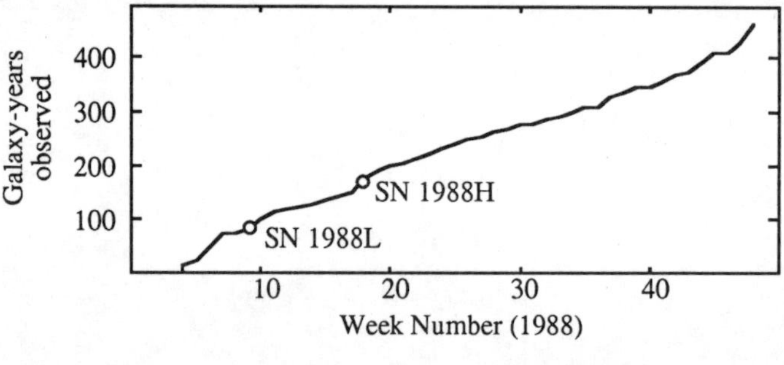

SENSITIVITY. The detection sensitivity of the search depends on the sky brightness, atmospheric transmission and seeing during the real-time observation of the galaxy and during the older reference observation that is subtracted. Our threshold is currently set to find supernovae at $m_V = 17$ on a typical night. This would find a Type 1a supernova at maximum light at a redshift of 7500 km/sec, or 5 magnitudes before maximum light if it were as close as the Virgo cluster. As a check of this sensitivity, we observed Supernova 1988A (discovered elsewhere before this search started running) as it faded. It was "re-discovered" in every one of the twenty observations before it fell below 17th magnitude, and then missed in 2/3 of the observations in the following month.

TIME SINCE PREVIOUS OBSERVATION. We aim to find as many nearby supernovae as possible as early as possible, and therefore attempt to observe all of the nearby galaxies ($z < 1500$ km/sec) in our reference catalog every day. The more distant galaxies, about 80% of our catalog, are scheduled for observations once a week during the remaining observation time. The weather at Leuschner Observatory in Lafayette, California, of course dictates how close we come to these goals. In practice, 25% of our observations are within a day or two of the previous observation of the same galaxy, while another 35% are within about 10 days.

SEARCH RESULTS. The two supernovae indicated on the figure were discovered at approximate magnitudes 16.0 and 16.5. The dearth of supernovae since then is apparently real, although still within statistical expectations for a supernova rate of one per hundred galaxy-years; no other researchers have discovered a supernova brighter than 17th magnitude in our catalog of galaxies during this period.

IMPROVEMENTS. We are currently working on improvements of this search. We have been speeding up the observation and image processing cycle time. Up through the beginning of December the average cycle time was about 136 seconds per image. This yielded typically 130 images per night, with a peak performance of 272 images on one long night. As of the second week in December we have almost double the rate, averaging 70 seconds per cycle, and yielding a peak performance of 577 images on one long night. Most of this speed is gained by better parallelizing of the computer tasks. We expect to run even faster than this soon.

To take advantage of this increase in speed, we have begun to observe reference images of new galaxies to add to our catalog. We are also installing a new TI CCD camera, which will allow us to set our detection threshold lower, since it contributes only 10 photoelectrons RMS to the noise, a factor of 10 improvement over our previous RCA CCD. Lastly, we have ordered a new telescope to be used at a better site. The sites we are considering have 21 mag/arcsec^2 sky brightness and 1.5 arcsec seeing, as opposed to 16.5 mag/arcsec^2 sky brightness and 5 arcsec seeing at Leuschner Observatory. These "hardware" changes should give a detection sensitivity of 20th magnitude for a 20 second observation.

UPCOMING SCIENCE. These improvements will result in more supernovae discovered per year, but we also want to get more science per supernova. We are adding an automated filter wheel to the Leuschner camera, and we intend to perform automated BVRI photometry on all supernovae that are bright enough. This will generate a database of complete, comparable light curves. We are currently calibrating our system for detection threshold and photometry, and expect soon to be calculating a supernova rate from our search results.

REFERENCES

1. Perlmutter, S., F. S. Crawford, R. A. Muller, C. R. Pennypacker, T. P. Sasseen, C. K. Smith, R. Treffers and R. Williams, "The status of Berkeley's real-time supernova search." In *Instrumentation for Ground-Based Optical Astronomy*, ed. L. B. Robinson, Springer-Verlag, New York, (1988) .
2. Huchra, J., ZCAT redshift catalog of galaxies, Harvard Smithsonian Astrophysical Observatory, 1985.

ON THE VARIATION OF TYPE IA SUPERNOVAE

Ken'ichi Nomoto and Toshikazu Shigeyama
Department of Earth Science and Astronomy, University of Tokyo

1. ABSTRACT

Variations among Type Ia supernovae (SN Ia) are seen in the bolometric light curves of SN 1972E and SN 1981B[5)] and the visual light curve of SN 1986G[11)] in Figure 1. SN 1986G shows significantly faster decline than SN 1972E and SN 1981B. We suggest that such a variation would mostly stem from the difference in the distributions of ^{56}Ni. Evolutionary origin of such a difference in the abundance distribution is discussed as well as the effect of the initial metallicity (i.e., population) of the white dwarf.

2. OPTICAL LIGHT CURVE AND DISTRIBUTION OF ^{56}Ni

We adopt the current standard model W7[10)] for our hydrodynamical model. In this model, a carbon deflagration occurs in the accreting white dwarf of mass 1.38 $M_\odot$ and 0.58 $M_\odot$ ^{56}Ni is synthesized. The star explodes completely with the kinetic energy of $E = 1.3 \times 10^{51}$ erg.

We calculate the light curves for three different distributions of ^{56}Ni as seen in Figure 1. Here we assume the optical opacity of 0.1 cm^2 g^{-1} and the γ-ray opacity of 0.03 cm^2 g^{-1}.

1) W71 (dashed): ^{56}Ni is confined in the central region of $M_r = 0$ - $0.58\ M_\odot$.

2) W72 (solid): ^{56}Ni is located at the intermediate region of $M_r = 0.28$ - $0.86\ M_\odot$ as in the original W7. At $M_r < 0.28\ M_\odot$, more neutron-rich ^{58}Ni, ^{54}Fe and ^{56}Fe are produced.

3) W73 (dash-dotted): ^{56}Ni is distributed in the outermost layers of $M_r = 0.80$ - $1.38\ M_\odot$.

The calculated light curves are compared with observations in Figure 1. The slow light curves of SN 1972E and SN 1981B are in good agreement with W72 (solid, i.e., the original W7), but W71 (dashed) is a little too slow[5)]. The fast decline of SN 1986G is well reproduced by W73 (dash-dotted) because γ-rays from the outer layer escape earlier. In order to reproduce SN 1986G with the same ^{56}Ni distribution as in W72, E/M should be 3 times larger than that of W7, which is impossible for carbon deflagration models. In fact, the expansion velocities of SN 1986G is even lower than SN 1981B.[3)]

This indicates that the light curve shape is more sensitive to the distribution of ^{56}Ni than E/M. This is consistent with the fact that the light curve of SN 1984A is similar to SN 1981B though its velocities are significantly higher.[3)]

3. VARIATIONS OF CARBON DEFLAGRATION MODELS

The distribution of ^{56}Ni depends on 1) the size of the neutronized core[6] if the carbon deflagration starts from the center and 2) the site of carbon ignition if an off-center deflagration occurs.

The neutronized core could not be very large, because it would lead either to the ejection of too much neutron-rich ^{58}Ni and ^{54}Fe or to the collapse due to rapid electron capture at high central density.[9]. Thus as far as the carbon deflagration starts from the center, variation of the light curve would not be so large.

If the off-center carbon deflagration frequently occurs, larger variations are expected. This would take place if a solid oxygen core has formed in the white dwarf[4] or the convective Urca neutrino process efficiently cools down the inner core after carbon ignition.[7] However, the former possibility has been ruled out by the recent findings that chemical separation between carbon and oxygen does not occur.[2],[8]

The convective Urca process will not occur if the carbon ignition density, ρ_{ig}, is higher than the threshold densities of Urca process, i.e., $\rho_{th} = 1.3 \times 10^9$ g cm^{-3} for ^{25}Na - ^{25}Mg pair and 1.8×10^9 g cm^{-3} for ^{23}Ne - ^{23}Na pair. If the accretion rate is as high as $\dot{M} > \sim 10^{-6}$ $M_\odot$ yr^{-1}, ρ_{ig} is lower than ρ_{th} for the A = 25 pair. Even for lower $\dot{M}$, the A = 25 Urca shell cooling would not operate if the white dwarf belongs to the old population and has low metallicity. In this case, $\rho_{ig} \sim 1.5 \times 10^9$ g cm^{-3} if $\dot{M}$ corresponds to the steady hydrogen burning as occurs in the AGB star or common envelope ($\sim 8 \times 10^{-7}$ $M_\odot$ yr^{-1})[10]; this would induce the central deflagration without undergoing convective Urca cooling. For still lower $\dot{M}$, the convective Urca process would operate; however, nova-like explosions might reduce the occurrence frequency of this case.

Our speculation is that the majority of the SN Ia progenitor undergo rapid accretion to avoid nova-like explosions[9]. Then the central deflagration is induced without undergoing convective Urca cooling. Resulting SN Ia form quite a uniform class of light curves. This would be more likely the case for the white dwarfs with low metallicity. The variations of the light curves would result from the off-center carbon deflagration that occurs for slower accretion as well as high metallicity white dwarfs.

4. REFERENCES

1) Barkat, Z., and Wheeler, J.C. 1989, private communication.
2) Barrat, J.L., Hansen, J.P., and Mochkovitch, R. 1988, *Astr. Ap.*, **199**, L15.
3) Branch, D., Drucker, W., and Jeffery, D.J. 1988, *Ap. J. (Letters)*, **330**, L117.
4) Canal, R., Isern, J., and Lopez, R. 1988, *Ap. J. (Letters)*, **330**, L113.
5) Graham, J. 1987, *Ap. J.*, **315**, 588.
6) Graham, J. 1987, *Ap. J. (Letters)*, **318**, L47.
7) Iben, I.Jr. 1982, *Ap. J.*, **253**, 248.
8) Ichimaru, S., Iyetomi, H., and Ogata, S. 1988, *Ap. J. (Letters)*, **334**, L17.
9) Nomoto, K. 1986, *Prog. Part. Nucl. Phys.*, **17**, 249.
10) Nomoto, K., Thielemann, F.-K., and Yokoi, K. 1984, *Ap. J.*, **286**, 644.
11) Phillips, M.M. *et al.* 1987, *P.A.S.P.*, **99**, 592.

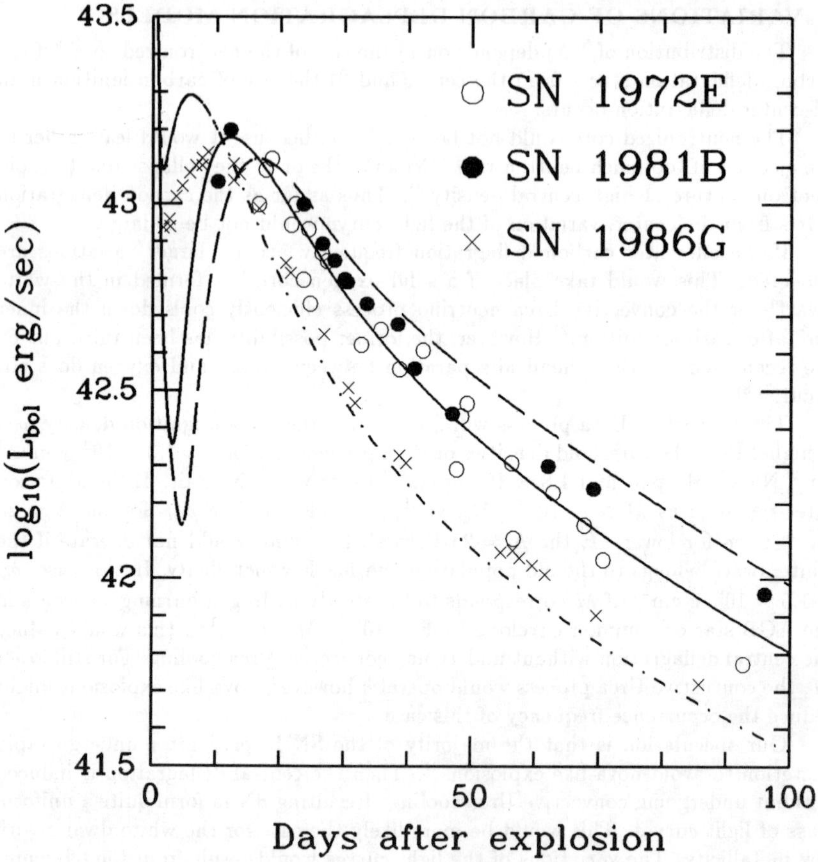

Figure 1: Bolometric light curves for the different distribution of ^{56}Ni in W7 as compared with the bolometric light curves of SN 1972E and 1981B[5)] and the visual light curve of SN 1986G.[11)] 1) W71 (dashed): ^{56}Ni is confined in the central region of M_r = 0 - 0.58 $M_\odot$. 2) W72 (solid): ^{56}Ni is located at the intermediate region of M_r = 0.28 - 0.86 $M_\odot$ as in the original W7. 3) W73 (dash-dotted): ^{56}Ni is distributed in the outermost layers of M_r = 0.80 - 1.38 $M_\odot$. The optical opacity of 0.1 cm^2 g^{-1} and the γ-ray opacity of 0.03 cm^2 g^{-1} are assumed.

Measurements of Light Curves of Supernovae Type Ia

Bruno Leibundgut
Astronomical Institute Basel
Venusstr. 7
CH-4102 Binningen
Switzerland

ABSTRACT: Light curves in six filters of supernovae of Type Ia show very little variations among different objects of this class. The good agreement in absolute luminosity in all these filters for a set of reasonably well observed supernovae Ia endorses the standard candle property. Using a technique of standard curves predictions of shapes of light curves at non-zero redshifts are presented. Furthermore a test to probe the nature of cosmological redshifts is proposed.

INTRODUCTION: The wealth of information in/light curves of supernovae (SNe) is still not fully exploited. With the extension to new filter bands, complementing the former studies which mostly treated the B band (Barbon et al. 1973, Pskovskii 1984, Doggett and Branch 1985), is it possible to describe the evolution of their energy output over a wide range in the electromagnetic spectrum. In the case of SNe Ia, which exhibit a very small scatter around a standard light curve in B (Cadonau et al. 1985), this extension to more filters probes the uniformity of the class. Reliable photometry of SNe Ia still rests on the well observed SNe early this decade (SNe 1980N, 1981B, 1981D) and the uniquely bright SN 1972E. Despite of spectral differences between individual SNeIa (Filippenko 1989, Branch et al. 1988) the light curves are very uniform. Table 1 lists the number

Table 1: Standard deviations of well observed supernovae

SNe	N_U	N_B	N_V	N_J	N_H	N_K	σ_U	σ_B	σ_V	σ_J	σ_H	σ_K
1972E	47	47	55	5	9	19	0.12	0.08	0.09	0.24	0.06	0.26
1980N	32	44	45	13	14	10	0.18	0.06	0.07	0.20	0.08	0.11
1981B	31	82	85	15	15	14	0.19	0.09	0.11	0.10	0.05	0.13
1981D	9	9	9	7	7	7	0.19	0.13	0.08	0.10	0.04	0.03
All	119	182	194	40	45	50	0.16	0.08	0.10	0.16	0.06	0.18

of observations and standard deviations of the four SNe which have well defined light curves in all investigated filters. The scatter is comparable to what has to be expected from photometric errors. With these SNe it is possible to establish a set of standard light curves similar to those given by Cadonau (1986, optical) and Elias et al. (1985, infrared). The fact that many SNe exhibit very similar light curves demands a narrow explosion scenario, as e.g. white dwarfs.

THE BOLOMETRIC CURVE: The curve integrated over all filters is bolometric in the sense that it includes the wide range from ~ 3500A (U) to 2.2μm (K). A comparison of this curve with theoretical predictions for a three zone model of an exploding white dwarf (Graham 1987) reveals an excellent agreement assuming a time of ~ 20 days from explosion to maximum light in B. This implies that only little energy is radiated outside the studied wavelength range.

ABSOLUTE LUMINOSITY: The given standard curves also allow the investigation of absolute brightness at maximum. The data were corrected for foreground absorption according to Sandage (1973). The distances were taken from Kraan-Korteweg (1986). Five SNe I in the Virgo cluster show a scatter at maximum of $\sigma(M_B^{max}) = 0.^m2$. For the 17 best observed SNe Ia in the field the corresponding value is $0.^m35$. This figure is an upper limit because of observational error, peculiar velocities and the neglect of intrinsic absorption in the parent galaxy. The infrared filters, which presumably are not affected by absorption, exhibit an even smaller scatter of $\sigma(M_H^{max}) = 0.^m15$, but this sample consists of only 5 SNe Ia. Thus SNe Ia show only a narrow range of absolute luminosities at maximum, which makes them very good standard candles.

THE EFFECT OF REDSHIFT: The standard curves also allow a detailed study of the expected light curves at non-zero redshifts. The K-corrections (Oke and Sandage 1968) for spectra of SN1981B (Branch et al. 1983) applied to the V light curve reveal a narrowing of the peak of this curve for $0 \leq z \leq 0.3$. The reason is that only the corrections at maximum light are negative (hence brightening the supernova), whereas for later epoches they are positive. This effect is stronger than the expected time dilation factor (1+z). Thus SNe Ia at these redshifts will have a steeper decline after maximum in V. The final decay, however, is shallower than for local SNe as time dilation overcomes the then small corrections.
The K-corrections for maximum light define new lines for SNe Ia in the Hubble diagram for different world models (i.e. q_0). The SNe are shifted towards brighter V magnitudes for $z \leq 0.6$, but become fainter at higher redshifts. Therefore supernova searches will be favored for redshifts of $z \approx 0.3$ as SNe Ia are relatively brightest at such distances.

CONCLUSIONS: The value of SNe Ia as standard candles is based on the uniform evolution of the light curves of these objects in six filters and their similar colours. Their very small scatter around a mean absolute luminosity makes them the most promising candidates for standard candles. Nevertheless a few exceptional cases are known (e.g. SNe 1986G and 1984A). Such objects may represent a small contamination to the class of ordinary SNe Ia. As standard candles SNe Ia are useful for the determination of distances to individual galaxies and the global value of H_0. To achieve this many more SNe must be discovered and followed in as many filters as possible. For the determination of q_0 they represent the most promising candidates. With the observation of a few SNe Ia at $z \approx 0.5$ it should be possible to distinguish between the cases of the empty ($q_0 = 0$) and the flat ($q_0 = \frac{1}{2}$) Friedmann universe. Such observations also would provide the data for the fundamental test of the nature of cosmological redshifts due to the differences of the shapes of the light curves in expanding or notexpanding universes.

ACKNOWLEDGEMENT: I am very grateful to Prof.Dr. G.A. Tammann for many discussions and suggestions which made this work possible.

References:

- Barbon,R.,Ciatti,F.,Rosino,L.: 1973, Astron. Astrophys. 25,241
- Branch,D.,Drucker,W.,Jeffery,D.J.: 1988, Astrophys.J. 330,L117
- Branch,D.,Lacy,C.H.,McCall,M.L.,Sutherland,P.G.,Uomoto,A.,Wheeler,J.C.,Wills, B.J.: 1983, Astrophys.J. 270,123
- Cadonau,R.: 1986, Ph.D. Thesis, University of Basel
- Cadonau,R.,Sandage,A.,Tammann,G.A.: 1985, in *Supernovae as Distance Indicators*, ed. N.Bartel, Berlin:Springer, p.151
- Doggett,J.B.,Branch,D.: 1985, Astron.J. 90,2303
- Elias,J.H.,Matthews,K.,Neugebauer,G.,Persson,S.E.: 1985, Astrophys.J. 296,379
- Filippenko,A.: 1985, this volume
- Graham,J.R.: 1987, Astrophys.J. 315,588
- Kraan-Korteweg,R.C.: 1986, Astron. Astrophys. Suppl. Ser. 66,255
- Oke,J.B.,Sandage,A.: 1968, Astrophys.J. 154,21
- Pskovskii,Y.P.: 1984, Sov. Astron. 28,658
- Sandage,A.: 1973, Astrophys.J. 183,711

TYPE II SUPERNOVAE AS DISTANCE INDICATORS: AN OVERVIEW

Robert V. Wagoner

Dept. of Physics, Stanford University
Stanford, California 94305-4060

The opportune explosion of SN 1987A produced an explosion of information about (at least initially compact) Type II supernovae.[1] In this note we will summarize what we feel these developments imply for the use of SNII in the Baade method of direct distance determination.[2] The points of view of the other groups active in this field are presented in the following talks by P. Hoflich, R. Eastman, and R. Russell. Our working philosophy is that the Baade method should be implemented in a manner that depends as little as possible on both the structure of the presupernova star (which will be unknown for all supernovae of cosmological interest) and the amount of intervening absorption (A_ν).

With respect to the first point, we believe that application of the method during the phase when the photosphere is recombining will produce the most accurate distance. Recall the equations from which one obtains the angular size θ and the radius R of any spherically symmetric sharp photosphere:

$$2\log\theta = \log(f_\nu/F_\nu) + 0.4A_\nu + 3\log(1+Z) , \quad R = v(t-t_0) ; \qquad (1a,b)$$

assuming $R \gg R_0$, giving the (proper motion) distance $D_m = R(t)/\theta(t)$ of a source with redshift Z. There are essentially four reasons for this belief: (1) The steep gradient in electron density at recombination produces a sharper photosphere, allowing the use of fewer model parameters in the computation of the emitted flux F_ν and a more constrained value of θ. (Our model parameters have been effective temperature, density scale height, and heavy element abundance.) (2) The weaker line profiles (such as FeII λ5018,5169) have sharper minima, making the determination of the velocity v at the photosphere more accurate. (3) The recombination phase typically begins after a few weeks, making the computation of R less dependent on an accurate knowledge of the time t_0 when the matter was last accelerated. (4) The method is less sensitive to our incomplete knowledge of the line opacity since there is less flux at wavelengths where it is important (λ < 4500 Å) when the temperature has dropped below 6000 K.

With respect to the second point, we have adopted the use of three-color(frequency) indices in order to make our comparison of computed (F_ν) and observed (f_ν) spectra independent of extinction. Of course, we have had to assume a universal dependence on wavelength: $A_\nu = E_{B-V}\Psi(\lambda)$, but only at wavelengths λ > 2600 Å where it is consistent with observations. After obtaining our best-fitting model photosphere by employing a variety of such indices, we can obtain the extinction-independent angular size θ by applying equation (1a) at a wavelength λ > 3 microns where $A_\nu = 0$.

In an earlier investigation[3], we obtained a preliminary distance to SN 1987A of 43.3 ± 4 kpc from observations on March 1, 1987. [Other measures of distance to other parts of the LMC (whose thickness is not negligible compared to its distance) range from 44-58 kpc.] We found the thickness of the photosphere to be less than 7% of the radius R = 4.3×10^{14} cm then; however, we still plan to relax our assumption of a plane-parallel atmosphere (as well as improve the convergence of our code and the computation of the line opacity). More recently, we have extrapolated our model results and compared them with observations on March 6, in particular SAAO and IUE photometry[4,5]. The preliminary change in the distance over this time (during which the velocity v dropped from 8551 to 5808 km s^{-1}) is -1.0 ± 3 kpc, providing an important check of our model.

A major potential difficulty for this method has emerged from observations which appear to indicate deviations from spherical symmetry. The most relevant for the recombination epoch of interest (March 1-14, 1987) is the March 7 polarization scan across the Hα line[6]. The only feature, representing a decrease in polarization of 0.2%, is seen at velocities approximately twice that of the matter at the photosphere. Since the otherwise constant level of polarization could be interstellar, as indicated by nearby stars, the amount of asymmetry at this epoch could be small. The next scan, on May 5, 1987, shows much more polarization. The other evidence of asymmetry is the resolution of the supernova image by speckle interferometry[7], giving an axis ratio of roughly 5/4 during February and April, 1988. However, since this refers to matter which was well below the photosphere in March, 1987, its relevance is not clear.

In any case, as Chevalier and Soker[8] have pointed out, there are at least three reasons why the more luminous supernovae (which arise from red supergiants, and will be used to determine the Hubble constant, etc.) should be more spherical: (1) Red supergiants have a flatter density profile in their mantle, which tends to decrease any (small) asymmetry in the shock wave as it propagates outward. (2) The probability of a companion close enough to affect the flow or tidally distort the mantle is small. (3) Red supergiants, being more extended, should have smaller rotational distortions.

This work has been supported in part by NSF (grant PHY 86-03273) and by NASA (grant NAGW-299). Important contributions have been made by Manorama Chilukuri, Stephen Hershkowitz, and Alan Karp.

1) Arnett, W.D., Bahcall, J.N., Kirshner, R.P., and Woosley, S.E., Ann. Rev. Astron. Ap., **20**, in press (1989).
2) Wagoner, R.V., in Theory and Observational Limits in Cosmology, ed. W.R. Stoeger, pp. 345-378 (Vatican Observatory, 1987).
3) Chilukuri, M. and Wagoner, R.V., in Atmospheric Diagnostics of Stellar Evolution (I.A.U. Colloquium 108), ed. K. Nomoto, pp. 295-304 (Springer-Verlag, 1988).
4) Menzies, J.W. et al., M.N.R.A.S., **227**, 39P (1987).
5) Kirshner, R.P., Sonneborn, G., Crenshaw, D.M., and Nassiopoulos, G.E., Astrophys. J., **320**, 602 (1987).
6) Cropper, M. et al., M.N.R.A.S., **231**, 3 (1988).
7) Karovska, M. et al., I.A.U. Circular No. 4604 (1988).
8) Chevalier, R.A. and Soker, N., Astrophys. J., submitted (1988).

SPECTRAL DIAGNOSTICS OF TYPE II SUPERNOVAE BY ATMOSPHERIC MODELS

P.Höflich

Max-Planck-Institut für Physik und Astrophysik
Institut für Astrophysik
Karl Schwarzschild Str. 1, D-8046 Garching, FRG

ABSTRACT: Spherical-NLTE calculations for scattering dominated photospheres are presented. Extension and NLTE effects are essential as well for the line as the continuum forming region if the density gradients are low or before the recombination phase of hydrogen. The sensitivity of the UV and optical spectra on the free model parameters are discussed. The absorption minima of weak lines are good first order indicators for the velocity fields in the region where the Thomson optical depths equals one. However, the continuum fluxes in the optical wavelength range are systematicaly lower by about 15 to 45 % than a corresponding black body would imply. Type II SN are very suitable as distance indicators. This is demonstrated by SN1987A as an example. Applications to other type II SN indicate a Hubble constant of about $65 \pm 10\ km/(s\ Mpc)$.

1. INTRODUCTION

Type II supernovae (SN) are one of the most luminous single objects which may be used as standard candles for extragalactic distance scales. However, the total luminosities show a wide range of individual variations. Therefore, detailed atmospheric models are needed for the determination of the absolute fluxes. Extension effects are substantial for the continuum forming regions, at least before the recombination phase of hydrogen, as has been shown by continuum models (Shaviv et al., 1984; Höflich et al., 1986). However, line blanketing affects are very important for the optical and the UV wavelength region. This can even be true if the U-B and B-V ratio is close to the expected value for a black body (Höflich, 1987). In order to investigate scattering dominated type II SN, we have extended our investigation of SN1987A to non-recombination bounded situations.

2. MODEL CALCULATIONS

Spherical-NLTE calculations have been carried out. Power law density profiles ($N(r) \propto r^{-n}$) are used. The expansion velocity v_{ex} of the matter is assumed to be a linear function of the distance r. The rate equations of several elements (H, He, C, N, O, Ne, Na, Mg, S, K, Ca, Ba) are solved consistently with the radiation transport. Our models are characterized by the following free parameters: a) the photospheric radius R_{ph}, b) the effective temperature T_{eff}, c) the exponent n, $v_{ex}(R_{ph})$ and d) a statistical velocity field. Here, models before the recombination phase of hydrogen are mainly discussed (see table 1).

The importance of a consistent treatment of the physical parameters should be noted. E.g. although the continuum forming regions are dominated by electron scattering, effects due to line blocking and departures from LTE of the radiation field have to be taken into account for the temperature profile (figure 1) because it strongly influences the emitted continuum radiation.

Table 1: The distance R_{ph} at which the optical depth is 1 for true absorption at 5000 Å, the effective temperature T_{eff}, the particle density N_o, and the value of n of the assumed density profile ($N(r) \propto r^{-n}$), and the expansion velocity v_o are given (cgs units are used). In addition, the Thomson optical depth $\tau_{sc}(R_{ph})$, the relative extension E of the continuum forming region (i.e. $r(\tau_{sc} = 1.6)/R_{ph}$) and the distance R_{HII}, up to which hydrogen is mainly ionized, are given. Solar abundances and a turbulence velocity of 200 km/s are assumed.

No.	R_{ph}	T_{eff}	N_o	n	v_o	E	τ_{sc}	R_{HII}
I	1.E15	7000.	2.5E11	5	3500.	2.3	46.	2.9E15
II	1.E15	7000.	2.5E11	5	7000.	2.3	45.	2.7E15
III	5.E14	10000.	5.3E11	5	3500.	2.3	50.	(2.7E15)
IV	1.E15	10000.	3.5E11	5	3500.	2.5	69.	...
V	1.E15	10000.	3.5E11	5	7000.	2.5	68.	...
VI	3.E15	10000.	1.6E11	5	3500.	2.9	99.	...
VII	5.E14	10000.	7.5E11	8	3500.	1.6	43.	...

Figure 1: Calculated temperature profile of model IV in comparison with the gray extended solution.

As can be seen (table 1), v_{ex} has only minor influence on the needed particle density, but it affects the ionization structure (i.e. τ_{sc}, R_{HII}). The scattering domination of the continuum forming region (i.e. τ_{sc}) increases with T_{eff}. Effects due to sphericity (i.e. E of table 1) become substantial in models before the recombination phase. However, they are not negligible for small density gradients (n $\leq$ 5...6) even after the recombination phase of H. The spectra show the following tendencies: a) The H line profiles are sensitive to v_{ex}, T_{eff} and the exponent n of the density profile. b) Generally, the line shift of the absorption feature decreases inside the Balmer series of H. The absorption minima of H_β and of weak lines of heavier elements occure at R (τ_{sc}) $\approx$ 1. c) The H line profiles are less sensitive to R_{ph}. d) However, the UV-excess increases with increasing radius as long as H is highly ionized. e) An UV-flux deficit is caused by low luminosities if v_{ex} is still high (e.g. SN1987A) because of the recombination of H in the UV line forming region. f) The occurance of strong optical lines due to heavy elements indicate recombination of H in the line forming region (e.g. SN1987A). g) The colour temperature in the optical wavelength range results in an overestimate of the predicted fluxes in the order of 15 to 45 % (table 2). (See Höflich (1988b) for more details).

Table 2: Comparison of the calculated fluxes (see table 1) δF in the V-filter band (at 5500 Å) with those as inferred by a black body of the temperature T_{BB} for the same colour index B-V.

No.	R_{ph}	T_{eff}	n	B-V	T_{BB}	δF
I	1.E15	7000.	5	0.50	6500	0.85
III	5.E14	10000.	5	0.09	10750	0.65
IV	1.E15	10000.	5	0.08	11000	0.6
VI	3.E15	10000.	5	0.06	11250	0.55
VII	1.E15	10000.	8	0.12	10250	0.6

3. SN TYPE II AS DISTANCE INDICATORS

In order to test the absolute fluxes of our models, we have used the spectral analysis of SN1987A in the Large Margellanic Cloud (LMC). The model series has been taken that fits the observed spectra best (Höflich, 1987, 1988a). On the basis of the monochromatic fluxes, a distance of 48±4 Kpc has been derived (table 3). This value is in good agreement with the those as derived for the LMC by other distance indicators (RR-Lyrae: 48.7 ±1.1 kpc, 48.3 ±1.3 kpc, Walker and Mack (1988); δ Cep.: 49.4 ±3.3 kpc; Walker (1987); 54.9 kpc, Martin et al., (1979); 49.0 kpc, Graham and Nemec (1984): 49.0 kpc) but it clearly tends to the lower values for the LMC. Note, that our new distance is somewhat larger than the value given before (46 kpc; Höflich, 1987). This correction is due to the incooperation of the line blanketing due to elements others than hydrogen in the new models. This causes a small increase of the radiation field and of the temperature in the continuum forming region and it results in an increase of the optical continuum flux. Note that the temperature change also influences the needed reddening correction as indicated by the models (Höflich, 1988a).

Table 3: The distance in Kpc of SN1987A as derived by comparison between the calculated and observed absolute fluxes (see table 1 of Höflich, 1988a) as a function of wavelength (mean value ±30 Å). A reddening of $E_{B-V} = 0.15^m$ is assumed.

No.	Date	5000	5300	5500	5700	6000	6300	at H_α
V	March 23	53.	44	49	46	52	52	46.5
VII	May 14	48.	50.	48.	49.	51.	50.	45.
VIII	June 02	48.	52.	46.	46.5	45.	50.	52.

In the last few years several distance determinations have been done by using the Baade-Wesseling method and black body fits (e.g. SN 1969L: $12 \pm 4kpc$ and SN 1970G: $7 \pm 2kpc$ Kirshner and Kwan (1974); SN1979C: $23 \pm 3kpc$, Branch et al. (1981)) implying a Hubble constant H_o of $57 \pm 4kms^{-1}Mpc^{-1}$ (Branch, 1988). Our calculations have shown that the line shift of the absorption features of weak line transitions (see above) are good first order indicators for v_{ex}. However, the absolute visual flux is systematicaly overestimated by about 15 to 45 % indicating a correction of H_o by about 10 - 20 % ($65 \pm 10kms^{-1}Mpc^{-1}$). Note, however, that detailed analyses of different dates are needed to get a reliable distance of a specific SN. In particular, the application of the Baade-Wesselink method may be very questionable in the recombination phase, because the correlation of the expansion velocity of the photosphere and v_{ex} of the matter, which is detected by the line shifts, depends very sensitively on the model parameters, e.g. R_{ph} may stay constant or may even decrease although blue shifted absorption features of the line spectra are observable (e.g. SN1987A; Höflich, 1988a).

3. REFERENCES

Branch,D., Proc. of the Symp. on Extragalactic Distance Scale., ed. van den Bergh, June 88, in press (1988)

Branch,D.; Falk,S.; McCall,M.; Rybski,P.; Uomoto,A.K.; Will,B.J., Astrophys.J. 244, 780 (1981)

Graham,J.A.; Nemec,J.M., IAU Symp. 108, 37 (1984)

Höflich,P., in preparation (1988b)

Höflich,P., Proc.Astron.Soc.Austr., Canberra, June 1988 , in press (1988a)

Höflich,P., Proc. of the ESO-workshop on SN1987A, Garching, July 1987, 447 (1987)

Höflich,P.; Wehrse,R.; Shaviv,G., Astron. Astrophys., 163, 105 (1986)

Kirshner,R., Kwan,J., Astrophys. J. 193, 27 (1974)

Shaviv,G., Wehrse,R., Wagoner,R.V., Astrophys. J., 289, 198 (1984)

Walker,A.R., Mon.Not.Roy.Astr.Soc., 225 , 627 (1987)

Walker,A.R.; Mack,P., Astron.J., 96, 872 (1988)

TYPE II SUPERNOVA DYNAMICS

S. A. Bludman[+]

Department of Physics, University of Pennsylvania
Philadelphia, Pennsylvania 19104, U. S. A.

ABSTRACT

In a realistic description of neutrino transport, neutrino-electron downscattering leads to so much neutrino leakage that the shock forms so far inside the iron core that it cannot traverse the large overlay mass, without being arrested by nuclear dissociation and neutrino radiation. To see whether the stalled shock may later be revived on the neutrino diffusion time scale, we developed an implicit general relativistic Lagrangian hydrodynamics on polar-sliced time that, because it is regular everywhere outside the apparent horizon, completely covers the space-time domain of outer communication, gives maximum coverage of material zones, and allows calculation through many time-steps of dynamic and of quasi-static evolution. In the same polar-sliced comoving gauge, we wrote down the general relativistic transport equation and moment equations and solved for the variable Eddington factor by a generalization of the tangent-ray method used in non-relativistic transport theory. We calculated the stationary transport in a weak and strong static gravitational field and found that gravity (1) Enhances the optical thickness and leads to a moderate reduction in the forward-peaking of the radiation field in an extended spherical atmosphere: (2) Red-shifts the luminosity and mean intensity, as expected.

I. PROBLEMS WITH THE PROMPT EXPLOSION MECHANISM

A. Nuclear Equation of State. In order to see whether a prompt explosion could be obtained from stellar collapse under the apparently most favorable stellar conditions, we[1,2] studied the relativistic collapse of a 1.17 $M_\odot$ iron core[3], assuming the low value K_o (N=Z) = 180 MeV for the incompressibility of symmetric nuclear matter, corresponding to K_o (N=2Z) = 140 MeV for neutron-rich nuclear matter. Both these assumptions may indeed be overly optimistic: Interpretations[4] of the

---

[+]Partially supported by DOE contract EY-76-D-02-3071. Work done in collaboration with E. S. Myra and P. J. Schinder.

Supernova 1987A light curve show that it came from a progenitor star of main sequence mass (15-20) $M_\odot$ and iron core mass (1.3-1.4) $M_\odot$. Recent determinations of the incompressibility of symmetric nuclear matter, derived from giant monopole measurements, show[5] $K_o(N=Z) = 290 \pm 20$ MeV. Nevertheless, in order to compare with Baron et al. (BBBCK)[6] who had reported strong $(1 - 2) \times 10^{51}$ ergs prompt explosions, we joined them in these two optimistic assumptions and in using the Cooperstein[7] nuclear equations of state, with the Baron-Cooperstein-Kahana (BCK) extrapolation to densities above saturation density ρ_o $(N=2Z) = 2.4 \times 10^{14}$ g cm^{-3}, $P = (K_o\rho_o/9\gamma)[(\rho/\rho_o)^\gamma - 1]$, $\gamma = 2.5$. The Cooperstein equation of state also contains a bulk-symmetry coefficient W_s, which directly controls the difference between neutron and proton chemical potentials and, therefore, the free proton fraction and the infall deleptonization ΔY_L.

B. Neutrino Radiation Hydrodynamics. The collapse of an iron core to a cold remnant neutron star releases $(3 - 5) \times 10^{53}$ ergs of gravitational binding energy, of which only 1% needs to be transferred to matter at the edge of the iron core in order to make a Type II supernova explosion. This makes the copious neutrino flux easy to estimate, but leaves the small neutrino-matter energy exchange as a delicate problem in neutrino radiation hydrodynamics. The neutrino neutral current scattering off nucleons and nuclei is almost energy conserving, while emission-absorption by charged currents and neutrino-electron scattering by charged and neutral currents (NES) thermalize the neutrinos by exchanging energy with nucleons and electrons.

The crucial weak, energy-dependent neutrino-matter coupling transpires in a density regime where the neutrino transport is neither simply diffusive nor simply ballistic. Various approximations to the neutrino transport have been used: (1) BCK and BBBCK describe the neutrino transport by a simple leakage scheme tuned to give average neutrino trapping at 1.0×10^{12} g cm^{-3}; (2) We use a multi-energy group flux-limited diffusion approximation including neutrino-electron scattering (NES) in a modified Focker-Planck approximation; (3) Bruenn[8] uses flux-limited diffusion without approximating NES.

None of these approximations is completely satisfactory in describing the delicate neutrino-matter coupling responsible for deleptonization during the infall and for shock wave dampening and possible matter ejection after bounce. Our multi-group flux-limited diffusion approximation is only approximately generally relativistic and becomes inaccurate in the low density regime where delayed shock revival is supposed to take place. Nevertheless, our transport and Bruenn's transport contain more physics than the BCK-BBBCK leakage scheme, which neglects NES and the energy-dependence of neutrino trapping.

C. Generally Relativistic Explicit Hydrodynamics in Comoving Synchronous Gauge. All calculations of spherical collapse use Lagrangian hydrodynamics in which the coordinates move locally with matter. This (1) Allows the matter equations to be directly integrated into conservation laws for energy, momentum, entropy and baryon number along matter trajectories; (2) Gives maximal matter coverage, because the comoving mass zones are concentrated where the matter is, not wasted where matter is absent or exceptionally diffuse.

In the prompt shock calculations[1,2], we used explicit hydrodynamics, in which numerical stability requires that the difference equations advance by time steps smaller than the Courant time for a sound wave to traverse a mass zone. We also used (May-White) "synchronous" comoving space-time coordinates, which become singular as soon as any mass zone reaches a general relativity trapped surface.

We found that (1) The fate of the prompt shock is already determined by the trapped lepton fraction Y_{Lf} at bounce: For $Y_{Lf} > 0.37$, the edge of homology where the shock forms is far out, the overlay mass is small, and a prompt explosion is obtained. For $Y_{Lf} \leq 0.37$, either because of NES or because of a more reasonable higher proton fraction, the prompt shock stalls; (2) The infall deleptonization down to Y_{Lf} depends, of course, on NES downscatter and on the number of free protons capable of capturing electrons. In the Hashimoto-Nomoto presupernova configuration, the electron-baryon ratio $Y_{ec} = 0.41$ at the center, so that electron capture by nuclei is already suppressed by neutron shell blockage over the bulk of the inner core; (3) A soft nuclear equation of state with general relativity hydrodynamics does create an initially energetic shock wave: For $\gamma = 2.5$, nuclear bounce took place at $\rho_b \sim 9 \times 10^{14}$ g cm^{-3}; (4) If we omit NES from out neutrino transport, we can obtain strong explosions of 2×10^{51} ergs, as do BBBCK. When NES is properly included, however, we find, even underestimating the free proton fraction, $Y_{Lf} = 0.37$ and a stalled shock; (5) NES is fatal to the prompt shock because it leads to infall deleptonization down to $Y_{Lf} = 0.37$. (Because our Focker-Planck approximation may underestimate neutrino downscattering in energy, NES may be even more harmful than we calculated.); (6) The neutrino radiation loss is always about (3.6-4.0) x 10^{51} ergs in our calculations, irrespective of the success or failure of the prompt explosion.

Even with a low mass iron core (1.17 $M_\odot$), a soft nuclear equation of state K_o (N = Z) = 180 MeV, an underestimated free proton fraction (W_s = 36 MeV), and general relativity hydrodynamics, the prompt shock mechanism fails. This happens because, in a realistic description of neutrino transport, neutrino-electron downscattering leads to so much neutrino leakage ($Y_{Lf} \leq 0.37$) that the shock forms so far inside the iron core, that it cannot traverse the large overlay mass, without being arrested by nuclear dissociation and neutrino radiation.

II. PROSPECTS FOR LATER REVIVAL OF THE STALLED SHOCK

A. Implicit Hydrodynamics. If the shock stalls on the hydrodynamic time scale (tens of milliseconds), it may[9] still be revived by mass accretion and neutrino diffusive heating on a time scale (seconds) for neutrinos of all flavors to diffuse out of the hot core. In order to economically follow the core evolution through this quasistatic stage of shock revival and later neutrino cooling, we must take much longer numerical time steps than the Courant time step allowed in explicit hydrodynamics. An implicit code is always numerically stable, but escapes the Courant time step restriction at the cost of solving a set of non-linear equations self-consistently at each time step. Because these large matrices of differential coefficients are inverted numerically, we also calculated the many differential coefficients numerically.

B. Improved General Relativity: Polar Slicing of Space-Time. Almost all previous general relativistic hydrodynamic calculations have used the comoving "synchronous" coordinates of May and White. These coordinates become singular everywhere as soon as a horizon is reached anywhere. We want to study stellar core collapses right up to black hole formation, because these produce initial shocks of maximum energy. In fact we also want to consider protracted collapses into a black hole, where the infalling matter lingers just outside the horizon, before slowly accreting enough mass to be drawn across the horizon. If the lingering is protracted enough, before being swallowed up by the black hole, we expect some neutrino radiation to escape to infinity. Neutrino-signalled stellar collapses may therefore be more frequent than neutron-star formation.

For these reasons, we have also shifted from the Schwarzschild areal radius coordinate R to a new comoving polar-sliced coordinate r(R,t) by the coordinate shift $dR = R'dr + \dot{R}dt = R'(dr + \beta^r dt)$, where $\cdot = \partial/\partial t$, $' = \partial/\partial r$ and $\dot{R} = R'\beta^r$. In the two gauges

$$ds^2 = -\alpha^2 dt + (dR/\Gamma)^2 + R^2 d\Omega^2 = -(\alpha^2 - \beta_r\beta^r)dt^2 + 2\beta_r drdt + (R'/\Gamma)^2 dr^2 + R^2(r,t)d\Omega^2 .$$

Both space-time coordinates are polar-sliced, meaning that the traces of the transverse part of the intrinsic curvature of the space-like surfaces vanishes. Schwarzschild coordinates are not co-moving but are Gaussian normal, meaning that the slicings (space-like surfaces t = constant) are everywhere normal to the velocity field of Eulerian observers. In our new comoving polar-sliced gauge, from one slicing to the next, the constant coordinate lines are shifted from the normal by $\beta^r = \alpha U$, depending on the material velocity U as well as the lapse α. The lapse vanishes exponentially near the apparent horizon (outermost trapped surface) so that evolution freezes near the apparent horizon while continuing elsewhere, until as $t \to \infty$, matter everywhere reaches the apparent horizon. Our polar-sliced coordinates therefore avoid coordinate singularities while allowing maximum coverage of the domain of outer communication outside of the trapped surfaces. Because they are comoving, these coordinates also allow maximum coverage of material zones.

Our code passed[10)] two crucial tests: (1) We allowed a nearly equilibrium polytrope to relax and found our implicit code taking longer and longer time steps as it reached and maintained equilibrium; (2) We collapsed an initially homogeneous dust cloud and found, after a relatively few time steps, the areal radius of each mass zero approaching limiting values and the cloud's outer radius approaching $R_s = 2M$, thus demonstrating the horizon-avoidance of polar-slicing and excellent agreement with the analytic Oppenheimer-Snyder solution.

C. Radiation Transport in Polar-Sliced Comoving Gauge. Almost all previous calculations of radiation transport during stellar collapse, including our own, assumed some form of diffusion approximation

In ref. 10, 11 we wrote down the general radiation transport equation in polar-sliced comoving gauge. For stationary transport in a static gravitational field, this partial differential equation can be integrated by the method of characteristics[12,11] or tangent ray method.

We are generally most interested in the mean intensity J and flux H, which are for stationary transport in a static gravitational field, given by the zeroth and first angular moments of the transport equation

$$\frac{\partial(\alpha R^2 H)}{\partial R} = R^2\alpha(\eta-\chi^J),$$

$$\frac{\Gamma\alpha R^2}{q}\frac{\partial(qfK^J)}{\partial R} = R^2\alpha(-\chi^H).$$

Here q(R) is the sphericity factor defined by

$$\frac{d\ln q}{dR} = \frac{(3f_K - 1)}{Rf_K} + \frac{1}{f_K}\frac{d\ln\alpha}{dR} \quad .$$

and $f_K(R)\ J(R) \equiv K(R)$ is the radiation pressure. These two moment equations are subject to flux boundary conditions $H(0) = 0$, $H(R_s) \equiv f_H(R_s)\ J(R_s)$. The transport equation must still be solved to know $f_K(R)$, $f_H(R_s)$, but this is the only information needed, in some approximation, from the transport equation, after which the integration of the moment equations is straightforward.

III. TEST CALCULATIONS USING THE GENERALIZED TANGENT-RAY METHOD FOR THE VARIABLE EDDINGTON FACTORS.

To study the effects of a strong static gravitational field on stationary radiation transport, we considered two stars of (1.64-1.65) $M_\odot$ in stable equilibrium under the n = 1 polytropic relation $P = K\rho^2$. In the weak gravity case, the constant K was chosen to make the central density $\rho_c = 10^9$ g cm^{-3} and the stellar radius $R_s = 1.37 \times 10^3$ km, values suggestive of a white dwarf star. This star was very non-relativistic, as indicated by the relativity index $\sigma \equiv (P/\rho c^2)_c = 8.9 \times 10^{-4}$, the central lapse $\alpha_c = 0.996$ and the surface red-shift $(GM/Rc^2)_s = 0.0018$. In the strong gravity case, K was chosen to make

$\rho_c = 4 \times 10^{15}$ g cm^{-3}, $R_s = 9.08$ km and $\sigma = 0.45$, $\alpha_c = 0.326$, $(GM/Rc^2)_s$ - 0.269, values characteristic of a neutron star. This case is practically maximally relativistic for a stable causal polytrope.

We tested our code for two source and opacity distributions: (A) Isothermal sources $S = \kappa B(T)/\chi E^3$ = constant, with opacity $\chi = R_s^{-1}$; (B) Isotropic coherent scattering in radiative equilibrium S = J, with varying opacity $\chi(R) = R_o/R_s^2$ and photospheric radius $R_o = 0.1\ R_s$. This is, therefore a star with a small core of infinite optical thickness surrounded by an extended atmosphere.

In both cases, we found that strong gravity decreased the Eddington factor f_K by about 20% near the edge of the extended atmosphere and reduced the luminosity $4\pi R^2 H$ and the mean intensity J by the red-shift factor $\alpha(R)$. In case (B), we found that gravity had no perceptible effect on any of the physical variables inside the small diffusive core. In the extended atmosphere, strong gravity enhanced the sphericity q(R) by about a factor two.

REFERENCES

1. Myra, G. S. and Bludman, S. A., Neutrino Transport and the Prompt Mechanism for Type II Supernova, UPR-0352T, submitted to Ap. J.
2. Bludman, S. A., Phys. Reports 163, 47 (1988).
3. Hashimoto, M. and Nomoto, K., Physics Reports 163, 13 (1988).
4. Arnett, W. D., Ap. J. 319, 136 (1987); Shigeyama, T., Nomoto, K. and Hashimoto, M., Astr. Ap. (1987); Woosley, S. E., Pinto, P. A. and Ensman, L., Ap. J. 324, 466 (1988).
5. Sharma, M. M., Borghols, W. T. A., Brandenberg, S., Crona, S., Meier, J. H., van der Woode, A. and Harakek, M. N. (preprint KVI-689, Groningen).
6. Baron, E. A., Bethe, H. A., Brown, G. E., Cooperstein, J. and Kahana, S., Phys. Rev. Lett. 59, 736 (1987).
7. Cooperstein, J., Nucl. Phys. A429, 527 (1985).
8. Bruenn, S. W., Ap. J. Suppl. 58, 771 (1985); Ap. J. 62, 331 (1986); Private communication (1988).
9. Wilson, J. R. in Numerical Astrophysics, ed. J. Centrella, J. LeBlanc, and R. Bowers (Jones and Bartlett, Boston, 1985); Bethe, H. A. and Wilson, J. R., Ap. J. 295, 14 (1985).
10. Schinder, P. J., Phys. Rev. D38, 1673 (1988).
11. Schinder, P. J. and Bludman, S. A., submitted to Ap. J. 1988.
12. Schmid-Burgk, J., Astrophys. Space Science 56, 191 (1978).

Nuclear and Astrophysical Evidence on the Equation of State

Norman K. Glendenning

Nuclear Science Division
Lawrence Berkeley Laboratory
University of California
Berkeley, California 94720

Abstract

The current situation concerning the equation of state is examined with particular reference to supernova simulations.

NUCLEAR EVIDENCE. It has been claimed that supernovae, by their occurance tell us that the nuclear equation of state must be soft [1,2]. In Fig. 1 we review a large body of data from nuclear evidence and neutron stars, which disagrees with this [3]. We examine below the particular path that led to the claim, show how the same equation of state employed in the supernova simulations is incompatible with neutron star masses, and conclude that there is an opportunity for discovering new physics possibly in the presupernova simulations or in the late stages of the collapse that will reconcile supernova with the existing knowledge of the equation of state

PROMPT SUPERNOVA EXPLOSIONS. Type II supernovae are those whose progenitor masses are greater than about 10 $M_{\odot}$. After evolving for about 10^7 years during which a core of neutron rich iron of mass of the order of the Chandrasekhar mass is built, stability against gravitational collapse is lost. The mass of the iron core is is crucial to the success or failure of the prompt bounce scenario. As the star collapses the core attains supernuclear densities, and a shock wave is formed which moves outward through the core toward the infalling mantle. For each 1/10 $M_{\odot}$ of iron core that the shock wave traverses, it dissipates about 2×10^{51} ergs due to the dissociation of the iron nuclei in its path. This is comparable to the entire energy output of a supernova. It is generally accepted that the shock cannot promptly eject matter if the iron core is more massive than about 1.35 $M_{\odot}$. Otherwise it stalls at about 100 km and becomes an accretion shock. So there is a very narrow window of iron cores in which the prompt scenario has a chance of working. Within this narrow window the Brookhaven-Stony Brook group found that *if* they chose a sufficiently soft equation of state, they could simulate a prompt bounce scenario for a supernova explosion, with a resultant neutron star residue[1,2]. If this were the whole story, then a tentative conclusion could be reached that the equation of state must be sufficiently soft at high density because of the occurrence of type II supernovae. This is indeed the conclusion that the authors drew, and which has been much advertised.

However there are many uncertainties in the physics of the simulation besides the equation of state. One of these has been mentioned, the iron core mass. Another is the specific entropy of the core just prior to collapse following an evolution of $\sim 10^7$ years! Both can easily change the amount of iron that the shock must dissociate before reaching the mantle by the crucial $1/10\ M_{\odot}$, less than a ten percent effect. Can such an evolution be modeled with such certainty?

Nevertheless taking all such effects in their favor, the soft equation of state is still found necessary to achieve the prompt ejection *with* the desired explosion energy[2] for SN1987A, the supernova event of early 1987. Using precisely their equation of state described in ref.[1], and the neutron fraction ($x \equiv Z/A = 1/3$) employed by those authors we have solved the Oppenheimer-Volkoff equations of star structure. The maximum mass for their equation of state is too small compared to the masses of two known neutron stars as shown in Fig. 2. One is the very accurately measured mass of PSR1913+16, and the other is the less well known mass of 4U0900-40.

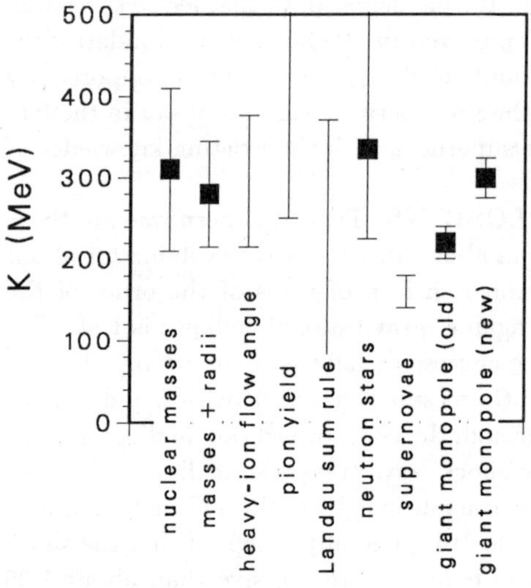

Fig. 1.

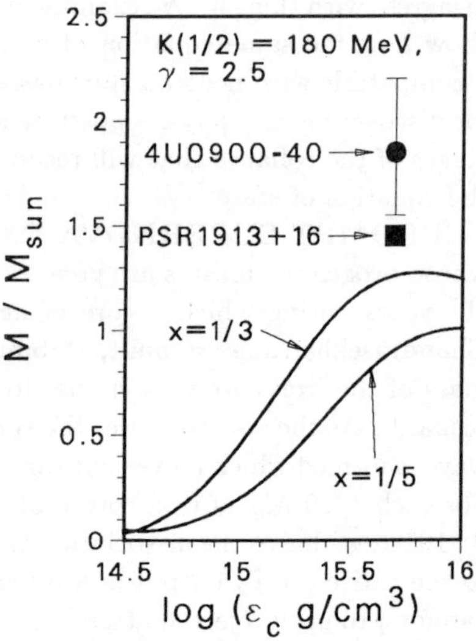

Fig. 2

One may be tempted to argue that the supernova matter is less dense than that of a neutron star, and that it is less neutron rich. However there are numerous physical effects at higher density that soften the equation of state. Neutronization is but one of them. If it were not energetically favorable it wouldn't happen, so it *does* soften the equation of state and its effect (the curve labeled $x = 1/5$) for the BCK equation of state is also shown.

Therefore it is unproven that the prompt-bounce mechanism can be made to work in supernova simulations, *given the present estimates of the iron core masses,* with equations of state that are soft enough to release enough energy for prompt ejection of the mantle, and stiff enough to be compatible with certain measured neutron star masses and with the preponderance of evidence from nuclear physics[3] shown in Fig. 1. Consequently it is premature to conclude as has been done, that supernovae, by their occurrence, imply a soft equation of state. This is all the more reinforced by the observation that there exists an alternative mechanism that does not impose this restriction on the equation of state, discovered by Wilson[4].

NEUTRON STARS. We have seen that with the present knowledge of supernova physics, the prompt bounce mechanism requires an equation of state that is too soft to support measured neutron star masses. How stiff do neutron stars require it to be? First we must realize that neutron stars are not pure in neutron. As the density of neutron matter is increased, the Fermi energy soon surpasses the threshold for decay first to proton and lepton and then to hyperons[5]. We have emphasized[6] how this hyperonization can soften the equation of state at the *higher* densities at which it occurs ($\approx 2\rho_0$) and *reduce the limiting neutron star mass* by as much as 3/4. The net effect therefore is that the equation of state must be rather stiff near saturation, so that with this softening at higher density, observed neutron star masses can be accounted for.

We analyse neutron stars in the framework of relativistic nuclear field theory, taking into account all higher baryon states to convergence. The full Lagrangian including leptons is given in ref. [5] together with the equation of state. When the field equations are solved subject to the subsidiary condition of isospin symmetry, the solution corresponds to symmetric nuclear matter. When this same theory is solved with subsidiary conditions of charge neutrality and beta equilibrium, we get the solution for neutron star matter. These solutions, by convention, will be denoted always by the properties of the corresponding solutions of symmetric matter. We vary the stiffness of the equation of state of neutron star matter at high density. This is accomplished through a variation of the coupling constants of the theory which leaves the bulk properties of cold symmetric matter fixed at saturation, with the exception of K. In this way we are able to place a lower bound on K.

The largest measured mass is is $1.85^{+0.35}_{-0.30}M_{\odot}$ for 4U0900-40. The most probable value of the mass of the above star is found to place a lower bound of $K \geq 335MeV$. If the lower bound on the mass measurement of 4U0900-40 is used, then the lower bound on K becomes about 225 MeV.

HYPERONIZATION MAY POWER SUPERNOVAE. We have seen that, taking all known factors to their advantage, the Brookhaven-StonyBrook group still found it necessary to adopt an equation of state that is so soft that it is incompatible with measured neutron star masses, and this after all is the most direct not to say only evidence we have on the high density equation of state of neutron rich matter. Moreover we have seen that taking into account the hyperonization of the neutron star core the equation of state is softened at high density, and that this softening is exploited by gravity very effectively as registered in the reduction of neutron star masses in comparison with calculations where hyperonization is neglected. *The time scale of star collapse ($\sim$ 100 milliseconds), is long compared to the weak interaction lifetime, so that the hyperonization of supernovae matter occurs during the collapse and the evolution of the proto-neutron star.* Hyperonization can effect the outcome of the collapse not only through the softened equation of state, but also through the various hadron interactions and decays. The associated kaons, neutral and charged according to the particular reactions, decay, producing photons and neutrinos. If shock formation occurs *after* the hyperon thresholds are reached these processes will help the shock by heating the matter through absorption in the iron exterior to the shock, thus reducing the dissociation energy. If the shock rebounds *before*, the same processes will help power the reignition of the stalled shock. So far simulations of both the prompt and late-time mechanism have not accounted for the hyperonization effects, which we propose as an additional source of explosion energy. We point out that in the core collapse it is assumed that about 10 % of the neutrons convert to protons, whereas we find that 20 % of the nucleons convert to hyperons in the limiting neutron star, so the neglected processes are as large or larger a source of neutrinos.

Acknowledgements: This work was supported by the Director, Office of Energy Research, Division of Nuclear Physics of the Office of High Energy and Nuclear Physics of the U.S. Department of Energy under Contract DE-AC03-76SF00098.

1) E. Baron, J. Cooperstein and S. Kahana, Phys. Rev. Lett. **55** (1985) 126.

2) E. Baron, H. A. Bethe, G. E. Brown, J. Cooperstein and S. Kahana, Phys. Rev. Lett. **59** (1987) 736.

3) N. K. Glendenning, Phys. Rev. C **37** (1988) 2733.

4) J. R. Wilson, R. Mayle, S. E. Woosley and T. Weaver, Ann. New York Acadamy of Sciences **470** (1986) 267.

5) N. K. Glendenning, Astrophys. J. **293** (1985) 470.

6) N. K. Glendenning, Z. Phys. A, **327** (1987) 295.

ATTACKING H_o AND Ω WITH SUPERNOVAE

ROBERT P. KIRSHNER

Harvard-Smithsonian Center for Astrophysics

Cambridge, MA 02138

U.S.A.

ABSTRACT Supernovae may provide powerful tools for direct estimates of both the Hubble constant, H_o, and the global value of the cosmic density Ω. The estimate of H_o comes from distances using the Expanding Photosphere Method as applied to Type II supernovae (SN II). The density estimate may result from the use of Type Ia supernovae (SN Ia) as standard candles in the Hubble diagram. In both cases, recent work gives reason for optimism, since the analysis of SN 1987A provides a stringent test of the accuracy of the expanding photospheres and pioneering work on faint supernovae in distant clusters is beginning to populate the Hubble diagram for $z > 0.2$. Increased understanding of the supernova events and a vigorous observational effort to find and study supernovae are important ingredients in realizing the promise of this approach to particle astrophysics. A powerful observing facility with 4 meter aperture and a commitment to the study of supernovae are needed to use supernovae to solve the problems of cosmology.

1. WHAT'S THE USE OF SUPERNOVAE?

Although supernovae mark the sharp and violent transition from a fusion-powered star through gravitational collapse to a neutron star, those dull souls who are not captivated by the physics of this evolution, and not dazzled by its utility in creating the chemical elements of the world we inhabit, may ask "What are supernovae good for?" At this conference, we have examined the nature of the supernova explosion and the observational evidence on supernovae with the goal of applying supernovae to cosmology. This is essential, since the notion of non-baryonic dark matter which informs much of the theoretical and experimental work reported at this meeting is based on the idea that the gap between the low values of Ω required for helium synthesis, or from a census of galaxies and their associated dark matter (Kirshner et al [1]), and the value of Ω = 1.00000000... prescribed by belief in inflationary models, demands that the Universe is dominated not just by matter that we cannot see, but of a kind that has never been detected.

To move this discussion from the realm of debate to the realm of observation and experiment we need a reliable value of the Hubble constant to know the age of the universe and the present value of the critical density, and we require a direct geometrical measurement of the space curvature to be sure that we are correctly assessing the effects of mass of all forms, visible, invisible, baryonic or not. The current status of this enterprise is unsatisfactory, with serious candidate values for H_0 that vary by a factor of two, and no fully credible measurement of Ω.

Supernovae have some special advantages over other astronomical objects in this pursuit. First, although the fact that supernovae emit 10^{53} ergs in neutrinos and only 10^{49} ergs in photons has lead some misguided spirits to regard supernovae as fundamentally sources of neutrinos, they are exceedingly bright optical objects: 10^6 times brighter than the Cepheid variable stars that form the backbone of the otherwise invertebrate extragalactic distance scale. A technique that employs supernovae has the potential to establish extragalactic distances to 1000 times the depth that Cepheids can be used.

Second, although supernovae are by no means completely understood, especially at the bouncing heart of core collapse (as detailed in the papers of Bludman, Barron, Glendenning, and Mayle in this volume), the observational facts indicate that Type II supernovae (which have hydrogen in their spectra) have relatively simple atmospheres so that even a wide range in the properties of individual objects (as graphically illustrated by Filippenko in this volume) can be analyzed individually to give distances, or at least sorted to select those which are suitable for detailed modelling. This is the "custom yardstick" approach outlined in this volume by Eastman, by Höflich, and by Wagoner. Conversely, the Type Ia supernovae show strong empirical evidence of being a very homogeneous class, and the theoretical picture for SN Ia favors the deflagration of a white dwarf of a well-defined mass and energy release (as discussed by Woosley in this volume). The advantage of a well-defined physical event over other possible standard candles is that even over large look-back times, we might expect to see the <u>same</u> events at large redshift as we do at low redshift. In our own Galaxy, the supernovae seen by the monks of St. Gallen in 1006 and by Tycho Brahe in 1572 may be essentially identical to those detected now in distant clusters. Unlike quasars, which have a wide spread in luminosity, or galaxies, which show definite signs of evolution, the SN Ia at large redshift may be the same as those nearby.

In any case, both these methods can be tested by observation: what is needed is a well-observed sample of SN II for application of the expanding photosphere method, and a sample of faint supernovae for the standard candle approach to the Hubble diagram for SN Ia. In this conference, the Berkeley Supernova Search is described by Pennypacker and the success of the Danish/Australian collaboration in finding and studying faint SN Ia's is described in the paper by Couch. These efforts are the beginning of what is needed to bring supernovae from the wings to the center of the stage in the observational determination of H_o and Ω.

2. THE CUSTOM YARDSTICK

Supernova 1987A has provided a strong stimulus to develop more detailed understanding of the stellar evolution that precedes a supernova explosion (see, for example, the review by Arnett, Bahcall, Kirshner, and Woosley [2]). In this case, we have the special advantage that the identity of the stellar progenitor is known: it was the hot blue supergiant star Sanduleak -69° 202. As described in the contribution by Nomoto, clues from the ultraviolet, X-rays, and optical can be combined with stellar evolution calculations and a schematic version of the explosion physics to calculate the hydrodynamics of the explosion, to account for the emission from the surface, and to estimate the yield of heavy elements. One surprising result for the Sanduleak -69° 202 star is that, although it has the mass expected for the progenitor of a SN II, of order 20 solar masses, its radius at the time of explosion was much smaller than for the SN II that are detected in most supernova searches. One effect of this is that when the shock arrives at the stellar surface, it produces a flash of ionizing radiation. The presentation by Fransson analyzed the effects of this UV flash on the surrounding material previously shed from the Sanduleak star. Observations with the IUE satellite demonstrate this fluorescence, and provide interesting constraints on the size and composition of the circumstellar matter.

One benefit of insight into the stellar progenitor is that it establishes the initial conditions for a detailed treatment of the radiation passing through the supernova atmosphere, as outlined in the contributions by Eastman, Höflich, Russel, and Wagoner. These calculations have many uses: they can check whether the observed run of density in the atmosphere agrees with the results of hydrodynamical calculations for the atmosphere of a blue supergiant, they can be used to improve the surface abundance estimates of elements, such as helium, which provide useful constraints on the stellar evolution, and best of all, they can provide distance estimates to supernovae in a way that is independent of all the treacherous steps on the extragalactic distance ladder.

The basic idea was developed in the context of SN II by Kirshner and Kwan [3], and summarized by Kirshner [4]: the angular size of the expanding photosphere of the supernova can be measured using the continuum flux and its temperature, while the corresponding velocity at the photosphere can be determined from the shapes of the absorption lines. What is new for SN 1987A is the superb quality of the observations, the superior knowledge of the initial conditions, and the application of much more sophisticated methods for solving the radiative transfer problem in the expanding medium. The results are very encouraging: for example Eastman and Kirshner[5] derive a distance to the LMC by this custom yardstick of 49+/-6 kiloparsecs, in excellent accord with recent results for RR Lyrae stars and for Cepheids. This stringent empirical test of the Expanding Photosphere Method provides the basis for hope that it may make a significant contribution to the extragalactic distance scale at cosmologically interesting distances which are beyond the reach of RR Lyrae stars or Cepheid variables.

Type II supernovae are demonstrated to have a wide range of spectra and luminosities, as vividly illustrated by the contributions of Wheeler, Filippenko, and Harkness to this workshop. One outstanding feature of the Expanding Photosphere Method is that it does not assume that all SN II are standard candles, but instead is a custom yardstick that makes a distance measurement to each individual object.

The Hubble constant can be measured well if it becomes routinely possible to make 10% or 20% distance measurements of SN II in galaxies at the distance of the Virgo Cluster and beyond. As described by Pennypacker and by Perlmutter in this volume, the Berkeley Supernova Search is aimed at the discovery and rapid reporting of supernova in this range. If spectra and photometry of the supernovae can be obtained by an instrument for which this is a high priority, then supernova contribute an important element to the question of the distance scale. At Harvard, we have discussed the construction of the HARVEST 4 meter telescope, which would have as one of its goals the prompt study of supernovae.

3. THE STANDARD CANDLE

Unlike the unruly zoo of SN II, the Type Ia supernovae appear to form a more nearly homogeneous class of events. The SN Ia recently became considerably more homogenous through the definition of the SN Ib subclass (Porter and Filippenko[6], Uomoto and Kirshner[6], Schlegel and Kirshner[7]). As described by Woosley, by Nomoto, and by Pinto in this Workshop, the conventional picture is that SN Ia result from the explosive burning of a Carbon-Oxygen white dwarf with a mass at the Chandrasekhar limit, while the SN Ib may result from massive stars that have suffered extensive mass loss. So, in their hearts, the SN Ib may be core collapse supernovae that lack a hydrogen envelope, more akin to the SN II. Leibundgut presented evidence at this workshop that a carefully selected sample of very well-observed SN Ia have a small internal scatter in luminosity of only 20%.

One way to calibrate the SN Ia standard candle is to observe some SN Ia in the same clusters of galaxies for which SN II have measured the distances. The Steward Observatory CCD Transit Instrument survey described in this workshop by McGraw will provide the data for supernovae in the Coma cluster, and may provide an important link between the detailed observations of nearby objects and the observations at the limit of technique. Even if the calibration of the SN Ia is incomplete, a small dispersion is the only requirement for their utility as standard candles.

If SN Ia have a small dispersion and are identical at large redshift, they can provide a powerful standard candle to illuminate the global effects of dark matter on the geometry of the universe through their redshift-magnitude diagram. The difference in apparent magnitude between $\Omega = 1$ and $\Omega = 0$ is about 0.13 magnitudes at a redshift of $z = 0.3$. Since this is only a little smaller than the scatter in intrinsic luminosity, in principle it would be possible to accumulate enough (redshift, magnitude) pairs by observing SN Ia to distinguish between these values of Ω, and determine this elusive cosmological parameter in a way that includes all the effects of dark matter, no matter what its composition.

Since the apparent magnitude for a SN Ia at a redshift of $z = 0.3$ is expected to be approximately $m = 22$, finding and studying these supernovae is a challenging technical task. A very promising beginning has been made by Hansen et al.[8] and by Norgaard-Nielsen et al.[9] as described by Couch in this workshop. Working at the Danish 1.5 meter telescope at the European Southern Observatory in Chile, this team has patrolled distant clusters for supernovae. In the recent case, a supernova which may be a SN Ia was detected at $m = 22.05$ in the cluster AC 118 at a redshift of $z = 0.31$. The evidence that the supernova is really a SN Ia comes from a spectrum obtained at the Anglo-Australian Telescope. While the spectrum of this painfully faint object is noisy, the features which are seen do resemble the spectrum of a SN Ia. The light curve for this object also resembles that of a SN Ia, after making suitable corrections for the time dilation, and including the K-corrections for the wavelength difference between the emitted frame and the observer's frame. While it would imprudent (or perhaps impudent) to derive a value of Ω from a single observation, these pioneering observations point out the possibility of making progress on the cosmological problem from diligent observation.

One sobering conclusion from the ESO/AAO survey so far is that the rate of supernova detections is lower than the most optimistic predictions, so beating down the uncertainty by observing many cases will not be an easy task. It is hard to see how this will be carried out without the substantial commitment of observing time to carry out the survey and to obtain the follow-up observations as needed on a telescope of 4 meters aperture. The HARVEST project under discussion at Harvard would be operated in a way that would make these observations possible. Supernovae are interesting events in their own right and should be studied even if they had nothing to contribute to determining H_o and Ω. Since they have the potential to provide accurate and independent estimates of these quantities that could I>H OGA >H I GW OA >I G AC> OHCCI C N A I W BACIC I Ω really is 1) or direct the energies of talented people elsewhere if it is not (for $\Omega < 0.3$), the study of supernovae deserves special attention. It is a fortunate circumstance that the brightest supernova in 383 years is available for our study and inspiration while these questions are on the table.

RPK's research on supernovae is supported by the NSF through grant AST-8516537 and by NASA through grants NAG5-645 and NAG5-841.

REFERENCES

1] Kirshner, R.P., Oemler, A., Schechter, P.L. and Shectman, S.A., Astron.J. **88**, 1285 (1983).

2] Arnett, D.W., Bahcall, J.N., Kirshner, R.P., and Woosley, S.E., Ann. Rev. Astron. Astrophys. **27**, 000 (1989)

3] Kirshner, R.P. and Kwan, J., Astrophys. J. **193**, 27.

4] Kirshner, R.P. in Supernovae as Distance Indicators, (N. Bartel, ed.) Lecture Notes in Physics #244, Springer-Verlag: Berlin (1985).

5] Porter, A.C. and Filippenko, A.V., Astron. J. **93**, 1372 (1987).

6] Uomoto, A.K. and Kirshner, R.P., Astron. Astrophys. **149**, L7 (1985).

7] Schlegel, E.M. and Kirshner, R.P., Astron. J., August (1989).

8] Hansen, L., Norgaard-Neilsen, H.U., Jorgensen, H.E., Ellis, R.S., and Couch, W.J., Astron. Astrophys. **211**, L9 (1989).

9] Norgaard-Nielsen, H.U., Hansen, L., Jorgensen, H.E., Salamanaca, A.A., Ellis, R.S., and Couch, W.J., Nature, June (1989).

X RAYS, GAMMA RAYS, and COSMIC RAYS

IN CHICAGO

COLD DARK MATTER ANNIHILATIONS: A SOURCE OF GAMMA RAYS AND ANTIPROTONS

Joseph Silk
Departments of Astronomy and Physics
Center for Particle Astrophysics
University of California
Berkeley, CA 94720

ABSTRACT

Massive particle candidates for dark matter provide a potentially observable signature by virtue of their annihilations in the galactic halo at a known but model-dependent rate. Possible signatures are described, including cosmic ray antiproton and gamma-ray production.

I. INTRODUCTION

Cold dark matter is the generic name for any non-baryonic particles that dominate the mass density of the universe and have negligible velocity dispersion during the matter-dominated epoch of the universe when structure formation occurs. Natural candidates for cold dark matter include massive weakly interacting particles such as the lightest supersymmetric particle (photino, higgisino, gravitino...), which is believed to be stable because of the new quantum number (R-parity) associated with supersymmetry.[1)] These exotic fermions differ by spin $\frac{1}{2}$ from their partner bosons that are either known (photon, graviton, W) or predicted (higgs) to exist. There are also known particles that may exist in a sufficiently massive flavor, most notably a neutrino with mass of order 1 GeV.

II. ANNIHILATION CROSS-SECTIONS

These massive particles are all non-relativistic when they thermally decouple from the universe, and this results in a generic property of cold dark matter. Particles that decouple at relativistic energies must be as numerous as photons or (standard) neutrinos, and thereby are light, providing a possible source of hot or warm dark matter. Massive non-relativistic decouplers are suppressed in number density relative to the relic photons by a Boltzmann factor, $\exp(-m_x c^2/kT_f)$, where the freeze-out temperature is defined by setting the annihilation rate equal to the Hubble expansion rate. The value of T_f is only logarithmically sensitive to the annihilation cross-section, and for typical weak-interaction strengths, one finds that $kT_f \approx m_x c^2/20$. The freeze-out condition consequently specifies the annihilation cross-section at freeze-out $\langle\sigma V\rangle^f_{ann}$ in terms of the relic density Ω_x (expressed in units of the critical density for closure, $3H_0^2/8\pi G$):

$$\langle\sigma V\rangle_{\mathrm{ann}}^{\mathrm{f}} = \frac{0.96\times10^{-26}\mathrm{cm^3s^{-1}}}{\Omega_x(h/0.5)^2}\left(\frac{1/20}{kT_f/m_xc^2}\right)\times$$

$$\times\left(\frac{N_F}{25.6}\right)^{\frac{1}{2}}\frac{a+bx_f}{a+\frac{b}{2}x_f}\left(\frac{T_x}{0.42T_\gamma}\right)^3 \qquad (1)$$

where $x_f = kT_f/m_xc^2$, $h = H_o/100\mathrm{km\ s^{-1}Mpc^{-1}}$, N_f is the number of relativistic species present at T_f, $(1+z)T_\gamma$ is the radiation temperature after freeze-out, $(1+z)T_x$ is the temperature prior to freeze-out, and the annihilation cross-section is approximated by $\langle\sigma V\rangle_{\mathrm{ann}} = a + bx$; $x = kT/m_xC^2$. The quantities a and b are specified for particular models: in general, note that equation (1) specifies $a + bx_f$, whereas the low-temperature annihilation rate, relevant for dark matter detection today, is $\langle\sigma V\rangle_{\mathrm{ann}} \approx a$.

For a massive Dirac neutrino, b=0, and

$$\langle\sigma V\rangle = A(m_f)m_{\nu_D}^2 \quad \text{with} \quad \Omega_{\nu_D}h^2 = (2\mathrm{GeV}/m_{\nu_D})^2. \qquad (2)$$

However for a majorana-mass dark matter candidate $b \neq 0$ and the low temperature cross-section is reduced below the freeze-out value. This reduction is considerable: for example, for a photino, one of the popular majorana-mass dark matter candidates, the reduction factor in the cross-section away from any threshold is proportional to $m_{\tilde\gamma}^{-2}$. This is because $a \sim \Sigma Q_f^4 m_f^2/m_{sf}^4$ and $b \sim m_{\tilde\gamma}^2 \Sigma Q_f^4/m_{sf}^4$. If $m_b < m_{\tilde\gamma} < m_t$ (where m_b is the bottom quark mass ~ 5GeV, and m_t the unknown but $\gtrsim 40$GeV top quark mass), one has in the low temperature limit that (with $m_{\tilde\gamma}$ in GeV)[2)]

$$\langle\sigma V\rangle_{\mathrm{ann}} \approx \frac{0.88\times10^{-26}}{1+0.025m_{\tilde\gamma}^2}\left(\frac{1/4}{\Omega_{\tilde\gamma}h^2}\right)\mathrm{cm^3\ s^{-1}}. \qquad (3)$$

This means that any annihilation flux from halo dark matter would be proportional to $n_{\tilde\gamma}^2\langle\sigma V\rangle$ and therefore, for $m_{\tilde\gamma} > 5$GeV, suppressed as $m_{\tilde\gamma}^{-4}$, since $n_{\tilde\gamma} \propto \rho_{\mathrm{halo}}\, m_{\tilde\gamma}^{-1}$, provided that the halo dark matter density ρ_{halo} is known.

There are two possibilities for boosting the cross-section. One can imagine that $\Omega_{\tilde\gamma} < 1$. In order for the halo to consist of photinos, one requires $\Omega_{\tilde\gamma} \gtrsim 0.02$. In fact, there is a minimum value of $\Omega_{\tilde\gamma}$ that arises from the requirement that $\Omega_{\tilde\gamma} \propto \langle\sigma V\rangle^{-1} \propto m_{sf}^4$: lower bounds on the selectron (and other chargino) masses from accelerator experiments limit generic neutralinos to $\Omega_x \gtrsim 0.04$. This still allows a possible cross-section boost by a factor ~ 25 at a specified $m_{\tilde\gamma}$. Moreover when $m_{\tilde\gamma}$ exceeds the top-quark mass, the cross-section is enhanced: a crude estimate[4)] yields

$$\langle\sigma V\rangle_{\mathrm{ann}} \approx \frac{0.88\times10^{-26}}{1+0.22(m_{\tilde\gamma}/m_t)^2}\frac{1/4}{\Omega_{\tilde\gamma}h^2}\mathrm{cm^3\ s^{-1}} \qquad (4)$$

III. HALO ANNIHILATIONS

The dark matter density in the galactic halo may be inferred by modelling the galactic rotation curve. To within an uncertainty factor of about 3, the halo density is given by

$$\rho_{\mathrm{halo}}(\mathrm{r}) = 0.3\left(\frac{\mathrm{a}^2+\mathrm{r}_\odot^2}{\mathrm{a}^2+\mathrm{r}^2}\right)\frac{\mathrm{a}}{\mathrm{b}}\mathrm{GeV\ cm}^{-3} \tag{5}$$

Here $\mathrm{r}_\odot = 8\mathrm{kpc}$ is the galactocentric distance, the halo core semi-major axis $\mathrm{a} = (3-10)\mathrm{kpc}$ and the semi-minor axis a satisfies $1/3 \lesssim \mathrm{b/a} \lesssim 1$. More generally, the halo may be triaxial, but this effort will be ignored here.

Annihilations in the halo lead to potentially observable fluxes of γ-rays, cosmic ray antiprotons, and cosmic ray positrons. The galactic background dominates any signal, and one has to carefully evaluate the characteristics of a possible halo feature.

a. *Antiprotons*

The expected antiproton signal from galactic cosmic ray interactions with interstellar matter is kinematically suppressed below several GeV. In particular, a $\bar{\mathrm{p}}$ signal below 1 GeV would require a reasonably exotic explanation. Photino annihilations provided a simple explanation[5)] for the original observation of low energy $\bar{\mathrm{p}}$'s with the choice $\Omega_{\tilde{\gamma}}\mathrm{h}^2 \approx 1/4$ although the annihilation model became rather more contrived with an attempt to simultaneously fit the higher energy $\bar{\mathrm{p}}$ data above 1 GeV. This is because the annihilation flux is proportional to $\mathrm{n}_{\tilde{\gamma}}^2\langle\sigma\mathrm{V}\rangle$, or form (3), $\mathrm{m}_{\tilde{\gamma}}^{-4}(\Omega_{\tilde{\gamma}}\mathrm{h}^2)^{-1}$ for $\mathrm{m}_{\tilde{\gamma}} > 5\mathrm{GeV}$. Hence since the annihilating particle mass must be $\sim 15-20\mathrm{GeV}$ to produce the 1–2GeV $\bar{\mathrm{p}}$ flux, one has to reduce $\Omega_{\tilde{\gamma}}\mathrm{h}^2$ by two orders of magnitude. The implied photino density[6)] is too low to account for dark halos ($\Omega_{\tilde{\gamma}} \sim 0.03$). Choice of an alternative SUSY relic, the higgsino[7)], allows a larger cross section in the required mass range, since the low temperature annihilation cross-section is approximately constant and one has the annihilation flux proportional to $\mathrm{m}_{\tilde{\mathrm{h}}}^{-2}$, with $\Omega_{\tilde{\mathrm{h}}}\mathrm{h}^2 = (0.05\mathrm{m}_{\tilde{\mathrm{h}}}/15\mathrm{GeV})^2$ for $\mathrm{m}_{\tilde{\mathrm{h}}} \gtrsim 15\mathrm{GeV}$. Two recent experiments[8),9)], however, demonstrate conclusively that there are no sub-GeV $\bar{\mathrm{p}}$'s, setting limits on $\bar{\mathrm{p}}/\mathrm{p}$ of 3.5×10^{-5} and 4.6×10^{-5} over 120 to 360 MeV and 205 to 640 MeV, respectively.

These energies are evaluated at the top of the atmosphere, and it is necessary to modulate the predicted interstellar $\bar{\mathrm{p}}$ fluxes to make a direct comparison. This negative result does not unduly constrain dark halo models because of the large uncertainty in any predicted $\bar{\mathrm{p}}$ flux due to the uncertainties in dark matter density and in halo confinement time.[26)] A further complication arises because adiabatic deceleration in dense clouds[10)] could account for a $\bar{\mathrm{p}}$ signal near the present sub-GeV upper limit, provided that 10 percent of cosmic rays were injected in this manner. Confirmation of the $\bar{\mathrm{p}}$'s observed over 1–15GeV would be of interest for the dark matter problem, however: a spallation origin requires high grammage, $\sim 100\text{gcm}^{-2}$ for a substantial fraction ($\gtrsim 10$ percent) of galactic cosmic rays, which implies a long halo confinement time, thereby boosting any potential annihilation signal at lower energy.

b) γ-rays

Dark matter annihilations generate hadronic cascades that produce γ-rays primarily through the production of neutral pions.[11),5)] About 10 percent of the decays are through rarer mesons[12)], including η's and B^*'s. The galactic background dominates the diffuse signal, and the annihilation flux is typically a ~ 1 percent modulation above 100 MeV. Several strategies are possible to enhance the signal-to-noise ratio.

i) Search for a particle with optimum cross-section for annihilation. Such particles have a low temperature cross-section approximately equal to the freeze-out value, that is to say, the temperature dependence in the annihilation cross-section is negligible, as is the case for a Dirac neutrino. Direct searches have eliminated $m_{\nu_D} > 12\text{GeV}$, and high energy neutrino searches for annihilations in the sun reduce the allowable window to $1.9(\Omega_{\nu_D}h^2)^{-\frac{1}{2}}\text{GeV} < m_{\nu_D} \lesssim 4\text{GeV}$. Below 4 GeV, particles evaporate from the sun, so that a unique opportunity exists for γ-ray astronomy to constrain (or detect!) a Dirac neutrino of precisely the mass (2–4 GeV) required for closure. The predicted signal[13)] in this case at 200 MeV amounts to about 35 percent ($m_{\nu_D} = 3.7\text{GeV}$) or 20 percent ($m_{\nu_D} = 2\text{GeV}$) of the mean high latitude diffuse galactic background.

ii) Use coarse spectral information. The predicted annihilation signal peaks at 70 MeV and has a characteristic exponential distribution of width $\sim 0.1 m_x$. This is distinct from the estimated galactic background[14)] which has a power-law spectrum, with differential photon flux approximately proportional to $E^{-2.7}$.

iii) Use angular resolution. Above 100 MeV, the galactic background is predominantly from π° decays induced by cosmic ray collisions with interstellar nuclei. There is also a contribution from cosmic ray electron bremsstrahlung. The inverse Compton component is inferred to be small, by successfully modelling the variation of galactic γ-ray flux with hydrogen column density. Analysis of COS-B data[15)] yields an empirical diffuse galactic γ-ray emissivity of $4 \times 10^{-27}\ \text{s}^{-1}\text{sr}^{-1}(\text{H}_{\text{atom}})^{-1}$ $(300 - 800\text{MeV})$ and $2 \times 10^{-27}\ \text{s}^{-1}\text{sr}^{-1}(\text{H}_{\text{atom}})^{-1}$ $(800\text{MeV} - 6\text{GeV})$, with a spectral dependence as $E^{-2.7}$. There are "holes" in the galactic HI distribution where the column density is $\sim 1 \times 10^{20}\text{cm}^{-2}$. The associated HII (and H_2) contribution is unknown, but should not exceed $\sim 0.5 \times 10^{20}\text{cm}^{-2}$. One infers a minimum background, accessible with subdegree resolution of $\sim 6 \times 10^{-7}\text{cm}^{-2}\text{s}^{-1}\text{sr}^{-1}$ over 300 MeV to 800 MeV. This would allow a very favorable signal-to-background ratio for Dirac neutrinos. Even for a

5 GeV photino, the predicted signal is 10 percent of galactic background near 1 GeV.

iv) Optimize the annihilation cross-section. The dark matter need not be of critical density. Suppose $\Omega_{\tilde{\gamma}}$ is the minimum needed to account for the astronomical constraints: large-scale dynamics (clusters and superclusters) requires $\Omega_{\tilde{\gamma}} \sim 0.1$, although one could conceivably tolerate an even smaller value $\Omega_{\tilde{\gamma}} \sim 0.05$, which would provide enough dark matter to account for observations of dark halos. This boosts the annihilation cross-section, and associated γ-ray fluxes, by up to an order-of-magnitude. Although there is a minimal Ω_χ for generic neutralinos[3)] of about 0.01 (and corresponding maximum $\langle\sigma V\rangle$) it does not seem possible for any astrophysically plausible candidate with density below $\Omega_\chi \sim 0.05$ to dominate the galactic halo.

IV. THE GALACTIC SPHEROID

Gamma ray emission is potentially observable from a centrally concentrated core of dark matter. Non-baryonic dark matter is concentrated into an approximately spherical distribution, according to at least some galaxy formation theories, identified with the galactic halo. The study of galactic rotation curves determines the dark matter distribution in this component of disk galaxies. However the galactic spheroid, defined by the distribution of population II stars, which is also nearly spherical, may also have a dark matter "problem." Mass models of our galaxy[16)] find that up to fifty percent of the inner spheroid cannot consist of a mix of stars similar to those in the solar neighborhood. This is an exceedingly weak argument for non-baryonic dark matter for at least two reasons: possible non-circular orbits in the inner disk complicate any simple modelling of the mass distribution and, indeed, conspire to yield overestimates of the mass density in naive models that assume circular motions, and the initial stellar mass function may well have been different when the inner spheroid formed. Nevertheless, one cannot rule out a significant dark matter component.

Theoretical models of galaxy formation in the context of cold dark matter suggest that, at some level, it is hard to avoid contamination of the spheroid with non-baryonic dark matter. There are two contributions: the initial dark matter content of the seed cloud around which the galaxy accreted, and dark matter entrained during the ensuing collapse. Studies of entrainment[17)] suggest that it may have been an important effect, necessary to explain the so called "disk-halo conspiracy" or the flat rotation curve within the disk-dominated inner galaxy. Ipser and Sikivie[18)] have calculated how much dark matter is required to reproduce the inner peak near ~ 300pc in the galactic rotation curve, which for a spherically symmetric mass distribution (and circular orbits) requires a central mass of $\sim 7 \times 10^9 M_\odot$. If the dark matter distribution is isothermal, these authors find that the core emission measure, defined as the enhancement within 150 pc of $\int \rho_x^2 dV$, is enhanced relative to its value in the absence of such a dark matter core by up to three orders of magnitude. One would expect an isothermal distribution to result as a consequence of the usual violent relaxation associated with core formation. An adiabatic model does, however, lead to a much smaller enhancement: this would have required a slow collapse of the core dark matter, over longer than a free-fall time of $\sim 10^8$yr.

The initial dark matter core parameters are uncertain, in the context of the cold dark matter theory. The rarity of the fluctuations from which our galaxy formed determines the epoch of collapse, and therefore the initial density of the core. This could have been as early as $z \sim 30$, thereby suggesting a virialized core density of $\sim 5 \times 10^{-23}$g cm^{-3} or $0.7 M_\odot pc^{-3}$. Coincidentally, this is precisely the range to which COS-B should have been sensitive for ~ 100pc core, if the dark matter consists of weakly interacting majorana-mass particles. Moreover, the observed spheroid core density is roughly $10 M_\odot pc^{-3}$, so the dark matter fraction need only be of order 10 percent to be observable, provided the particle mass is in the 1–10 GeV range.[19)] However uncertainties in galaxy formation modelling hitherto do not lead to my definitive prediction: high spatial resolution time-dependent collapse calculations of dark and baryonic matter are needed to address this question.

V. GALACTIC NUCLEI

Galaxy formation models in which there is sufficiently early collapse can lead to much higher density galactic cores. Such cores provide an environment that can have observable effects: for example, Salati and Silk[20)] note that stellar evolution could be drastically modified in such circumstances. A natural origin for dense dark matter cores arises in two classes of theories: isocurvature density fluctuation and cosmic string-plus-cold dark matter scenarios. In the former case, one normally assumes a baryon-dominated universe; however a non-baryonic component may be necessary to reconcile the model with the isotropy of the cosmic microwave background, as well as providing a dark matter source with minimal assumptions. The natural seed scale is the post-recombination Jeans mass $\sim 10^6 M_\odot$ at $z \sim 1000$, and dark matter cores should form in this mass range with internal density $\sim 10^5 M_\odot pc^{-3}$ and core ~ 2pc. Even with generic SUSY relic masses of ~ 100GeV, one would still expect an annihilation γ-ray luminosity of $\sim 10^{37}$erg s^{-1} for such objects, detectable as point-like γ-ray sources in nearby galactic nuclei.

The cosmic string-plus-cold dark matter scenario provides an even more dramatic signature of annihilating cold dark matter. One generically expects halos in this model to be clumpy, at least initially. Halos form via accretion around cosmic string loops. Accretion of cold dark matter becomes effective once the universe is matter-dominated at $z < 3 \times 10^4 \Omega h^2$ The loops are born (at epoch t) with a high velocity ($\sim$ c) and accrete an elongated wake of length ($\sim$ ct) but of width comparable to the loop radius, typically 0.01 ct. The loops oscillate, eventually decaying by gravitational radiation, and this varying potential produces large density fluctuations in the accreting material. The wakes are unstable to fragmentation: since the loops fill most of space, this results in a highly inhomogeneous distribution of cold dark matter. The details are model-dependent, depending on, for example, presently controversial numerical simulations to resolve the number of loops per galaxy. Moreover, long strings also drive wakes that are unstable to fragmentation in a cold dark matter-dominated universe, and the relative roles of loops and long strings will affect the mass distribution of the fragments. The galaxy formation process will most likely erase much of the substructure, but a considerable amount should remain in intergalactic space. The initial lumpiness is also critical in determining the maximum core density attained in dark matter collapse. While estimates of the detailed mass spectrum of cold dark matter clumps are in progress[21)], it seems clear that a substantial diffuse γ-ray background could result from annihilation in a nonhomogeneous intergalactic medium of cold dark matter. Rare, massive clumps might survive fragmentation: extreme parameters for the expected $\sim 10^7 M_\odot$ of cold dark matter entrained at $z = 3 \times 10^4$ around a loop destined to accrete a $\sim 3 \times 10^{11} M_\odot$ halo are core density $\sim 10^9 M_\odot pc^{-3}$, core size $\sim 1 - 3pc$, γ-ray luminosity $\sim 10^{10} m_{100}^{-1} L_\odot$, and annihilation time-scale $\sim 10^{10} m_{100} yr$ where $m_{100} = m_x/100GeV$.

VI. LINE γ-RAYS

Gamma rays from direct two-photon annihilations by halo dark matter particles with $v/c \sim 10^{-3}$, rather than via hadronic cascades and the ensuing π°-channel, produce almost monoenergetic lines, with $\Delta E/E \sim 10^{-3}$. The potential of this signature for exploring the nature of halo dark matter was originally proposed[22)] for $\lambda\lambda \to q\bar{q}\gamma$, but the branching ratio was found[23)] to drop approximately as $m_{\tilde{\gamma}}^{-2}$, falling below 10^{-4} for $m_{\tilde{\gamma}} > 5GeV$. This branching ratio estimate was confirmed by Bergstrom and Snellman[24)], who also proposed that $\lambda\lambda \to \gamma\gamma$ would have a much more favorable branching ratio. For $\tilde{\gamma}\tilde{\gamma} \to \gamma\gamma$, Rudaz[25)] finds a branching ratio $1.1 \times 10^{-3}(m_{\tilde{\gamma}}/10GeV)^2$. The 2-photon production cross-section therefore is independent of $m_{\tilde{\gamma}}$, only scaling as $(\Omega_{\tilde{\gamma}}h^2)^{-1}$; according to reference 27, $\langle\sigma V\rangle_{2\gamma} \approx 3 \times 10^{-31}(1/\Omega_{\tilde{\gamma}}h^2) cm^3\ s^{-1}$. However at fixed $\Omega_{\tilde{\gamma}}h^2$, the predicted flux of line photons decreases as $m_{\tilde{\gamma}}^{-2}$ or E_γ^{-2}. For comparison, the expected background due to cosmic ray interactions with interstellar nuclei falls off steeply (the differential spectrum declining as $E_\gamma^{-2.7}$). At some specified resolution, such as the $\sim 15\%$ for EGRET or $\sim 1\%$ eventually for ASTROGAM, the signal-to-noise ratio for a possible line signature declines as $E_\gamma^{-0.3}$, and allows a potentially detectable flux for $m_{\tilde{\gamma}} \lesssim 30GeV$.

Acknowledgements

I gratefully acknowledge discussions with A. Bouquet, K. Freese, S. Rudaz, P. Salati and A. Stebbins. This research has been supported in part by DOE grant DE-FG03-84ER40161 at the University of California, Berkeley.

References

1. J. Ellis, *et al.*, *Nucl. Phys. B.*, **238** (1984), 453.
2. M. Srednicki, M., K.A. Olive, and J. Silk, *Nucl. Phys. B.*, **274** (1987), 804.
3. K. Griest, *Phys. Rev. D.*, **38** (1988), 2357.
4. A. Bouquet, P. Salati and J. Silk, (1989), in preparation.
5. J. Silk and M. Srednicki, *Phys. Rev. Lett.*, **53** (1984), 624.
6. F.W. Stecker, S. Rudaz and T.F. Walsh, *Phys. Rev. Lett.* **55** (1985), 2622.
7. S. Rudaz and F.W. Stecker, *Astrophys. J.*, **325** (1988), 16.
8. S.P. Ahler *et al.*, *Phys. Rev. Lett*, **61** (1988), 145.
9. R.E. Streitmatter *et al.*, *Adv. Sp. Sci.* (1989), in press.
10. B.G. Mauger and S.A. Stephens, *Proc. 18th Int. Cosmic Ray Conf.*, Bangalore, **9** (1983), 171.
11. J.E. Gunn *et al.*, *Astrophys. J.*, **723** (1978), 1015; F.W,. Stecker, *Astrophys. J.*, **223** (1978), 1032.
12. F.W. Stecker and A.J. Tylka (1988), preprint LHFA/Th-88-46.
13. K. Freese and J. Silk, *Astrophys. J.*, (1988), submitted.
14. C. Dermer, *Astron. Astrophys.*, **157** (1986), 223.
15. H. Bloemen, *Ann. Revs. Astron. Astrophys.*, **27** (1989), in press.
16. J. Caldwell and J. Ostriker, *Astrophys. J.*, **251** (1981), 61; J. Bahcall, M. Schmidt and R. Soneira *Astrophy. J.*, **265** (1983), 730.
17. G. Blumenthal *et al.*, *Astrophys J. Lett.*, **301** (1986), 27; J. Gunn and B. Ryden, *Astrophys. J.*, **318** (1988), 15.
18. J. R. Ipser and P. Sikivie, *Phys. Rev. D*, **35** (1987), 3695.
19. J. Silk and H. Bloemen, *Astrophys. J. Lett*, **313** (1987), L47.
20. P. Salati and J. Silk, *Astrophys. J.* (1989), in press.
21. A. Stebbins and J. Silk (1989), in preparation.
22. M. Srednicki, S. Theisen and J. Silk, *Phys. Rev. Lett.*, **56** (1986), 236, 1883.
23. S. Rudaz, *Phys. Rev. Lett.*, **56** (1986), 2128.
24. L. Bergstrom and H. Snellman, *Phys. Rev. D.*, **37** (1988), 3737.
25. S. Rudaz, preprint UMM-TH-716/89 (1989).
26. J. Ellis *et al.*, *Phys. Lett. B*, (1988), submitted.
27. A. Bouquet, P. Salati and J. Silk, preprint LBL-26749 (1989).

GASEOUS HOT HALOS AND DARK MATTER: AN OBSERVATIONAL VIEWPOINT

G. Fabbiano

Harvard/Smithsonian Center for Astrophysics

ABSTRACT.

X-ray observations with the *Einstein* satellite have been used to measure the mass of early-type galaxies. These measurements are based on the assumption that the X-ray emission is thermal bremsstrahlung from hot gaseous halos, and that the gas traces the potential, while not contributing to it significantly. This approach has demonstrated the presence of a large dark halo in M87. When applied to 'normal' early-type galaxies, which are $\sim$ 100 times less X-ray luminous, this method cannot constrain usefully the mass to light ratios, given the uncertainties of the present X-ray data. Future X-ray observations will be essential for constraining these measurements.

X-ray observations with the *Einstein* satellite have been used to measure the mass of early-type galaxies. These measurements are based on the assumption that the X-ray emission is thermal bremsstrahlung from hot gaseous halos, and that the gas traces the potential, while not contributing to it significantly. Three approaches have been used: the first has been to measure the binding mass within a certain radius, assuming hydrostatic equilibrium and spherical symmetry. This method was first applied to M87, and has more recently been used to estimate the mass of less X-ray luminous galaxies. The second approach has made use of more detailed modeling to match theoretical predictions to the observed X-ray surface brightness profiles. This method has not given unequivocal results so far, and I will not discuss it here. Finally, Fabian et al (1986) devised a method to estimate lower limits to the *total* mass of a galaxy. In the following I will discuss the results of these analyses, with an eye to the assumptions made and to the limitations of the current data. This paper is based closely on a review that I have recently written for *Ann. Rev. Astr. Ap.* (Fabbiano 1989, in press). A more complete discussion can be found there.

The discovery of the extended thermal X-ray source in the Virgo cluster centered on M87 prompted the first attempts at estimating the binding mass of an early-type galaxy within radii well away from the optical core, to which optical measurements have been restricted. The first estimates, based on the assumption of the gas being in hydrostatic equilibrium in the galaxy potential at a temperature T$\sim$ $3 \times 10^7 K$, suggested very large binding masses, between 10^{13} and $\geq 10^{14} M_{\odot}$ (Bahcall & Sarazin 1977; Mathews 1978). However, Binney & Cowie (1981) showed that the X-ray data then available could also be fitted with a model of pressure confined cooling atmosphere surrounding a low mass galaxy. The observations of M87 with the *Einstein* IPC, producing an image and spatially resolved spectral information, have led to a more accurate measurement of the binding mass, which confirmed the earlier suggestions of the presence of a massive dark halo (Fabricant et al 1980; Fabricant & Gorenstein 1983). In particular, Fabricant & Gorenstein (1983) could exclude a radial dependence of the temperature of the gaseous halo consistent with Binney and Cowie's model, and found that the integral mass to light ratio (M/L) of M87 in solar units must increase from 5-15 at a radius of 1' ($\sim$ 4.4 kpc; from the optical data of Sargent et al 1978), to over 180 at 20' ($\sim$ 87 kpc), implying the presence of a dark massive halo. They derived a total mass M$\sim 3 - 6 \times 10^{13} M_{\odot}$, within 60' ($\sim$ 260 kpc) from the core.

The equation used to estimate the binding mass of M87 by Fabricant et al (1980) was derived from the equation of hydrostatic equilibrium, combined with the ideal gas law, in the assumption of spherical symmetry, and is given below:

$$M(< r) = -(r_{gas} k T_{gas} / G \mu m_H)(dlog\ \rho_{gas} / dlog\ r + dlog\ T_{gas} / dlog\ r) \qquad (1)$$

Equation (1) shows that the measurement of the binding mass depends on four variables: the radius within which the mass is estimated, the gas temperature at this radius, and the logarithmic gradients of both temperature and gas density at the same radius. Each of these variables is a potential source of uncertainty. While the observations of M87 yielded enough signal to noise to allow a relatively well constrained mass estimate, the same unfortunately is not true when this method is applied to early-type galaxies not at the center of clusters. These galaxies typically are $\sim$100 times less luminous than and at least as distant as M87, and therefore their radial profiles and especially temperature information are not as well constrained. Forman et al (1985) pointed out that the assumption of hydrostatic equilibrium for these galaxies is reasonable even in the presence of cooling flows, because these would be largely subsonic. They derived M/L ratios ranging from $\sim$10 to 90 for a sample of 13 galaxies, assuming that the X-ray emission is due to an isothermal gas at T=1.2 $\times 10^7 K$, adopting as r_{gas} the outermost radius at which X-ray emission could be detected, and using an average fitted value for the density gradient. Taken at face value, this result would suggest that dark extended halos are a common feature of early-type galaxies. A supporting argument given by these authors for the widespread presence of massive dark halos is the suppression of galactic winds. However, as discussed above in the case of M87, radiative wind suppression in the inner regions and an external pressure from a surrounding intracluster or intergalactic medium could also be responsible for the retention of the hot interstellar medium. The temperature in this case should increase at the outer radii. Temperature profiles are needed to exclude this possibility (Sarazin & White 1987; Vedder et al 1988). The present data does not provide tight enough constraints (see Fabbiano 1989).

Subsequent work (Trinchieri et al 1986; see also Fabbiano 1985, 1986a and b; Canizares 1987) has shown that the uncertainties on this type of mass measurements can be large. In particular, detailed analysis of the *Einstein* data of 6 E and S0 galaxies, which are among those detected with higher signal to noise in the sample of Forman et al (1985), shows both departures from spherical symmetry, that can be ascribed either to interlopers in the field or to the interaction with the surrounding medium, and peculiar variations of the gas density gradient at the outermost radii. Even ignoring some of these effects, the uncertainties on the gas temperature and on its gradient are such that mass estimates differing by up to a factor of 10 are possible for a single galaxy (fig. 1). Within these uncertainties, the X-ray measurements are generally consistent with the mass estimates from optical velocity dispersion. Therefore, large dark halos are allowed by these data, but not required by this analysis. The use of less luminous galaxies or of Sa galaxies will introduce an additional uncertainty, which cannot be resolved with the present data. This is the possibility that most of the X-ray emission is due to the evolved stellar component and not to a hot gaseous medium (Fabbiano & Trinchieri 1985; Trinchieri & Fabbiano 1985; Canizares et al 1987; Fabbiano et al 1989), thus invalidating the whole approach.

Instead of relying on a measure of the binding mass within a certain observed radius, Fabian et al (1986) devised a method forestimating a lower limit to the *total* binding mass. Their method is based upon three assumptions: (1) that the gas within the observed radius is confined by a hydrostatic outer atmosphere; (2) that the halo is convectively stable; and (3) that the pressure gradient is always negative. With these assumptions they obtain the equation:

$$M(r_\infty) \geq 5r_okT_o[1-(P_\infty/P_o)^{2/5}]/2G\mu m_H(1-r_o/r_\infty) \qquad (2)$$

where r_o is the outermost detected radius of the halo, and P_o and T_o are the gas pressure and temperature at this radius. From this equation, and using the outermost radii (r_o) and best fit values of temperatures from the literature, they derive mass to light ratios for a sample of early-type galaxies, which lead them to conclude that there is overwhelming evidence for large massive halos in these galaxies. However, the same uncertainties apply here as in the calculation of the binding mass. Using the five galaxies for which the binding mass had been measured by Trinchieri et al (1986), and temperature estimates that take into account the uncertainties of the spectral fits, together with the values of r_o derived in that paper, I have calculated lower limits to the mass with equation (2). They are shown in fig. 1. For two of the galaxies (NGC 4649 and NGC 4472) these limits do not require particularly large M/L ratios, in excess of those inferred from optical measurements. For the other three galaxies (NGC 4636, NGC 1395 and NGC 720), however, $M/L > 20$ are clearly required. Therefore, there is evidence for large dark halos in some normal elliptical galaxies at least, unless the three basic assumptions of Fabian et al do not apply. This might be the case for galaxies experiencing considerable external forces, as for example M86, which could be ram pressure confined (Fabian et al 1980; see Fabian et al 1986). Also, as remarked by the latter, if there is a significant confining pressure due to intracluster or intragroup gas, the lower mass limit may not apply to the galaxy, but to the group as a whole.

Considering these results, I think that, purely on *observational* grounds, there is still insufficient evidence to prove that dark massive halos are a general feature of early-

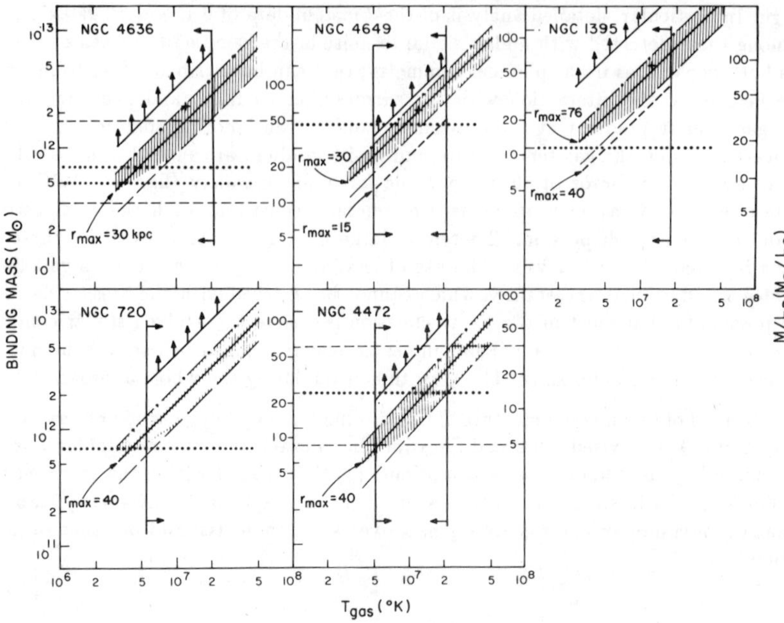

Figure 1: Mass estimates and related uncertainties for 5 early-type galaxies

(adapted from Trinchieri et al 1986). The shaded areas are the allowed regions from eq. 1, as in the above paper; the central solid line is the estimate for an isothermal halo, and the dashed and dot-dashed lines are the estimates for halos with moderate radial dependences of the temperature , $T \sim r^{0.5}$ and $T \sim r^{-0.5}$, respectively. The vertical lines restrict the temperature ranges to the values obtained by the spectral fits of the X-ray data (at the 90% confidence). The diagonal dashed lines are isothermal estimates of the binding mass for more conservative choices of the halo outermost radii, which exclude regions where the radial surface brightness profiles depart from a power law (see Trinchieri et al 1986). The horizontal dashed and dotted lines are mass to light ratio deteminations from optical velocity dispersions (see Trinchieri et al 1986, and references therein). The diagonal lines with upwards pointing arrows are the lower limits to the total binding mass (and mass to light ratios) calculated following Fabian et al (1986, eq. 2), for the less conservative estimates of the outer radii (larger r_{max}); using the smaller r_{max} values these estimates will be displaced downward.

type galaxies, although there is evidence of their existence in cluster dominant galaxies and perhaps in some of the most X-ray luminous normal ellipticals. The use of X-ray data to measure the mass of these galaxies is very promising, but future more sensitive X-ray observations are needed to get a more definite answer.

ACKNOWLEDGEMENTS

This work was supported by NASA contract NAS8-30751.

REFERENCES

Bahcall, J. N., Sarazin, C. L. 1977, *Ap.J.(Letters)* 213:L99.

Binney, J., Cowie, L. L. 1981, *Ap.J.* 247:464.

Canizares, C. R. 1987, *Proc. of IAU Symp. No. 117, Dark Matter in the Universe*, ed. J. Kormendy, G. R. Knapp, 165. Dordrecht: Reidel.

Canizares, C. R., Fabbiano, G., Trinchieri, G. 1987, *Ap.J.* 312:503.

Fabbiano, G. 1985, *Proc. ESA Workshop: Cosmic X-ray Spectroscopy Mission* 33. ESA SP-239.

Fabbiano, G. 1986a, *Pu.A.S.P.*, 98:525.

Fabbiano, G. 1986b, *Gaseous Halos of Galaxies*, ed. J. N. Bregman, F. J. Lockman, 203. NRAO/AUI.

Fabbiano, G. 1989, *Ann. Rev. Astr. Ap.*, in press.

Fabbiano, G., Gioia, I. M., Trinchieri, G. 1989, *Ap.J.*, submitted.

Fabbiano, G., Trinchieri, G. 1985, *Ap.J.* 296:430.

Fabian, A. C., Schwarz, J., Forman W. 1980, *MNRAS* 192:135.

Fabian, A. C., Thomas, P. A., Fall, S. M., White, R. E. III 1986, *MNRAS* 221:1049.

Fabricant, D., Gorenstein, P. 1983, *Ap.J.* 267:535.

Fabricant, D., Lecar, M., Gorenstein, P. 1980, *Ap.J.* 241:552.

Forman, W., Jones, C., Tucker, W. 1985, *Ap.J.* 293:102.

Mathews, W. G. 1978, *Ap.J.* 219:413.

Sarazin, C. L., White, R. E. III 1987, *Ap.J.* 320:32.

Sargent, W. L. W., Young, P. J., Boksenberg, A., Shortridge, K., Lynds, C. R., Hartwick, F. D. A. 1978, *Ap.J.* 221:731.

Trinchieri, G., Fabbiano, G. 1985, *Ap.J.* 296:447.

Trinchieri, G., Fabbiano, G., Canizares, C. R. 1986, *Ap.J.* 310:637.

Vedder, P. W., Trester, J. J., Canizares, C. R. 1988, *Ap.J.* 332:725.

THE COSMIC X-RAY BACKGROUND

Elihu Boldt
Laboratory for High Energy Astrophysics
NASA / Goddard Space Flight Center
Greenbelt, MD 20771

1. ABSTRACT

Basic measured characteristics of the cosmic X-ray background are reviewed, including recent results from autocorrelation studies of surface brightness fluctuations examined with HEAO-1 and the Einstein Observatory (HEAO-2). Prospects for addressing some key outstanding issues with future experiments are discussed as regards possible weak large-scale anisotropies (e.g., dipole) and a spectral test to help discriminate among candidate origins (i.e., diffuse radiation versus discrete sources).

2. OBSERVATIONS

The X-ray sky (at ≥3keV) is dominated by a remarkably isotropic extragalactic CXB (cosmic X-ray background) exhibiting an optically thin thermal-type spectrum corresponding to $T \approx 5 \times 10^8$K (kT=40keV), much as the sky in the microwave band is dominated by an isotropic thermal CMB (cosmic microwave background), albeit characteristic of a 3°K blackbody. Discrete X-ray sources resolvable with the imaging telescope of the Einstein Observatory (0.2-3.5keV) account for about 20% of the CXB at 3keV[1)]; they number about 0.01 per square arc minute. The upper limit to arc-minute scale fluctuations in the apparent surface brightness of the unresolved background observed with this telescope implies that most of this residual CXB is either diffuse or arises from a discrete source population having a number density on the sky exceeding one per square arc-minute[2)]. An upper limit to the autocorrelation function for surface brightness fluctuations on scales ≥5 arc-minutes has also been derived from Einstein Observatory data[3)]; it sets an upper bound of 10 h^{-1} Mpc on the correlation length for the possible dominant sources of the residual CXB at large redshifts (i.e., $4 \geq z > 1$).

The broadband (2-50keV) all-sky study of the background carried out with the proportional counters of the HEAO-1 A2 experiment resolved only the brightest foreground sources[1)]; in toto these resolved sources account for about one-percent of the CXB (at 3keV). Fluctuations in the surface brightness of the CXB on scales ≥3degrees observed with the A2 experiment are consistent with random variations in the expected number of unresolved foreground sources[4)]; the corresponding rms fluctuation in the apparent surface brightness of the CXB is ≈6% for 9 square degree pixels. The A2 derived upper limit to the autocorrelation function for CXB surface brightness fluctuations[5)] on scales ≥3 degrees sets an upper bound on the correlation length for Abell clusters of galaxies consistent with the correlation found in the optical[7)]. For the unresolved AGN (active galactic nuclei) within the present epoch (at $z \leq 0.1$) contributing to the foreground fluctuations in surface brightness of the CXB the A2 limit implies an upper bound on the correlation length[5,6)] ≈ 20 h^{-1} Mpc [where $h \equiv H_o / (100 \text{ km s}^{-1} \text{ Mpc}^{-1})$].

3. ISSUES

If the proper frame of the CXB were at rest relative to the proper frame of the CMB, then our own velocity (v) relative to this universal system (i.e., Compton-Getting effect) would induce a CXB dipole anisotropy in the direction of the CMB apex having an amplitude $(\alpha+3)v/c$ that varies from $\approx$0.4% at $\leq$10keV (where the effective CXB energy spectral index $\alpha\approx0.4$) to $\approx$0.6% at $\geq$100keV (where the effective $\alpha\approx1.6$, obtained from HEAO-1 A4 CXB data[8]). However, fluctuations in CXB surface brightness observed with the HEAO-1 A2 experiment amount to $\geq$0.3% rms for bands on the sky which are $\leq$ one steradian in solid angle. Within the uncertainty imposed by such fluctuations the CXB dipole anisotropy is consistent with that implied by the CMB[4]. If we could resolve out foreground sources at a level corresponding to $\geq$10 per square degree (i.e., comparable to the point source sensitivity level of the Einstein Observatory or better) over the whole sky then large-scale surface brightness fluctuations in the residual CXB might well be suppressed by at least an order of magnitude, thereby permitting a precise determination of the CXB Compton-Getting dipole anisotropy. Possible structure of the CXB on intermediate angular scales[9] could, however, constitute a fundamental complication that would still have to be addressed. At $\geq$100keV the best available all-sky data are from the HEAO-1 A4 scintillation counter experiment. The limitation in measuring weak large-scale CXB anisotropies with this experiment arises from variable radioactivity induced in the CsI/NaI scintillators by the radiation environment of the HEAO-1 orbit, in spite of the precautions employed[8]. Future high-energy studies of the CXB will have to avoid or further minimize this complication.

The spectrum of the CXB over the limited interval 3-10keV may be characterized by an energy index $\alpha\approx0.4$ which is significantly smaller than the indices that describe the steeper power-law spectra measured for AGN over the same band with HEAO-1 and EXOSAT[10]. If the *residual* CXB does indeed arise from faint discrete X-ray objects, then we should begin to observe point sources having spectra characterized by $\alpha<0.4$ over the 3-10keV band accessible to forthcoming broadband high-resolution imaging telescopes such as AXAF. On the other hand, if we fail to find such a population of point sources, then we can use "fast" moderate-resolution X-ray optics (e.g., such as employed for BBXRT) for arc-minute mapping of the weak surface brightness variations that could be characteristic of a diffuse CXB. The study of such variations might provide us with a direct indication of the large-scale distribution of all matter (non-baryonic[11,12] as well as baryonic) out to at least $z\approx8$ before scattering of X-rays by electrons (bound as well as free) begins to wash out CXB anisotropies[1].

4. REFERENCES

1. Boldt, E., "The Cosmic X-ray Background" Phys.Rept. 146, 215 (1987)
2. Hamilton, T., and Helfand, D., Ap. J. 318, 93 (1987)
3. Barcons, X., and Fabian, A., "The small-scale autocorrelation function of the X-ray background", preprint (1988)
4. Shafer, R., University of Maryland PhD thesis, NASA TM 85029 (1983)
5. Persic, M., De Zotti, G., Boldt, E., Marshall, F., Danese, L., Franceschini, A., and Palumbo, G., Ap. J. 336, L47 (1989)
6. De Zotti, G., Persic, M., Franceschini, A., Danese, L., Palumbo,G., Boldt, E., and Marshall. F., "X-ray Constraints on AGN Clustering", in IAU Symposium No. 134, in press (1988)
7. Bahcall, N., and Soneira, R., Ap. J. 270, 20 (1983)
8. Gruber, D., personal communication (1988)
9. Jahoda, K. and Mushotzky, R., personal communication (1988)
10. Turner, T. J., Leicester University PhD thesis (1988)
11. Daly, R., Ap. J. 322, 20 (1987).
12. Persic, M., and Salucci, P., NASA-GSFC/LHEA Report 87-014 (1987)

COMPACT SOURCE ORIGIN OF COSMIC RAY ANTIPROTONS

Charles D. Dermer
Physics Department, Lawrence Livermore National Laboratory
P.O. Box 808, L-297, Livermore, CA 94550

The flux of cosmic ray antiprotons with kinetic energies between ~1 and 15 GeV is ~5 times greater than the flux predicted on the basis of the leaky-box model.[1] This excess is attributed to secondary antineutron production in compact sources.[2,3] Because the antineutrons are not confined by the magnetic field of the compact source, they leave the interaction site, decay in interstellar space and account for the apparent excess cosmic ray antiproton flux. The escape and decay of neutrons produced in association with the antineutrons is a source of cosmic ray protons. Observations of the angular variation of the intensity and spectral shape of 100 MeV γ-rays produced by neutron-decay protons in the reaction $p+p \rightarrow \pi^o \rightarrow 2\gamma$ could reveal compact-source cosmic ray production sites. COS-B observations of spectral hardening near point sources,[4] and future high-resolution observations of galactic point sources by Gamma-1 and the Egret telescope onboard the *Gamma Ray Observatory* may provide supporting evidence for this model.

Measurements of the ratio of the cosmic ray antiproton (CR p-) flux to the cosmic ray proton flux are shown in Fig. 1. Also plotted is the predicted flux of secondary CR p- expected from the leaky-box model of galactic cosmic ray propagation.[5] This model, developed to explain the observed cosmic ray isotopic compositions, attributes all CR p- as secondary products of collisions between cosmic rays and diffuse galactic gas and dust. The deficiency in the integrated flux between 1 and 15 GeV, coupled with the lack of observations of CR antinuclei, is interpreted here as evidence for an additional CR secondary production source. This interpretation has the virtue that the chemical composition of the CR is unaffected by the existence of these sources, since the magnetic field at the interaction site inhibits the escape of charged nuclei but permits neutral species to freely escape.

We estimate the discrete source neutron production rate by normalizing to the CR p- excess, giving the fit through the data shown in Fig. 1. The normalization is carried out by noting that secondary antineutrons with E<12 GeV are produced primarily by energetic protons with $E \lesssim 250$ GeV in the reaction $p+p \rightarrow p+p+n+\bar{n}$. Neutron production proceeds predominantly through pion-associated reactions such as $p+p \rightarrow p+n+\pi^+$ and related multi-prong reactions. We approximate the inclusive differential cross section for neutron production in proton collisions by the expression $d\sigma_{p+p\rightarrow nX}/dE_n = 25 \text{ mb } \delta(E_n - E_p/2)$, where E_p and E_n refer to proton and neutron kinetic energy, respectively. For the neutron production rate per discrete source, we obtain

$$\frac{dN_n}{dE_n dt} = \begin{cases} 1.3\times10^{41}\ g\,(2E_n + m_p)^{-2.2}, & s=2.2 \\ 2.8\times10^{41}\ g\,(2E_n + m_p)^{-2.75}, & s=2.75 \end{cases}$$

where $m_p = 0.94$ GeV, and the spectral index of the assumed nonthermal ion flux in the compact sources is denoted by s. The term $g=n_{0.2}V_{67}f_5/N_{100}$, where the diffuse galactic disk hydrogen density $n_H=0.2n_{0.2}\text{cm}^{-3}$, the galactic disk volume $V=10^{67}V_{67}$ cm^3, the integrated CR p- excess is $5f_5$ times greater than the leaky box model prediction, and the number of discrete sources in the galaxy is given by $100N_{100}$.

Neutrons follow a straight-line escape from the compact sources unless they collide with intervening matter (within $\sim10^{-4}$ psc of the source for a 10 GeV neutron). After decaying, the neutron-decay protons diffuse in the galactic magnetic field. Except for extremely strong sources, the flux of neutron β-decay electrons and knock-on electrons from neutron-decay protons is small compared to the flux of ambient galactic CR electrons. This is also usually the case for secondary electrons and positrons resulting from pion production. Except possibly for the 2.2 MeV line formed by neutron capture on protons in the atmosphere of a binary stellar compaion,[3] the most pronounced signature of a compact source origin of antiprotons would be neutral pion production and decay from collisions of neutron-decay protons with galactic matter.

The intensity of the proton flux near a compact source can be obtained by solving the transport equation[6,7] for cosmic ray propagation and energy loss. Neglecting the energy loss term, and assuming a steady point source radiating at the origin of the radial (ρ) and z coordinates, the differential density of neutron-decay protons in a geometry with absorbing boundaries at $z=\pm a$ is given by

$$n(\rho,z,E_p) = \frac{3(dN_n/dE_n dt)}{4\pi c\, l_\perp^{1/2}\, l_\parallel} \sum_{n=-\infty}^{n=\infty} (r_n - t_n) \;,$$

where $l_\perp$ and $l_\parallel$ are the mean free paths of the neutron-decay protons perpendicular and parallel to the galactic plane, and

$$r_n^{-2} = \frac{\rho^2}{l_\parallel} + \frac{(z-4na)^2}{l_\perp} \quad ; \quad t_n^{-2} = \frac{\rho^2}{l_\parallel} + \frac{(z+4na-2a)^2}{l_\perp} \quad .$$

Fig. 2 shows model results for neutron-decay proton fluxes in the vicinty of a compact source, assuming s=2.2, a=300 psc, $l_\perp$=0.06 psc, and $l_\parallel = l_\perp/2$, $l_\perp$, and $2l_\perp$. These values pertain to protons with 2 GeV kinetic energies; a dependence of $l \propto E_p^{0.5}$ at higher energies is expected from energy-dependent diffusion. As can be seen, an enhanced CR proton flux is found in the vicinity of discrete sources, which would reveal itself through an enhancement in the neutral pion decay gamma radiation. Detailed information about the propagation characteristics of the galaxy and the time dependence and spectra of the injected nonthermal neutrons depend on high sensitivity imaging observations near point sources at photon energies ε~100 MeV. For example, angular resolution << 5° is required to map the presumed CR proton halo around Cyg X-3, if the halo is ~150 psc in radial extent. Both Gamma-1 and Egret onboard GRO have ~1.5° single-photon direction determination at ε=100 MeV, and are capable of mapping variations in the diffuse galactic gamma-ray emission on scales of a fraction of a degree.[8] Investigation of CR halos near magnetized, accreting neutron stars, such as the X-ray sources Cyg X-3, Her X-1, or Vela X-1, should be a high priority of these missions.

This work was performed under the auspices of the U.S. Department of Energy by the Lawrence Livermore National Laboratory under contract W-7405-ENG-48.

1]Stephens, S. A. and Golden, R. L., *Space Science Reviews*, **46**, 31 (1987).
2]Dermer, C. D. and Ramaty, R., *Nature*, **319**, 205 (1986).
3]Dermer, C. D. and Ramaty, R., in *Accretion Processes in Astrophysics*, eds. J. Audouze and J. Tran Thanh Van (Editions Frontieres), 85 (1986).
4]Ormes, J. F., Özel, M. E., and Morris, D. J., *Astrophysical Journal*, **334**, 722 (1988).
5]Tan, L. C. and Ng, L. K., *Journal of Physics G*, **9**, 227 (1983).
6]Syrovatskii, S. I., *Soviet Astronomy*, **3**, 22 (1959).
7]Ramaty, R. and Lingenfelter, R. E., in *Isotopic Composition of the Primary Cosmic Radiation*, ed. P.M. Dauber (Lyngby: Danish Space Research Institute), 203 (1971).
8]Thomson, D. J., *Nuclear Instruments and Methods in Physics Research*, **A251**, 390 (1986).
9]Streitmatter, R. E., et al., *Advances in Space Science*, in press (1989).
10]Ahlen, S. P., et al., *Physical Review Letters*, **61**, 145, (1988).
11]Bogomolov, E. A., et al., *Proc. 20th Int. Cosmic Ray Conference* (Moscow), **2**, 72 (1987).
12]Golden, R. L., Mauger, B.G., Nunn, S., and Horan, S., *Astrophysics Letters*, **24**, 75 (1984).

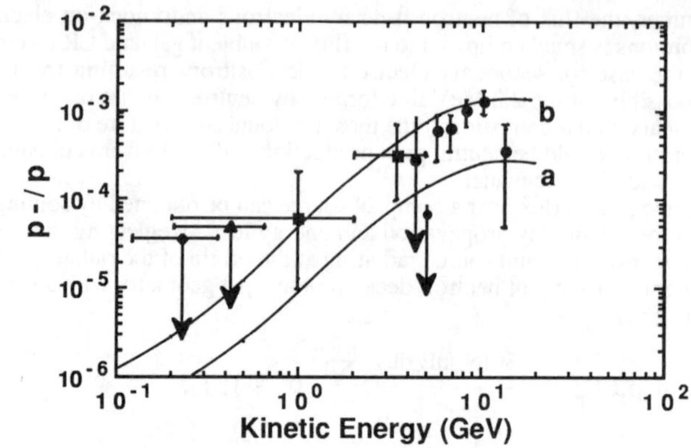

Fig. 1. Ratio of the fluxes of antiprotons to protons in the cosmic radiation. Data are from Ref. 9 (diamond), Ref. 10 (triangle), Ref. 11 (squares), and Ref. 12 (circles). Curve a is the leaky box model prediction of Ref. 5, and curve b, used to normalize the compact source neutron production rate, is 5 times the magnitude of curve a.

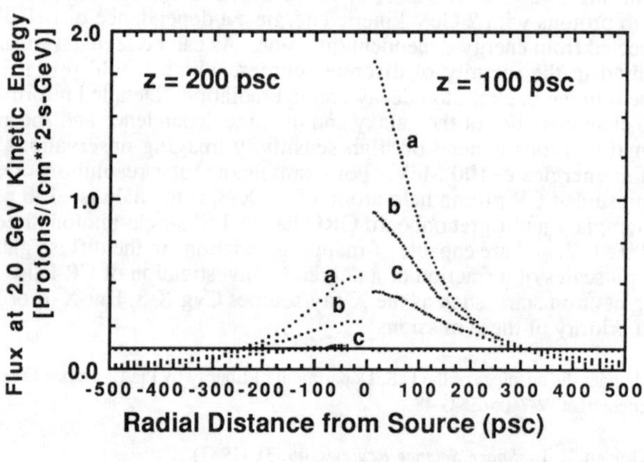

Fig. 2. Radial dependence of the flux of 2 GeV protons at 100 and 200 psc above the z=0 galactic plane, for a diffusive propagation model with absorbing boundaries at z=±300 psc. The source of nonthermal energetic neutrons, located midway between the boundaries, is assumed to emit neutrons continuously. The perpendicular diffusion mean free path is 0.06 psc, and the ratios of the parallel and perpendicular mean free paths are given by 0.5, 1.0, and 2.0 in curves a, b, and c, respectively. Solid line shows the local value of the cosmic ray proton flux at 2 GeV.

REPROCESSING OF GAMMA-RAYS AT COSMOLOGICAL REDSHIFTS

ANDRZEJ A. ZDZIARSKI

Space Telescope Science Institute, 3700 San Martin Dr., Baltimore, MD 21211, USA

AND

ROLAND SVENSSON

NORDITA, Blegdamsvej 17, 2100 Copenhagen Ø, Denmark

1. ABSTRACT

This abstract summarizes our current work on absorption and reprocessing of γ-rays at cosmological redshifts [1]–[4]. We study Compton scattering and pair production by γ-rays on the cosmic baryonic matter, and photon-photon scattering and photon-photon pair production by γ-rays and Compton scattering of relativistic pairs on the cosmic blackbody background. We point out the cosmological importance of photon-photon scattering (a process not previously considered in astrophysics). We derive analytical formulae for the optical depths of the universe to the processes listed above. We study in detail γ-ray reprocessing due to photon-photon pair cascades and photon-photon scattering. We determine the regions of dominance of the elementary processes on the photon-energy–redshift plane. The problem of cosmological γ-ray reprocessing is relevant, e.g., for observational γ-ray astronomy, for which the signatures of the cosmological γ-ray origin need to be determined, and for studies of the effects that pair cascades, caused by the decay of unstable particles from a hot Big Bang, have on the primordial nucleosynthesis.

2. REFERENCES

1] Svensson, R., and Zdziarski, A. A., in preparation (1989).
2] Zdziarski, A. A., *Ap. J.*, **335**, 786 (1988).
3] Zdziarski, A. A., and Svensson, R., *Ap. J.*, submitted (1989).
4] Zdziarski, A. A., and Svensson, R., *Nuclear Physics B*, in press (1989).

Cosmic Gamma-Ray Bursts from Galactic Neutron Stars

K. Hurley

University of California, Space Sciences Laboratory, Berkeley, CA 94720, and Centre d'Etude Spatiale des Rayonnements, Toulouse, France.

and

D. Hartmann

Board of Studies in Astronomy and Astrophysics University of California, Santa Cruz, CA 95064.

ABSTRACT

We review some of the essential observational aspects of γ-ray bursts. The arguments in favor of a neutron star origin and in favor of an extragalactic origin are summarized, and new statistical tests for the origin of bursters are discussed. We consider the implications of a galactic neutron star origin, particularly the information which bursts can give us on neutron star properties.

I. INTRODUCTION

Cosmic γ-ray bursts (hereafter GRB's) are short transient events, typically lasting from several tens of milliseconds to a minute.[22] GRB "light curves" display structure on a number of timescales. A typical rise time for an individual peak might be around 0.5 s. The time history of one exceptional burst (March 5, 1979) displayed an initial spike with a rise time of less than 0.2 ms, followed by a much fainter pulsation phase in which an 8 s periodicity was clearly observed.

Many theories of GRB's involve galactic neutron stars. This is partly because neutron stars are known to power some steady X-ray and γ-ray sources, and partly because the energetics and the number of sources required fit reasonably well with this idea. In addition, a 500 ms timescale is consistent with the free-fall time from the magnetopause of a magnetized neutron star[24], suggesting that perhaps GRB's are an accretion phenomenon, like some of the steady X-ray sources. Additional evidence for the neutron star hypothesis is found in the ≤ 0.2 ms rise time of the March 5, 1979 event, which may imply a source size ≤ 60 km, consistent with a neutron star. It is also in the range of dynamical time scales near the surface of neutron stars. An 8 s periodicity is consistent with those of X-ray pulsars, which are now well demonstrated to involve neutron stars.

Of great significance to the neutron star idea are two types of features observed in the energy spectra of some bursts. The first is single or double absorption lines in the $\sim$ 20-70 keV region,[7,27,33] and the second is an emission line around 400 keV.[27,43] The former might be due to the cyclotron resonance in a magnetic field of strength $\sim 2\text{-}5\times10^{12}$ gauss. The prevalent interpretation of the second feature is gravitationally redshifted electron-positron pair annihilation originating near the surface of slowly rotating neutron stars.[25,27] However, because of large uncertainties associated with the (unknown) emission process the observed spectral features, if real, cannot be used to prove a neutron star origin of GRB's or to constrain the nuclear matter equation of state.[23]

The spatial distribution of GRB's is apparently isotropic,[1,8,20] a fact which leaves the distance scale undetermined: it is possible that we are sampling a disk distribution, but that our instruments are only sensitive enough to record the nearby, apparently isotropic sources. Typical burst energies in that case are less than 10^{39} ergs. But it is also possible that we are sampling a truly isotropic extragalactic population, implying that the events are indeed extremely energetic with energies perhaps exceeding 10^{52} erg.

II. GAMMA-RAY BURST COUNTERPARTS

Several deep optical studies have been carried out for some of the smaller GRB error boxes, but no GRB source has yet been identified unambiguously with a source of quiescent radiation in any wavelength band.[14] The exception may be the association of the March 5, 1979 event with the supernova remnant N49 located in the LMC. Deep exposures of several GRB error boxes with instruments aboard the EINSTEIN satellite resulted in a marginal detection of one X-ray point source in the field of GB781119[9,36] but subsequent observations with the EXOSAT satellite did not confirm this detection.[2,3] The non-detection of a substantial X-ray flux places severe constraints on the neutron star surface temperature or alternatively the mass accretion rate.[2,3,9,19,36] In the context of the thermonuclear model[10,45–47] accretion rates less than about 10^{-15} $M_\odot$ y^{-1} imply source distances in excess of $\sim$ 500 pc.

GRB counterparts may also be detected during the outburst phase if they radiate a significant amount of energy at longer wavelengths. The intriguing discovery of three historical optical flashes coincident with modern GRB error boxes on a subset of the archival photographs at Harvard[39,40] suggests that optical emission may indeed accompany GRB's. Deep CCD surveys of the 1928-OT/1978-GRB field[35,41] revealed no quiescent counterpart brighter than about 23rd magnitude. Three optical flashes were observed from an identical position near, but not inside, the error box of GB790325b.[18] No optical counterpart brighter than $m_v \sim$ 24 was detected on deep CCD images of that position.[17,38] Although there is at present no compelling observational evidence that GRB's are indeed accompanied by optical flashes theoretical arguments lead us to believe that there should be emission at long wavelengths on some level. London and Cominsky[26] considered reprocessing of a small fraction of the hard radiation by a companion main sequence star, a degenerate dwarf, or an accretion disk[5] and concluded that the particular event 1928-OT/1978-GRB could not be explained within the *binary reprocessing scenario.* The problem stems from the fact that one has to explain both the bright optical display during outburst and the extremely low emission during quiescent times, as inferred from the deep CCD surveys. Rappaport and Joss[37] reconsidered this scenario and found it to be viable for source distances less than about 100 pc. The new twist in their work was the consideration of nearby binary systems with H-rich secondaries of mass less than about 0.06 $M_\odot$. Detailed calculations of the reprocessing of hard photons in the atmospheres of such low-mass stars[4,30] find only marginal agreement between the observed and calculated ratios of γ-ray fluence to optical fluence at the earth for source distances as small as 25 pc. Better prospects exist if the reprocessing occurs in an accretion disk around the neutron star.[5,31,32]

A different scenario involves comet impacts in white dwarf/neutron star binary systems.[44] Alternatively, optical flashes might be generated far above the surface of strongly magnetized neutron stars undergoing either a thermonuclear explosion of plasma accreted over long periods of time or equivalently a sudden impact of a large amount of plasma.[11,47] In this model the mechanism for the optical emission is cyclotron radiation in a region far above the surface of the neutron star. The radiation is self-absorbed for about 100 harmonics and the electrons are radiatively excited by Compton collisions with photons of the GRB. The strong surface fields required in this model are indicated by the observed low-energy features in GRB spectra.[7,27,33]

III. THE GREAT DEBATE OF THE 80's

Are GRB's galactic or extragalactic? The leading theories involving galactic neutron stars are thermonuclear explosions, comet or asteroid collisions with neutron stars, starquakes, and disk accretion instabilities. The inability of the observations to constrain the distance scale has led in recent years to a consideration of what might be termed "exotic" (i.e., extragalactic and cosmological) models. The current exotic theories involve superconducting cosmic strings and gravitational lensing.

The arguments in favor of neutron stars were given in the introduction. The arguments in favor of "exotic" theories may be summarized as follows. 1) "Hard" experimental evidence for neutron star origin (i.e., line features) exist in only a small fraction of the cases; thus any explanation for the remaining events may be acceptable, 2) The angular distribution of GRB's allows practically any distance scale, and 3) There has been no identification of a GRB source counterpart in any other energy band which would contradict an exotic origin.

Two statistical tools for determining the spatial distribution of GRB's are their frequency-brightness distribution and their angular distribution on the celestial sphere. Although the lack of precise burst localizations poses severe problems for our attempts to identify quiescent GRB counterparts, the body of localizations as a whole contains valuable information. Hartmann and Epstein[13] proposed a Monte Carlo method to obtain an objective estimate of the degree of isotropy of the burst distribution. The method takes the finite sizes of the burst error boxes into account. To quantitatively test if the distribution of GRB locations is consistent with isotropy one calculates the dipole and quadrupole moments of the GRB distribution on the celestial sphere and compares the moments obtained for the catalog under investigation with the statistical distribution of these moments for an ensemble of isotropic catalogs. Hartmann and Epstein[13] analyzed the localizations of 56 events observed by experiments aboard satellites of the interplanetary network[1] using an ensemble of 1000 Monte Carlo realizations of that catalog and found that particular set of data consistent with isotropy. In particular, the single positions in this catalog showed a remarkable degree of isotropy, only 1.5 % of all Monte Carlo simulations had a smaller dipole moment than the observed distribution. There is no evidence for any quadrupole moment with statistically significant amplitude.

Recently an independent quantitative evaluation of the uniformity of the radial distribution of GRB's means of a V/V_{max} test was suggested by Schmidt *et al.*[42] The basic idea of this test is to evaluate for each object in the sample its radial location within the volume available to it as determined by the sample detection limit. Schmidt *et al.* find an average V/V_{max} consistent with a uniform distribution of burst sources (i.e., a value near 0.5). Although experimental selection biases affect the V/V_{max} test[15] the data appear to be consistent with homogeneity.

The results obtained from the above methods and the cumulative brightnesss distribution[6,21,34] imply that GRB's are either very nearby (e.g., distances of order of a few disk scale heights) or at cosmological distances (i.e., well beyond the local supercluster). However, isotropy and homogeneity do not exclude possible clustering of the sources. Indeed it is the angular correlation of GRB sources that could reveal their association with the large scale structure of the universe. Hartmann and Blumenthal[12] have recently calculated the angular autocovariance function of the single positions in the catalogs of the interplanetary network[1] and the Konus experiment.[8,28,29] Both catalogs are consistent with zero correlation on angular scales between 1 and 100 degrees. For a spatial correlation resembling that of galaxies or clusters of galaxies their upper limit on the correlation magnitude suggests that GRB's are observed to a minimum distance of order 100 Mpc. However, the derived limits on the angular correlation are also useful to test local GRB models involving neutron stars. For example, if the sources are drawn from an infinite disk with an exponential scaling in one direction, the observed limits on the autocovariance function imply that present catalogs sample to less than about two disk scale heights.

IV TESTING THE PARADIGM

Let us assume that GRB's originate on galactic neutron stars. To utilize the statistical constraints discussed above a detailed dynamical simulation of the spatial distribution of galactic neutron stars has to be carried out.[16] We have calculated $\sim$ 3000 neutron star orbits assuming a constant birthrate. Without any specific assumption about the GRB mechanism or occurence rates we are interested in the question whether or not the kinematic characteristics of the neutron star population on the whole are consistent with GRB observations. Going outwards from the solar position the spatial distribution of the neutron stars can be determined as a function of sampling depth. Out to distances of order of 8 kpc we obtain a power law for the neutron star density $\rho \propto D^{-\beta}$ with exponent $\beta \sim 0.4$. If this shallow distribution is representative of the GRB population, then the expected deviations from homogeneity can be determined analytically: $V/V_{max} = (3-\beta)/(6-\beta) \sim 0.465$ (the corresponding Log N–Log S index is $-1.5 + \beta/2 \sim -1.3$). These results are not very sensitive to the assumed neutron star birth properties. If all presently available observations are utilized the predicted deviations from homogeneity could perhaps be tested. Larger deviations, expected if we sample the sources out to more than $\sim$ 8 kpc, would be quite surprising because we also find that for sampling depths exceeding more than $\sim$

5 kpc the galactic distribution becomes extremely anisotropic and correlated, in conflict with the observed angular distribution.

K. H. acknowledges support under NASA grant NGL 05-003-497. The work of D. H. was supported in part by the National Science Foundation (AST 88-13649) and by the Cal Space Institute under grant CS 32-88.

REFERENCES

1. Atteia J.-L. *et al.* 1987, *Astrophys. J. Suppl.*, **64**, 305.
2. Boer, M., *et al.* 1986, *Adv. Space Res.*, **6**, (4), 65.
3. Boer, M., *et al.* 1988, *Astr. Ap.*, **202**, 117.
4. Cominsky, L. R., London, R. A., and Klein, R. I. 1987, *Ap. J.*, **315**, 162.
5. Epstein, R. I. 1985, *Ap. J.*, **291**, 822.
6. Epstein, R. I., and Hurley, K 1988, *Astrophys. Letters*, in press
7. Fenimore, E. E., *et al.* 1988, *Ap. J.(Letters)*, **335**, L71.
8. Golenetskii, S. V. 1988, *Adv. Space Res.*, **8**, (2)653.
9. Grindlay, J. E., *et al.*, 1982, *Nature*, **300**, 730.
10. Hameury, J. M., Lasota, J. P., Bonazzola, S., and Heyvaerts, J. 1985, *Ap. J.*, **293**, 56.
11. Hartmann, D., Woosley, S. E., and Arons, J. 1987, *Ap. J.*, **332**, 777.
12. Hartmann, D., and Blumenthal, G. R. 1988, *Ap. J.*, in press
13. Hartmann, D., and Epstein, R. I. 1989, *Ap. J.*, submitted.
14. Hartmann, D., and Woosley, S. E. 1988, in *Multiwavelength Astrophysics*, ed. F. Córdova, Cambridge Univ. Press., pg. 189.
15. Hartmann, D., Gonzalez, J. J., and Hurley, K. 1988, in preparation.
16. Hartmann, D., Epstein, R. I., and Woosley, S. E 1989, *Nucl. Phys. B*, in press
17. Hartmann, D., Pogge, R. W., Hurley, K., Vrba, F., and Jennings, M. 1989, *Ap. J.*, in press.
18. Hudec, R., *et al.* 1988, *Adv. Space Res.*, **8**, (2-3), 665.
19. Hurley, K. 1987, in *Origin and evolution of neutron stars*, Proceedings of the 125th Symposium of the IAU, Nanjing, China, May 26-30, 1986, pg. 489.
20. Hurley, K. 1988, Erice School on *Cosmic γ-rays and cosmic neutrinos*, in press
21. Jennings, M. C. 1988, *Ap. J.*, **333**, 700.
22. Klebesadel, R. W., Strong, I. B., and Olson, R. A. 1973, *Ap. J. (Letters)*, **182**, L85.
23. Kluzniak, W. 1989, *Ap. J.*, **336**, 367.
24. Lamb, D. Q., *et al.* 1973, *Nature Phys. Sci.*, **246**, 52.
25. Liang, E. P., 1986, *Ap. J.*, **304**, 682.
26. London, R. A., and Cominsky, L. R. 1982, *Bull. Am. Astr. Soc.*, **14**, 867.
27. Mazets, E. P., *et al.* 1981, *Nature*, **290**, 378.
28. Mazets, E. P., *et al.* 1981, *Astrophys. Space Sci.*, **80**, 3.
29. Mazets, E. P. 1988, *Adv. Space Res.*, **8**, (2)669.
30. Melia, F., Rappaport, S., and Joss, P. C. 1986, *Ap. J. (Letters)*, **305**, L51.
31. Melia, F. 1988, *Adv. Space Res.*, **8**, (2), 641.
32. Melia, F. 1988, *Ap. J. (Letters)*, **324**, L21.
33. Murakami, T., *et al.* 1988, *Nature*, **335**, 234.
34. Paczyński, B., and Long, K. 1988, *Ap. J.*, **333**, 694.
35. Pedersen, H., *et al.* 1983, *Ap. J. (Letters)*, **270**, L43.
36. Pizzichini, G., *et al.* 1986, *Ap. J.*, **301**, 641.
37. Rappaport, S. A., and Joss, P. C. 1985, *Nature*, **314**, 242.
38. Ricker, G. *et al.* 1989, *Ap. J.*, in press.
39. Schaefer, B. E. 1981, *Nature*, **294**, 722.
40. Schaefer, B. E., *et al.* 1984a, *Ap. J. (Letters)*, **286**, L1.
41. Schaefer, B. E., Seitzer, P., and Bradt, H. V. 1983, *Ap. J. (Letters)*, **270**, L49.
42. Schmidt, M., Higdon, J. C., and Hueter, G. 1988, *Ap. J. (Letters)*, **329**, L85.
43. Teegarden, B., and Cline, T. 1980, *Ap. J. (Letters)*, **236**, L67.
44. Tremaine, S., and Zytkow, A. 1986, *Ap. J.*, **301**, 155.
45. Woosley, S. E., and Wallace, R. 1982, *Ap. J.*, **258**, 716.
46. Woosley, S. E. 1982, MPE Report **177**, p. 189.
47. Woosley, S. E. 1984, AIP Conf. Proc. **115**, ed. S. E. Woosley. pg. 485.

HALO ANTIPROTONS AND GAMMA RAYS FROM COLD DARK MATTER ANNIHILATION

J. Ellis[1], R.A. Flores[2], K. Freese[3], S. Ritz[4], D. Seckel[5], and J. Silk[6]

[1]CERN – Geneva
[2]School of Physics and Astronomy, University of Minnesota, Minneapolis, MN 55455
[3]Dept. of Physics, MIT, Cambridge, MA 02139
[4]Dept. of Physics, Columbia University, New York, NY 10027
[5]Bartol Research Foundation, University of Delaware, Newark, DE 19716
[6]University of California, Berkeley, CA 94720

Reports in 1981 (Buffington *et al* [1]) of a high flux of low-energy cosmic ray antiprotons not easily explained as secondary $\bar{p}$ led to the speculation (Silk and Srednicki 1984[2]; Stecker, Rudaz, and Walsh 1985[3]; Rudaz and Stecker 1988[4]) that these observations might be due to annihilation of dark matter particles in the galactic halo, $\chi\bar{\chi} \to \bar{p} + X$. New measurements (Ahlen *et al* 1988[5]) failed to reproduce this high flux and placed new upper limits on the flux which are more than an order of magnitude below the previous claims (see the line in Figure 1). We find, however, that these upper limits do not rule out any supersymmetric particles as candidates for halo dark matter (Ellis *et al* 1988)[6].

The flux observable at the Earth of $\bar{p}$'s produced by dark matter annihilation in the halo is

$$\phi(E) = \frac{1}{4\pi}\,\langle\sigma V\rangle_A\; V_{\bar{p}}\;\frac{dN}{dE_{\bar{p}}}\;\langle n_\chi^2\rangle\;\tau_{\bar{p}}$$

where $\langle\sigma V\rangle_A$ is the velocity-averaged annihilation cross-section, $V_{\bar{p}}$ is the galactic $\bar{p}$ velocity, $\frac{dN}{dE}\Big|_{\bar{p}} = \frac{1}{\sigma_{\chi\bar{\chi}}}\frac{d\sigma}{dE_{\bar{p}}}$ is the differential cross-section for inclusive $\bar{p}$ production, suitably modulated by the solar wind, $\tau_{\bar{p}}$ is the $\bar{p}$ confinement time, and n_χ is the number density of dark matter particles in the halo.

The hadronization process in

$$\chi\bar{\chi} \to q\bar{q} \to \begin{pmatrix} \pi^0 \\ \text{or} \\ \eta \\ \text{or} \\ \vdots \end{pmatrix} \to \bar{p} + \cdots$$

cannot be calculated analytically; to obtain an estimate of the differential cross-section one can either use e^+e^- date directly or a Monte Carlo adapted to fit e^+e^- data. The problem with taking $\frac{dN}{dE_{\bar{p}}}$ from e^+e^- experiments directly is that the e^+e^- branching ratios to various quarks are different from the $\chi\bar{\chi}$ branching ratios. With a Monte Carlo one can compute and sum the quark contributions with branching ratios appropriate to the dark matter annihilation (although the extrapolation away from the regime of the e^+e^- data depends on modeling). Using a Lund Monte Carlo (Sjöstrand 1988)[7],[8] to obtain $\frac{dN}{dE_p}$, for a halo density $\rho_\chi = M_\chi n_\chi = 0.3$ GeV cm^{-3}, and $\tau_{\bar{p}} = 2 \times 10^{15}$ sec, we find the $\bar{p}/p$ flux ratio illustrated in Figure 1 for photino mass $M_{\tilde{\gamma}} =$ 3, 6, and 10 GeV and higgsino mass $M_{\tilde{h}} = 15$ GeV. Present upper limits on cosmic-ray $\bar{p}$ do not rule out any supersymmetric particles as candidates for halo dark matter.

Freese and Silk (1988)[9] have studied gamma-ray production from cold dark matter annihilation in the halo, and find that signals are potentially detectable.

REFERENCES

[1] Buffington, A., Schindler, S.M., and Pennypacker, C.R. 1981, Ap. J. 248, 1179 (1981).
[2] Silk, J. and Srednicki, M., Phys. Rev. Lett. 53, 624 (1984).
[3] Stecker, F.W., Rudaz, S., and Walsh, T., Phys. Rev. Lett. 55, 2622 (1985).
[4] Rudaz, S. and Stecker, F.W. Ap. J. 325, 16 (1988).
[5] Ahlen, S.P. *et al* Phys. Rev. Lett. 61, 145 (1988).
[6] Ellis, J, Flores, R.A., Freese, K., Ritz, S., Seckel, D., and Silk, J. CERN preprint CERN-TH.5062/88 (1988).
[7] Sjöstrand, T., University of Lund preprint LUTP 85-10 (1985).
[8] Sjöstrand, T., Int. J. Mod. Phys. A3, 751 and references therein (1988).
[9] Freese, K.,and Silk, J., 1988 preprint.

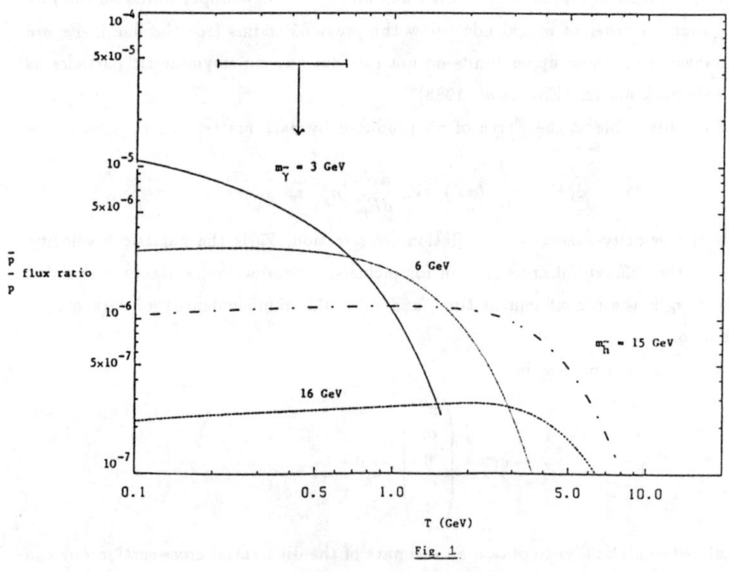

Fig. 1

Gamma Ray Fingerprints of Supersymmetric Dark Matter

Lars Bergström
Department of Physics, University of Stockholm
Vanadisvägen 9, S–113 46 Stockholm, Sweden

Here I wish to review some recently proposed mechanisms [1,2] that may produce observable high energy gamma rays from neutralino dark matter annihilation. The lightest supersymmetric particle (neutralino) λ is in most models a mixture of the supersymmetric partners of the photon, the neutral Higgs particles and of the Z^0. It is expected to be stable and has generically weak couplings to ordinary matter and annihilation cross sections of weak interaction strength. It is therefore an attractive candidate for the dark matter that seems to be needed in the Universe. If neutralinos constitute a sizeable fraction of our galactic halo, there should in principle exist many possibilities to detect their presence. The most direct proof of their existence would perhaps be the direct detection of their scattering in terrestial detectors. However, another very attractive possibility would be the detection of very sharp lines or other structures in the cosmic gamma ray spectrum coming from annihilations in the halo. The diffuse spectrum of gamma rays coming from the fragmentation of the main annihilation products (giving, e.g., π^0 mesons) is in general too low in rate and energy and tends to drown in the galactic and extragalactic gamma ray background [3] (unless this poorly known background has been considerably overestimated). It then seems that the only possibility to get an observable signal is to detect photons from direct (prompt) processes. The important point is that since the neutralinos are expected to have typical galactic velocities ($v/c \sim 10^{-3}$), the annihilations take place essentially at rest and therefore processes like $\lambda\lambda \to V\gamma$ (V: vector meson like J/Ψ, Υ etc) or $\lambda\lambda \to \gamma\gamma$ have the nice feature of producing nearly monoenergetic lines. Likewise, $\lambda\lambda \to b\bar{b}$ with b or $\bar{b} \to s\gamma$ would give a step in the γ distribution. The process $\lambda\lambda \to V\gamma$, suggested in [4] and elaborated in [5,1], unfortunately drops rapidly with increasing m_λ due to form factor effects. On the contrary, $\lambda\lambda \to \gamma\gamma$, suggested in [1,2] has the remarkable feature of having a branching ratio that is either constant or, generically, even growing with m_λ. As with all calculations in supersymmetric models there are many unknown parameters entering. However, we may make the following general observations. Using standard values for halo parameters etc [3], the integrated flux from $\lambda\lambda \to \gamma\gamma$ is appoximately, in the direction of the galactic pole

$$\Phi_\gamma = \frac{0.7 \cdot 10^{-4}}{m^2\, s\, sr} \times \left(\frac{1\, GeV^2}{m_\lambda^2}\right)\left(\frac{\sigma v_{rel}}{1.2 \cdot 10^{-26}\, cm^3 s^{-1}}\right)\left(\frac{B.R.(\lambda\lambda \to \gamma\gamma)}{10^{-3}}\right), \tag{1}$$

where typical values for σv_{rel} and the branching ratio for $\lambda\lambda \to \gamma\gamma$ are indicated. It must be realized, though, that the uncertainty of these values and of the astrophysical parameters related to the halo imply an uncertainty of a few orders of magnitude of the flux estimate of Eq. 1. Using this estimate anyway as a reasonable guess, we get for a λ mass of 5 GeV a flux of around $3 \cdot 10^{-6}\, m^{-2}\, s^{-1}\, sr^{-1}$. It is given in the form of a spike of relative width 10^{-3} at $E_\gamma = m_\lambda$. Therefore, in order to detect the signal a large area detector with very good energy resolution is needed. With 1 $m^2 sr$ acceptance and 1 % energy resolution, as anticipated in recent proposals for the Space Station [6], around 100 events would be collected with a signal to background ratio close to 1:1 for this example. It is worthwhile to note that the total event rate is roughly independent of m_λ up to several tens of GeV, whereas signal to background improves with increasing m_λ. An interesting but more speculative scenario is provided if there exist a fourth generation of quarks and leptons [2]. The loop-induced process $\lambda\lambda \to \gamma\gamma$ can then be even larger (in particular, there is a subprocess $\lambda\lambda \rightarrow H_3^0 \rightarrow \gamma\gamma$, with H_3^0 a neutral pseudoscalar Higgs particle, that may be very significant). Moreover, if $m_{b'} < m_t$, then $b' \rightarrow b\gamma$ could be very important [7]. In Fig. 1 we examplify the structures possible in this case for $m_{b'} = 35$ GeV and $m_\lambda = 40$ and 70 GeV. Also shown is the very small continuum part coming from the fragmentation of the main annihilation products (which in this case is b quarks and gluons). As can be seen, the edge from $b' \rightarrow b\gamma$ and the spike from $\lambda\lambda \rightarrow \gamma\gamma$ (assumed to have a branching ratio of $5 \cdot 10^{-4}$) should be clearly visible with an appropriate detector.

References

[1] L. Bergström and H. Snellman, Phys. Rev. D37 (1988) 3737.

[2] L. Bergström: Possible Structure in Cosmic Gamma Rays from Dark Matter Particle Annihilation, USITP-88-12 (1988).

[3] J. Ellis, R.A. Flores, K. Freese, S. Ritz, D. Seckel and J. Silk, Phys. Lett. 214B (1988) 403; F.W. Stecker, Phys. Lett. 201B (1988) 529.

[4] M. Srednicki, S. Theisen and J. Silk, Phys. Rev. Lett. 56 (1986) 263; 56 (1986) 1883 (E).

[5] S. Rudaz, Phys. Rev. Lett. 56 (1986) 2128.

[6] ASTROGAM proposal, J.H. Adams et al. (1988), unpublished; SUSY-SKY proposal, P. Carlson et al. (1988), unpublished.

[7] W.-S. Hou and R.G. Stuart, MPI-PAE/PTh 53/88 (1988), unpublished.

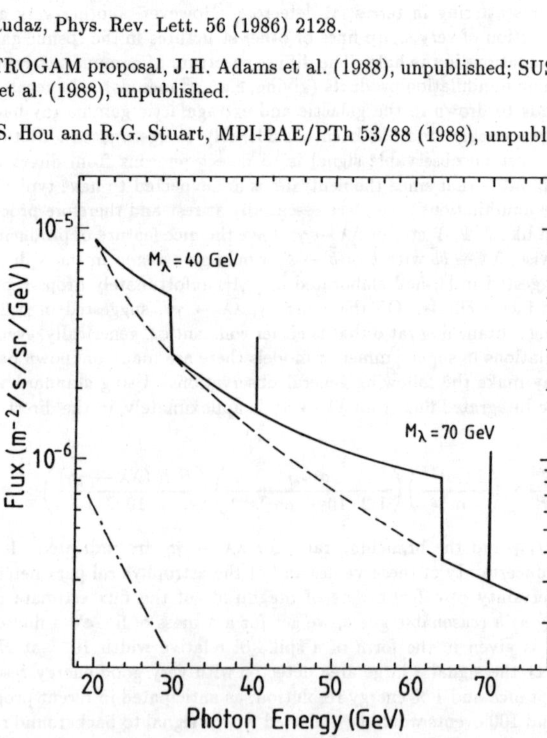

Fig. 1. Possible steps and lines coming from neutralino annihilations. The dashed line represents the estimated isotropic cosmic background and the dash-dotted line is the diffuse gamma radiation produced in the annihilations.

Status of the Chicago Air Shower Array

Leslie J Rosenberg
The University of Chicago
The Enrico Fermi Institute

representing the
CASA Collaboration*

ABSTRACT

The Chicago Air Shower Array (CASA) is a large-area surface scintillator array designed to study PeV sources. The complete detector will have more than 1000 detection stations, distributed on a square grid 15m apart. The first 49 of these stations are commissioned, and together with the University of Michigan underground muon detectors, constitute a powerful instrument for the study of point sources. We report our current status, and progress on construction of the full array.

The CASA detector, when fully instrumented, will consist of at least 1064 stations, distributed on a 15m grid, shown in Figure 1 by regular dots. The CASA detector surrounds the University of Utah Fly's Eye detector, shown in Figure 1 as the central cluster of dots. Buried beneath CASA is the University of Michigan muon array, shown in Figure 1 as the solid rectangles.

Four pieces of 1/2" acrylic scintillator comprise the detection elements of each station. Fast timing signals are exchanged between adjacent stations. We reconstruct the shower front directions from these station-to-station relative times, and thus eliminate the cabling and logic required with a common start. Each station contains its own analog, analog to digital, data buffer, microcomputer, and Ethernet communication circuitry. Each station is covered by $1X_0$ of lead, which lowers the effective shower energy threshold of the array.

The array trigger is digital in nature. It requires a minimum number of stations detecting signals within a time window. When a station detects a sufficient number of its four scintillators are struck (the exact condition is programmable–we nominally require a four of four coincidence), a current pulse is injected into a bussed coaxial cable. A trigger circuit near the center of the array sinks the currents and replies with a voltage level on another coaxial cable, bussed back to the array, when the current sum exceeds a certain threshold (this array trigger threshold is programmable– we nominally require that three stations are injecting current simultaneously). This trigger causes struck stations to digitize and save their data, which consists of relative times and pulse heights. Every ten seconds, the entire array is polled for data over a coaxial Ethernet link to a VAX located at the center of the array. These three coaxial cables are the only signals running to a central collection area.

As of the end of 1988, there were 49 stations operational and taking data. These stations are located over a muon detection bank. We are in the process of analyzing data from this initial installation. For example, Figure 2 shows a reconstructed timing front from a shower that struck all 49 of the stations. The vertical coordinate is the absolute arrival time of the shower front in units of nanoseconds (there is an arbitrary offset).The other two coordinates are the location of the boxes in the array.

We are now in the process of expanding the number of stations in the array. We plan to commission 500 more stations in 1989, and another 500 stations in 1990.

⋆ The members of the CASA Collaboration are J.W.Cronin, K.G.Gibbs, H.A.Krimm, N.C.Mascarenhas, T.A.McKay, D.Müller, B.N.Newport, R.A.Ong, L.J Rosenberg and M.E.Wiedenbeck.

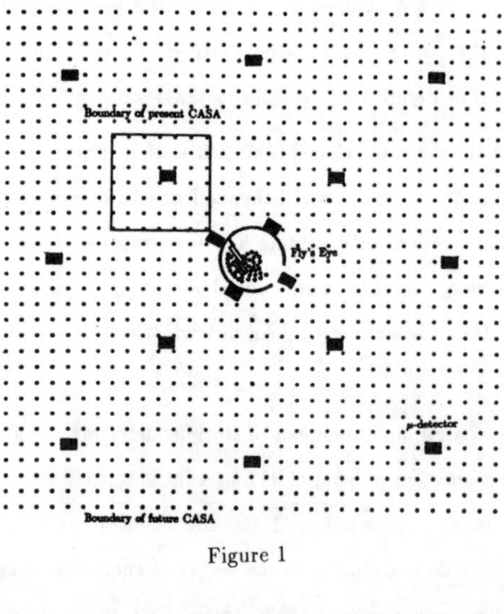

Figure 1

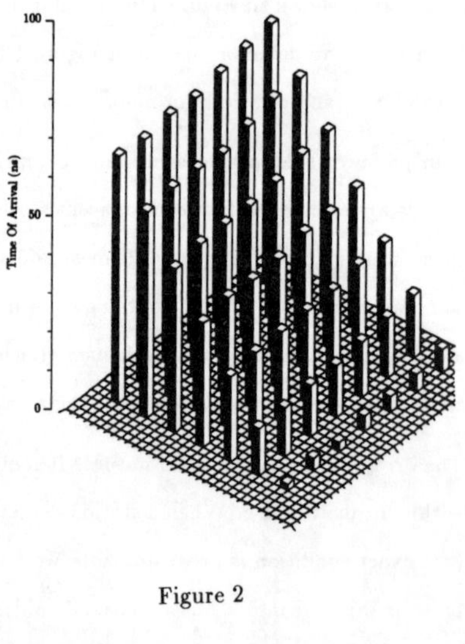

Figure 2

Neutrino Detection with the High Resolution Eye

G.L. Cassiday, R. Cooper, B.R. Dawson, J.W. Elbert, B.E. Fick, K.D. Green, S. Ko, E.C. Loh, M.H. Salamon, J.D. Smith, P. Sokolsky, P. Sommers, and S.B. Thomas

Physics Department, University of Utah

ABSTRACT: The High Resolution Eye, a second-generation Fly's Eye, is briefly described, and its neutrino aperture is presented. Expected numbers of upward- and downward-going neutrino events for various models are discussed.

THE HIGH RESOLUTION EYE:

The High Resolution (HiRes) Eye, presently in the stage of prototype development, will be a second-generation Fly's Eye, located near the present Fly's Eye [1] site in Utah. Like the Fly's Eye, it will detect ultra-high-energy (UHE) cosmic rays (> 0.1 EeV) by measuring the atmospheric N_2 fluorescence generated by the cosmic ray's air shower; measurement of the geometry and longitudinal profile of the shower allows the determination of its incidence direction, energy, and position of shower maximum. These in turn are used to obtain the energy spectrum [2], mass composition [3], and directional anisotropy [4] of the UHE cosmic rays.

This observatory will consist of three individual "eyes", each separated from the others by 15 km, ensuring that most events will be observed by at least two eyes, providing tight constraints on the showers' trajectories. The primary difference between the Fly's Eye and the HiRes Eye is in the segmentation of the sky; in the Fly's Eye, each photomultiplier tube (PMT) views $5.5^\circ \times 5.5^\circ$ of the sky, while in the HiRes Eye each PMT will view $1^\circ \times 1^\circ$. This reduction in angular aperture has two benefits: it increases signal-to-noise, thereby significantly increasing the aperture for UHE cosmic ray detection, and the increased number of samples along a shower track greatly enhances the quality of track reconstruction. A figure of merit is the number of events/yr of E > 10 EeV whose shower maxima are resolved to < 30 g/cm2 of atmospheric depth; for the present Fly's Eye this number is 1, while for the HiRes Eye it will be > 270.

NEUTRINO APERTURE:

The HiRes Eye will also have an increased aperture for neutrino detection. We have calculated this aperture as a function of neutrino energy using a Monte Carlo code that models the HiRes Eye, similar to the one used for the Fly's Eye [1]; the results are shown in Figure 1. These are the "effective" aperture values, in that the neutrino interaction probability is already included in these results; i.e., to obtain the number of observed events, simply multiply the neutrino flux by the aperture values in the figure.

The neutrino interaction cross-sections used are those of Reno and Quigg [5], which include the latest estimates of the effect of scaling violations in the low-x parton distribution. At HiRes energies, these effects cause the neutrino-nucleon cross-section to be as much as an order of magnitude higher than earlier standard model predictions [6]. This has two effects: first, the atmospheric interaction probability is proportionally higher, yielding an increase in anticipated event rate; second, the increased cross-section results in severe Earth-shadowing for upward-going neutrinos.

Deeply penetrating (or upward-going) events are a signature for a neutrino event. These are electron neutrinos that undergo a charged-current interaction deep in the atmosphere (or Earth), converting into an electron with a mean inelasticity of 0.2 [7]; this electron initiates an electromagnetic cascade which is measured by the HiRes detector. If the Reno and Quigg cross-sections are correct, it is extremely unlikely that we will ever see an upward-going event. Figure 2 shows the attenuation of neutrinos as a function of energy and nadir angle, given a model of the Earth in which the lithosphere is 100 km thick with a mean density of 2.8 g/cm3, and the mantle density is 5.4 g/cm3 (the core is neglected). It is seen that the attenuation is severe even for neutrinos just a few degrees below the horizon; thus the detection of any upward-going event would be of great significance. In our calculation of aperture, therefore, we have neglected the upward neutrino flux.

At HiRes energies, the major source of astrophysical neutrinos is via photopion production from the interaction of UHE cosmic ray protons with the cosmic background radiation [6] (we do not consider here the plethora of cosmic string models). The gamma-proton cross-section peaks at the delta(1232) resonance, corresponding to the collision of a 3°K photon with a ~100 EeV proton; this yields a flat integrated neutrino spectrum estimated by Stecker [6] to be $\sim 10^{-17}$/cm2-sr-s at all energies up to 10 EeV. With this flux, < 1 neutrino would be observed in 10 years of operation. However, recent calculations [8] that incorporate an early, "bright" phase of Galactic evolution [9] (that imply a concommitant increase in cosmic ray intensity at redshifts ranging up to z=10) suggest that the UHE neutrino flux may be as large as 10^{-14}/cm2-s-sr at E > 1 EeV; at this level, several neutrino events will be seen each year. Thus the HiRes Eye will be able to place significant limits on models of Galactic evolution after a few years of operation.

Although the number of expected neutrino events is small, there is much that can be learned from neutrino detection at the HiRes Eye, in that even the absence of a signal is astrophysically significant. Of course, should point sources of UHE neutrinos exist, such as cosmic strings, the possibility of their detection is an exciting one.

REFERENCES:

1) Baltrusaitis,R.M., et al., Nucl. Instr. Meth. A240,410 (1985).

2) Baltrusaitis,R.M., et al., Phys. Rev. Lett. 54, 1875 (1985).
3) Baltrusaitis,R.M., et al., Proc. 19th Inter. Cosmic Ray Conf. (La Jolla) 7, 159 (1985).
4) Cassiday,G.L., et al., Phys. Rev. Lett. 62, 383 (1989).
5) Reno,M.H. and Quigg,C., Fermilab-Pub-87/66-T (1987).
6) Stecker,F.W., Ap. J. 228, 919 (1979).
7) Quigg,C., Reno,M.H., and Walker,T.P., Fermilab-Pub-86/50-AT (1986).
8) Hill,C.T. and Schramm,D.N., Phys. Rev. D31, 564 (1985).
9) Partridge,R.B. and Peebles,P.J.E., Ap. J. 147, 868 (1967).

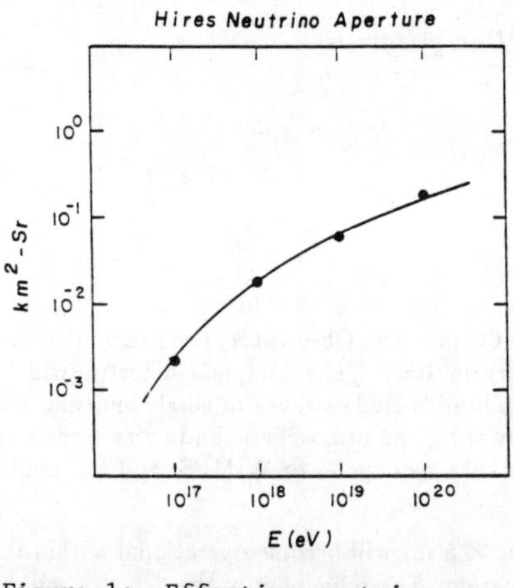

Figure 1: Effective neutrino aperture for the HiRes Eye.

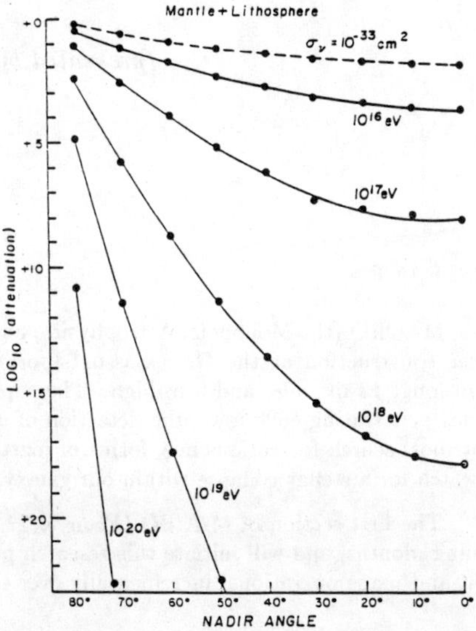

Figure 2: Attenuation of upward-going neutrino flux by Earth-shadowing. The dashed curve shows neutrino attenuation for the earlier standard model cross-section of 10^{-33} cm2 at 10 EeV.

Neutrino Astronomy with MACRO

MACRO Collaboration

Bari, Bologna, Boston, Caltech, Drexel, Frascati, Indiana, L'Aguila, Lecce, Michigan, Napoli, Pisa, Rome, Texas A & M, Torino

(presented by Barry Barish)

ABSTRACT

MACRO (the Monopole, Astrophysics, and Cosmic Ray Observatory) is a large detector under construction at the Gran Sasso Laboratory in Italy. The complete detector will be 72 m long, 12 m wide, and 9 m high. The capabilities include studies of single and multiple muons penetrating 4000 mwe, the detection of upward-going atmospheric and extra-terrestrial neutrinos, search for various new forms of matter, like monopoles or WIMPS, and the ability to watch for a stellar collapse within our galaxy.

The first section of MACRO (12 m × 12 m × 5 m) will become operational within the coming months, and will initiate this research program. The remainder of the detector will be built and made operational incrementally over the next 2-3 years.

MACRO AND THE INITIAL PROGRAM

The MACRO detector, as presently conceived, will contain three horizontal layers of scintillator tanks and 18 horizontal layers of streamer tubes. The streamer tubes are separated from each other by steel boxes filled with rock, reducing rates of correlated noise from natural radioactivity, and setting a minimum energy for muons of $E > 3$ GeV. The detector will be closed on the sides and ends by layers of vertical streamer tubes and scintillators to maximize the geometric acceptance. A horizontal sandwich consisting of several layers of track-etch detectors and absorber will be placed between the middle and bottom scintillator layers.

The total sensitive mass is 1000 tons of liquid scintillator. The scintillators (2 tons each) are 75 cm wide, 25 cm high and 12 m long. Pulse height, timing and wave form information is recorded. The tracking system consists of streamer tubes 3 cm × 3 cm × 12 m allowing an angular resolution of $0.2\,^{\circ}$ for particles penetrating the full thickness of MACRO. This is well below the $\sim 1^{\circ}$ multiple scattering angle for muons penetrating the overburden.

The complete detector is shown in figure 1.

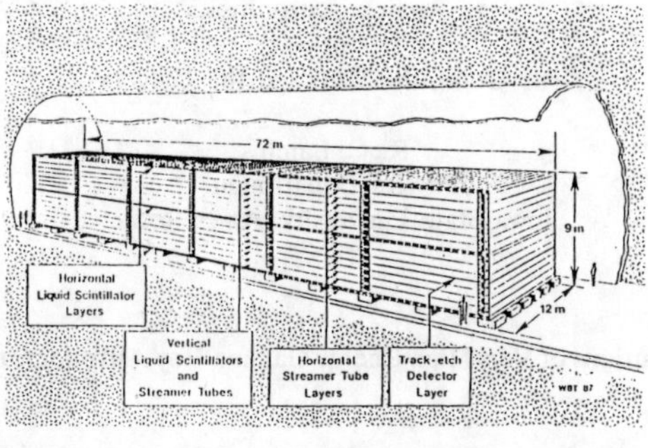

The first supermodule for MACRO is presently being turned-on for early physics studies. This section of the detector consists of the first "supermodule" which is 12 m × 12 m × 5 m representing ~ 10% of the total detector. An example of the recent detection of a multiple muon event in this supermodule is shown in figure 2.

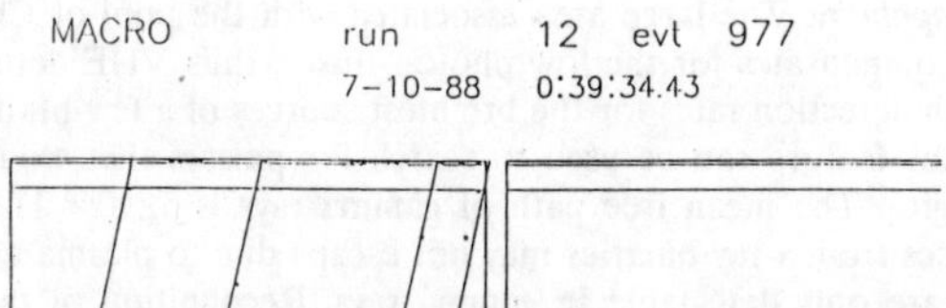

In the initial running (~ 3 months of data taking) to take place in Spring 1989, we will acquire

1) ~ 10^5 single muons for studies of the > 1 TeV muons penetrating the mountain to MACRO

2) ~ 3 10^3 multiple muons for studies of the consituents of the high energy primaries

3) ~ 10^3 muons "pointing-back" through the extensive air shower array at the top of the mountain for coincidence information

4) First detection of upward-going muons from neutrino interactions (we expect ~ 5 events)

The MACRO detector will grow incrementally both in size and electronics sophistication following this initial data run.

Contributions of Very High Energy Gamma-Ray Astronomy to Particle Astrophysics or "The Good, the Bad, and the Ugly(?)"

R. C. Lamb
Physics Department
Iowa State University, Ames, Iowa

ABSTRACT: The contributions of Very High Energy Gamma-Ray Astronomy to particle astrophysics are summarized.

The designation "Very High Energy" has come to mean 10^{11} to 10^{14} eV, roughly the region spanned by the atmospheric Cherenkov technique. Thus a discussion of the contributions of VHE astronomy to particle astrophysics is linked to this technique with its strengths (the "Good") and its weaknesses (the "Bad").

1. THE GOOD:

High-rate capability. The large area associated with the pool of Cherenkov light on the ground ($\sim 10^9$ cm^2) compensates for the low photon fluxes, thus VHE detectors are intrinsically high-rate devices, with detection rates for the brightest sources of a few photons/minute and burst rates even higher. This feature can be used to search for pulsars that would be hidden at x-ray and radio-wave energies. The mean free path of gamma rays is 5×10^{25} H-atoms/cm^2 vs $< 10^{22}$ for x-rays. Radio pulses from x-ray binaries may not escape due to plasma absorption. Thus there may be pulsars that are only detectable in gamma rays. Recognition of this fact has motivated several groups[1,2,3)] to search for a possible pulsar associated with Cygnus X-3. Establishing its presence would remove the uncertainty about the nature of the compact object in this system and measurements of possible arrival time variation in the pulses would determine the size of the projected binary orbit.

The high-rate capability can also be used to explore short time scale phenomena characteristics of acceleration mechanisms and/or changes in beam-target geometry. A number of the x-ray binary VHE sources appear to favor emission at certain orbital phases. The most heavily studied binary, Hercules X-1, appears to have bursts on a time scale of an hour or less. One episode coincides with the time of x-ray eclipse,[4)] clearly indicating that the site of gamma-ray production is distinct from that of x-rays. This episode, along with an episode[5)] at a similar time for Vela X-1, has suggested that in some cases the target material generating the gamma-rays is the limb of the companion star.

An important message for the experimentalists is that we should strive for increased rate. For the detection of a dc effect, an increase in rate, R, which preserves the percentage of gamma-ray signal to background cosmic rays, will improve the significance of a positive result as $\sqrt{R}$. However, if one is searching for a known or suspected periodic effect *one gains exponentially in the rate.* For example, if the overall rate during the episode by which the Durham group detected a possible 12.6 ms pulsation[1)] from Cygnus X-3 were doubled, then the probability that that episode was a statistical fluke would change from 10^{-2} to 10^{-9}. A further advantage of higher rate is that it allows one to explore possible source variability on even shorter time scales than the present limits. One way to achieve higher rates is by reducing the energy threshold through increases in detector aperture. A reflector of ~ 100 meters aperture could study flares on a time scale of seconds. Such a time scale for flares may be likely if mechanisms similar to those that produce solar flares are operative. A reflector of this size would also have an energy threshold of ~ 10 GeV, providing an important overlap with the energies covered by satellite detectors.

Primary identification. Simulations of air showers initiated by gamma rays and by cosmic rays indicate that there are substantial differences in the pattern of the Cherenkov emission for the two kinds of shower. These differences are most pronounced in the lateral distribution of the light and in the appearance of the Cherenkov images. Distributed arrays of Cherenkov receivers such as those at Pachmarhi and Themis can exploit the lateral differences. Imaging receivers are used at the Whipple Observatory and the Crimean Astrophysical Observatory. The Whipple Observatory Collaboration has recently reported[6] a weak flux from the Crab Nebula in which a raw 1-σ result was converted to a positive 9-σ detection by means of image selection criteria. This result was achieved with a single reflector and a relatively coarse-grained camera. The imaging technique appears to be extendable in a variety of ways which ultimately will lead to a virtually background-free detector. If there are new types of radiation amongst the gamma rays then such imaging detectors should also be able to determine its distinctive characteristics.

Ground-based. The technique is necessarily ground-based. This gives freedom to the experimentalist to develop new techniques at relatively low expense. It also means that one can think of establishing facilities with the necessary longevity for a full development of the field.

2. THE BAD:

The Cherenkov technique has the same limitations of ground-based optical astronomy, namely observations are restricted to good weather during dark sky conditions; for a good site this means ~10% of the time is useful. The observations are not all-sky such as those of the EAS arrays, but are pointed; thus a program of monitoring an established source competes with observations of a possible new source. The solution: multiple detectors. At this time none of the observations at TeV energies has extracted an energy spectrum from a pulse height spectrum. This will change in the next few years since both the Cherenkov arrays and imaging detectors have the capability to determine a total Cherenkov light yield that can, in turn, be converted to a primary energy measurement.

3. THE UGLY(?):

There is a bit of possible ugliness in the picture of VHE astronomy. The Whipple Observatory has some apparent detections of Hercules X-1.[4,7] When the same image selection criterion that was used to enhance the Crab Nebula signal was applied to the Hercules X-1 data, none of the apparent episodes of emission behaves as if it were predominately photonic. To see if this is a real "ugliness" - something that is statistically unlikely or a failure to understand some systematic error - or only the first glimpse at some beautiful "new physics," remains to be seen. Only greatly improved equipment will settle this issue decisively.

CONCLUSION: Two crucial experiments need to be done: (1) find where the Cosmic Department of Energy has located the accelerators; and (2) measure the energy and time spectra and the content of the long-lived neutral beams made by these sources.

REFERENCES

[1]Chadwick, P. M., et al., Nature 318, 642 (1985).
[2]Zyskin, Yu. L., et al., Astron. Circular, Bur. Astron. Info. USSR Acad. Sci. No. 1508, (1987).
[3]Fegan, D. J., et al., Astron. Ap. (in press).
[4]Gorham, P. W., et al., Ap. J. (Letters) 308, L11 (1986).
[5]North, A. R., et al., Nature 326, 567 (1987).
[6]Weekes, T. C., et al., Ap. J., July 1, 1989.
[7]Lewis, D. A. et al., Proc. of Vulcano Workshop 1988-Frontier Objects in Astrophysics and Particle Physics, in press (1989).

PeV GAMMA RAY ASTRONOMY

D. E. Nagle
Los Alamos National Laboratory

A. STATUS OF AIR SHOWER OBSERVATIONS

Cygnus X-3

Cygnus X-3 is a low-mass x-ray binary system with a 4.8-hour periodicity, usually attributed to the orbital motion. It is not seen at optical wavelengths, because of its distance (~10 kpc) and obscuration by intervening dust clouds. It is seen in radio, infrared, and VHE and UHE gamma rays. Observations of 4.8-h-phase since 1984 have generally shown a phase of 0.6–0.8, by many observers, in contrast to the Kiel observation of 0.2. The flux is generally reported to be at least an order of magnitude less than reported by Kiel (Samorski and Stamm) for the period 1976 to 1980. Conflicting claims of a 12.6-ms and a 9.22-ms pulsar have been made by VHE observers.

Kiel had reported an anomalously high muon content in the air showers from Cyg X-3: this was partially corroborated by the CYGNUS collaboration at Los Alamos from 1986 data on Cyg X-3 (Dingus et al.).

Recently, Cassiday et al. (the Fly's Eye group) reported a "dc" excess in 10^{18} eV gamma rays for the period November 1981 through May 1988. Weak evidence for a phase excess at ~0.2 is given. This is especially interesting, because an energy of ~10^{19} eV is implied for the accelerator in Cyg X-3.

Her X-1

This eclipsing binary is extensively studied over a wide range of energies. The distance is ~6 kpc. The x rays are pulsed at the neutron star period of ~1.24 s, and the orbital period is ~1.7 d. A 35-day cycle is usually attributed to a precession of the accretion disc or of the neutron star.

Recently two air Cerenkov observatories (Whipple and Haleakala) reported sightings during the summer of 1986 of a pulsar period which was blue shifted by ~0.16% from the x-ray period. (Another observation of the same shifted frequency was discussed by Morse of Haleakala at this meeting.) The same frequency has also been seen by the CYGNUS group in air showers at 10^{15} eV. Possible explanations in terms of a moving target in the accretion disc have been given by K. C. Cheng and M. Ruderman in a recent preprint. A very intense pulse in 1986 was seen by the workers at Pachmari, India, with a DC excess of ~40 standard deviations. The CYGNUS group also reported that in the showers from Her X-1 the muon content was about like that for ordinary cosmic ray showers.

Vela X-1

Vela X-1 is a massive x-ray binary system in the southern sky at a distance of ~1.4 kpc. The orbital period is ~8.9 days. Evidence was found at the 0.01% level by the Buckland Park (Adelaide) group for emission at an orbital phase of ~0.6 in showers above 3 PeV, from 1979 to 1981. Recently (Ciampa and Clay, Adelaide preprint 88-100/E52) reported a slight excess in the period 1986 through 1988, with a probability of ~10% after all trials. Potchefstroom found a 1.8% chance probability effect during approximately the same time.

Other Sources

A variety of other sources has been reported over recent years: Cen X-3, LMCX-4, the Crab Nebula, Cygnus X-1, Cen A, and others. Generally, the statistical significance is not high.

B. PHYSICS QUESTIONS

Reality of VHE Sources: Cyg X-3, Her X-1, and Vela X-1 are the best established as VHE sources. The others await further confirmation.

Muon Anomalies: The muon anomalies reported by Kiel and by CYGNUS require further confirmation because of their drastic implications. Are the gamma rays really gammas? Apparently related anomalies are reported in this meeting by R. Lamb of the Whipple collaboration. See also F. Halzen, this meeting, for an exposition of the idea that above ~1 TeV, the photon form factor has an abundant gluon component. Other authors have suggested a new neutral particle as the explanation.

Microwave Absorption: As of yet, there is no evidence of absorption in the Cyg X-3 "spectrum" owing to pair production off photons of the three-degree microwave background radiation. The scatter in the points is too large, because of the sporadic character of the emissions of Cyg X-3. A good broadband array might be able to see the effect in a single large burst from Cyg X-3. That would establish that the 10^{15}-eV emissions are photons, and would give perhaps a value for the distance to Cyg X-3, a number which bears on the dark matter problem (see Dickey).

Accelerators: Mechanisms for acceleration are a prime topic, point sources vs. distributed acceleration.

Extragalactic and Galactic Sources; Their Relative Contribution:

Other Sources: Other PeV sources need to be searched for. Bursters are a possibility.

Cosmic Ray Elemental Composition: The muon detectors make possible composition studies.

PeV Gamma Rays from SN1987a: None have yet been seen, but as the cloud disperses and its opacity drops, the pulsar may become detectable.

C. NEW ARRAYS AND NEW TECHNIQUES

Many arrays are under construction or are being upgraded. Table I is an incomplete listing of new arrays. New techniques are also proposed, some at this meeting. We may expect an order of magnitude improvement, at least, in the near future.

The Chicago CASA array at Dugway, when completed, will be the world's most sensitive. It is described elsewhere in this session.

The CYGNUS II array is an extension of the present one. When completed, about 6×10^4 m^2 will be sampled. The total number of detectors will be 204. In addition, about 70 m^2 of muon detectors will be added, bringing the total to 110 m^2. Completion is scheduled for 1989. The combined array will have sensitivity from about 5×10^{13} eV to 10^{17} eV.

D. RELATIONSHIP WITH OTHER FIELDS

Several of the problems mentioned above are multidisciplinary.

REFERENCES

For references mentioned in this talk, and others, I suggest the article by D. E. Nagle, T. K. Gaisser, and R. J. Protheroe, *Annual Review of Nuclear and Particle Science* **38**, 609–657 (1988).

Table I. UHE gamma-ray experiments under construction.[a]

Location	Country	Organization	Location (deg)	Elevation (km)	Energy PeV)	Angular Resolution (deg)	Year
Dugway	USA	Utah, Michigan, Chicago[b]	40N,112W	1.5	0.1	0.5	1990[b]
La Palma	Spain	Kiel	29N,18W	2.2	0.1	1	1988
Gran Sasso	Italy	Turin	42N,14E	2.0	0.01	1	1988
Mt. Norikura	Japan	Tokyo	36N,137E	2.8	0.1	1	1988
South Pole		Bartol, Leeds	90S,0	2.5	0.1	1	1988
Black Birch Mountain	New Zealand	JANZOS	42S	1.7	(0.01–10)		1987
Yanbajing	Tibet	Japan, China	30N,91E	4.3	0.03	0.2	1989

[a]For details, see *Nuclear Instruments and Methods* **A234** (1988).
[b]Completion date for the University of Chicago part of the installation is estimated as 1990.

GAMMA RAY ASTRONOMY: THE IMPORTANCE OF MUONS

FRANCIS HALZEN
Physics Department, University of Wisconsin, Madison, WI 53706 USA

ABSTRACT

We present a simple and compelling argument why one expects that photon showers contain fewer muons than proton showers. Too many muons means new physics. In this context we evaluate the relative performance of surface and underground detectors as gamma ray telescopes emphasizing the pivotal role of muons.

TeV/PeV photons can reveal important information about astrophysical beam dumps (such as cosmic accelerators and young supernovae), the origin and acceleration of cosmic rays and, possibly, particle physics. The low fluxes of high energy photons force us to use the atmosphere as a detector in the TeV/PeV energy regime. Air showers initiated by cosmic photons have to be identified in a background of hadronic cosmic ray showers. This is the challenge of high energy astronomy. Calculations agree that the relative multiplicity of muons and electrons is about 30 times smaller in a γ-initiated cascade than in a proton shower.[1] Measurement of the muon content of the shower can therefore be used to enhance photon signals above cosmic ray background. Evidence is accumulating, however, that γ-showers have at least ten times the muon content previously calculated.[2] If true, this has important implications for our choice of future detectors. Muon-rich γ-showers can indeed be observed with deep underground detectors which are less sensitive to cosmic rays than air shower arrays. We will quantify this statement further on.

Why calculations find low muon multiplicities in photon cascades is easy to demonstrate using Heitler's approximation to an electromagnetic cascade. In every radiation length (λ_R) a photon will produce an e^+e^- pair and electrons will radiate a photon. Pair production and bremsstrahlung have roughly equal high energy cross sections. The atmosphere is filled with sequential layers of depth λ_R (actually $\sqrt{2}\,\lambda_R$) in which electromagnetic particles subdivide in 2 particles with half the initial energy. We construct explicitly the cascade initiated by a $E_0 = 100$ TeV photon:

Layer (n)	Energy ($E_0/2^n$)	Multiplicity (2^n)
0	100 TeV	1 (γ)
1	50 TeV	2 (e^+e^-)
2	25 TeV	4 ($e^+\gamma e^-\gamma$)
⋮	⋮	⋮
6	1.6 TeV	64
7	0.8 TeV	128
⋮	⋮	⋮
12	25 GeV	4100
13	12 GeV	8200 ($\frac{1}{3}\gamma$, $\frac{2}{3}e^\pm$)

Layer $n = 13$ has about 8200 particles of which 1/3 are photons of average energy 12 GeV.

This is the last layer where one can produce muons of a few GeV energies relevant to air shower experiments. The number of muons is $R \times \frac{1}{3} \times 8400$, where R is the relative probability that a photon produces a π, which decays to a muon, rather than an e^+e^- pair. We estimate R as follows:

$$R = \frac{A\,\sigma(\gamma N \to \pi X)}{\sigma(\gamma \to e^+e^-)} \simeq 3 \times 10^{-3}. \tag{1}$$

Here A ($\simeq 14.5$) is the atomic number of air, $\sigma(\gamma N \to \pi X)$ ($\simeq 10^{-1}$mb) is the π-photoproduction cross section (measured at Fermilab up to about 10^{-1} TeV) and $\sigma(\gamma \to e^+e^-)$ ($\simeq 500$ mb) is the high energy Beth-Heitler cross section on air. Thus the number of GeV muons in a 100 TeV shower is roughly $2 \times R \times \frac{1}{3} \times 8400 \simeq 20$. The factor 2 allows for the fact that all other layers together contain as many photons as the last layer. The result is inescapable as photoproduction is almost exclusively by γ-rays with energies previously explored in accelerator experiments. Muon production by other processes such as $\gamma \to \mu^+\mu^-$ and $\gamma \to$ charm $\to \mu$ is negligible.[1] For a 100 TeV proton shower the number of muons would be 30 times larger as a result of the abundant hadronic production and decay of charged π's. Tagging muon-poor showers is a powerful tool for doing high energy γ-ray astronomy.

Several experiments have detected γ-rays from Hercules X-1 and Cygnus X-3 which produce an atmospheric muon flux exceeding calculations by at least one order of magnitude.[2] It is tempting to speculate that one is seeing a new form of radiation, carried by neutral particles produced in the 10^5 TeV (or more) cosmic beam dump. Given the 1.24 seconds pulsed nature of the Hercules γ-ray beam, the mass of these long-lived particles is less than 1 GeV and it is unlikely that they escaped the scrutiny of accelerator searches. The alternative is to photo-produce more pions by increasing the value of $\sigma(\gamma N \to \pi X)$ and therefore R. This cannot be done in the last radiation layer where muons are produced by 12 GeV photons. The cross section has been measured by accelerator experiments. A large value of $\sigma(\gamma N \to \pi X)$ at 10-10^2 TeV is useless because there are too few photons of that energy in the shower to make a difference. There is only one solution: a photoproduction threshold with $R \geq 1$ above $0.5 \sim 1$ TeV. Referring back to layer $n = 7$ in the cascade, we obtain $2 \times R \times \frac{1}{3} \times 128 \simeq$ 100 muons. It is interesting to notice that the anomalous muons are of high enough energy to accommodate claims of observation of point sources in underground detectors.[3] They are possibly of sufficiently low energy to deform the Cerenkov images in the Whipple observation[4] of Hercules X-1.

Rather than speculate, we would like to emphasize the importance for the future of TeV astronomy of settling the muon puzzle. To do this we compare the sensitivity to point sources of a 1 km^2 air shower array with that of a 0.1 km^2 underground detector observing 100 TeV muons. We assume 1° resolution for both. The background rate per year from cosmic rays is 10^6 showers in the array and 5 muons in the underground detector. The observation of a 5σ excess corresponds to 5000 showers and 15 muons per year. This translates into a flux sensitivity of 2×10^{-14} and 2×10^{-12}cm^{-2}s^{-1}, respectively. In deriving the last number we assumed that photon showers are muon-poor. The array wins (although signal-to-noise is far superior in the underground detector), especially as it has the possibility to further increase its sensitivity by 10^2 by selecting muon-poor showers. If the muon content of photon showers is similar to that of protons, air shower arrays lose this advantage and underground detectors have an increased sensitivity to photons. The underground experiment is competitive and, arguably, preferable as it can also detect neutrinos at the rate of 10 per year for a Cygnus X-3 type source.

REFERENCES

1. See *e.g.* Stanev, T. *et al.*, Phys. Rev. D32, 1244 (1985) and Phys. Rev. D34, 2061 (1986).
2. Yodh, G., Nagle, D., this volume.
3. Marshak, M., this volume.
4. Lamb, R., this volume.

Anomalous Periodicity Observed in TeV Gamma Rays from Her X-1

Robert Morse†

Haleakala Gamma Ray Observatory*

ABSTRACT

HGO reports a 900 second burst from Her X-1 observed on 23 May 1987 with an anomalous period consistent with the periods reported in 1986. The period found is 1.23578 ± .0002s or 0.16% shorter than the X-ray period.

INTRODUCTION

Additional evidence at the 2% *false alarm* level for an anomalous periodicity in TeV γ rays from Her X-1 that is approximately 0.16% shorter than the X-ray period has been observed again in 1987 by the HGO collaboration. Studies in 1986 of γ rays from Her X-1 in the TeV and PeV regions by the HGO collaboration, the Mt. Hopkins group and by the CYGNUS group at Los Alamos had suggested anomalous periodicities significantly different from that of the rotating neutron star (Resvanis *et al.* 1988, Lamb *et al.* 1988, and Dingus *et al.* 1988).

Earlier period measurements of Her X-1 at TeV energies yielded a large number of observations different from the neutron star rotation period as determined by X-ray measurements. Detection of 7 bursts by the Mt. Hopkins group during 1984 and 1985 also provided some early evidence for a VHE period different from the X-ray period (Gorham, 1986). We summarize this period information in Table 1.

Since measurement durations of VHE emissions are often short (about 10^3s), the determination of the period is limited by the large *Fourier-independent-period-*

† Presenter of this paper. Paper presented at the LBL Particle Astrophysics Workshop, Berkeley (1988). The 1987 Her X-1 results presented are primarily the work of Linda Kelley, Univ. of Hawaii.

* The Haleakala Gamma Observatory is a collaboration of: Univ. of Athens, Columbia Univ., Univ. of Hawaii, Purdue Univ., and the Univ. of Wisconsin.

spacing (FIP) of $\delta P = \frac{P^2}{T}$. Given this burst nature it is possible that anomalous periodicities from other binary objects exist as well, but the effects have either not yet been resolved, or have not been pursued by observers.

OBSERVATIONS

The 1987 result is taken from 102 hours of observation during March-June. The data were divided into 900 second intervals corresponding to a FIP resolution of 0.00171s, and searched at 19 periods which covered a range of 3 FIP's centered on the extrapolated X-ray period of 1.23778s. A total of 291 intervals were searched revealing one robust *on-source* instance of a Rayleigh power of 11.9 which has a false-alarm probability of 2%. The measured period is 1.23578±0.00020s or 1.16 FIP's away from the X-ray period, and is in agreement with the three anomalous 1986 results. The distribution of on-source Rayleigh powers is shown in Figure 1. The multiple peaks are the result of the aforementioned oversampling, and this effect has been noted in calculating the false-alarm probabilities.

This signal was observed at 40338.24s UTC on 23 May 1988, occurring at an orbital phase of 0.32 and at a 35 day phase of either 0.95 using the 35 day parameters by Staubert, *et al.* 1983, or phase 0.99 using the 35 day period and ephemeris from Ariel 5 data as given by Trumper *et al.* 1986. This burst thus occurs at the beginning of the *high-on* region of the 35 day period

The largest *off-source* Rayleigh power is 9.2, which corresponds to a probability of 26%. The off-source Rayleigh-power distribution is shown in Figure 1.

This signal shows a 1.3σ on-source excess when the data are corrected for zenith angle effects. This rate excess is lower than the 2.5σ which would be most probable from a signal with a Rayleigh power of 11.9, but the two results are still statistically compatible, and show the same excess level as the previous measurements.

DISCUSSION

Table 1 shows that none of the individual VHE results can be taken very seriously, since they all exhibit about a 1-2% probability of being spurious. However, when considered as three independent *wrong-in-just-the-right-way* results they merit consideration. Furthermore, episodes with period deviations this large may have been present in earlier observations, but were missed because the range of periods searched was too narrowly confined around the X-ray period. Such old data might be usefully reexamined.

These results suggest that the gamma rays and X-rays do not originate from the same location (Halzen, 1988). To explain anomalous period TeV radiation from binary pulsars, beam-target models have the target structure rotating at the observed VHE periods, and for the specific case of Her X-1, more rapidly than the neutron star (Slane, et al.,1988). Similar advances for the subpulses in radio pulsars have also been noted.

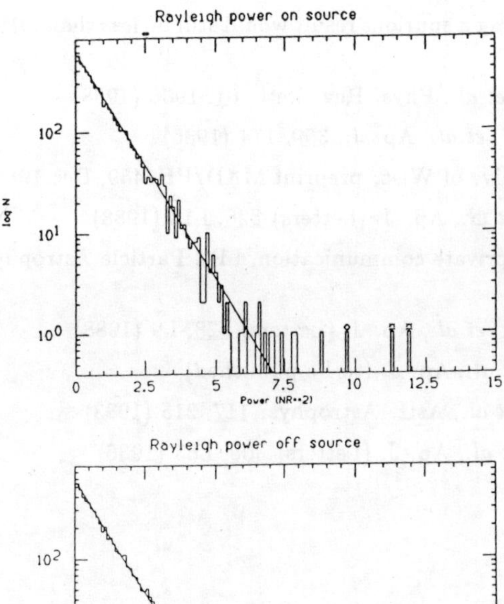

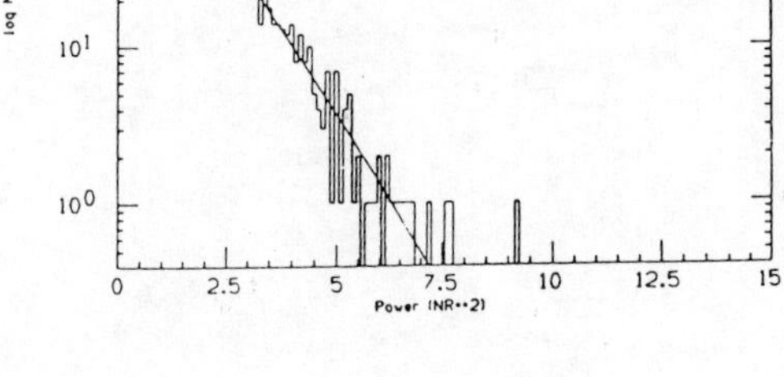

Table 1

Observatory	Period	Power	S/N	Prob.(%)	ϕ_{orb}	ϕ_{35}
X-ray	1.23778	–	–	–	–	–
HGO-I	1.23593±.0002	12.8	1.4σ	0.7%	.81	.22
HGO-II	1.23578±.0002	11.9	1.3σ	2.0%	.32	.95
Mt. Hopkins	1.23579±.0002	13.1	1.9σ	0.9%	.70	.04
Los Alamos-I	1.23572±.0004	(a)	–	.002%	.8	.23
Los Alamos-II	1.23575±.0003	(a)	–	.002%	.9	.23

(*a*) The Los Alamos group uses Protheroe Statistics to assess probability levels, but they feel if a Rayleigh analysis had been performed then the probability of their reporting a spurious result would still be less than .01% (Nagle 1988).

REFERENCES

1. Dingus, B. L. *et al.*, Phys. Rev. Lett. 61, 1906 (1988)
2. Gorham, P. W. *et al.*, Ap. J., 309, 114 (1986)
3. Halzen, F., Univ. of Wisc. preprint MAD/PH/459, Dec 1988
4. Lamb, R. C. *et al.*, Ap. J. (Letters) 328, L13 (1988)
5. Nagle, D. E., private communication, LBL Particle Astrophysics Workshop (1988)
6. Resvanis, L. K. *et al.*, Ap. J. (Letters) 328, L9 (1988)
7. Slane, P. O. *et al.*, Ap. J. (to be published)
8. Staubert, R. *et al.*, Astr. Astrophys. 117, 215 (1983)
9. Trumper, J. *et al.*, Ap. J. (Letters) 300, L63 (1986)

RELATIVISTIC PLASMAS IN COMPACT ASTROPHYSICAL OBJECTS

ROLAND SVENSSON

NORDITA, Blegdamsvej 17, 2100 Copenhagen Ø, Denmark

AND

ANDRZEJ A. ZDZIARSKI

Space Telescope Science Institute, 3700 San Martin Dr., Baltimore, MD 21218, USA

1. ABSTRACT

Recent work on the theory of compact relativistic plasmas and their applications to compact astrophysical objects are reviewed. Of particular relevance in such plasmas is the production of electron-positron pairs.

2. INTRODUCTION

The most important advance in the theory of relativistic plasmas during the 1980's may have been the final realization that electron positron pairs are produced in great abundance in certain circumstances affecting both the properties of the plasma and the spectral shape of the emerging radiation. The importance of photon-photon pair production in compact plasmas was first pointed out by Jelley [1] in 1966 and by Herterich in 1974 [2]. The crucial parameter determining whether pairs are produced or not is the compactness, i.e. the luminosity to size ratio, introduced by Cavaliére and Morrison [3] in 1981. Pioneering studies were performed already in 1971 by Bisnovatyi-Kogan, Zeldovich and Sunyaev [4] on thermal pair production and by Bonometto and Rees [5] on nonthermal pair production. Several reviews by Rees, McCray and others in the late 1970's called for detailed investigations into the properties of pair plasmas. Since 1982 several such studies have been made whose main results are summarized here. More detailed reviews of work prior to 1986 can be found in [6] and [7].

3. PHYSICAL PROCESSES IN RELATIVISTIC PLASMAS

Calculations of thermal and nonthermal rates [8]–[18], in particular at mildly relativistic energies, formed the necessary prerequisite for the recent work on compact relativistic plasmas. Of particular importance has been the developments regarding the synchrotron, Compton scattering, pair production, and pair annihilation processes.

4. THERMAL PAIR PRODUCTION

When the electron temperature of a plasma reaches approximately $m_e c^2/k \approx 10^9 - 10^{10}$K, i.e. when the kinetic energies become of the same order as the electron rest mass energy then pair production may become important. Detailed studies [19]–[27] showed that for plasmas in pair equilibrium, i.e. when the pair annihilation rate balances the pair production rate, there exists a maximum temperature above which the annihilation of pairs cannot keep up with the pair production. The compactness

$$\ell = \frac{L}{R}\frac{\sigma_T}{m_e c^3} \simeq 3000 \left(\frac{L}{L_{Edd}}\right)\left(\frac{3R_{Schw}}{R}\right) \tag{1}$$

is the important parameter, and in pair dominated plasmas the maximum temperature is a unique monotonically but slowly decreasing function of the compactness. For compactnesses $1 < \ell < 1000$, expected in compact objects, the maximum temperature is of the order 100-500 keV. Another important result [21],[24] is that an annihilation line is not expected from plasmas in pair equilibrium. At high temperatures other emission processes dominate the broad line and at smaller temperatures the Thomson depth is large enough to wipe out the line.

Using different assumptions pair dominated thermal pair equilibria have been shown to be both stable [28] and unstable [29]–[30] to small pertubations. Adding phenomenological accretion terms gives rise to a flaring limit cycle behavior of the luminosity for $\ell \approx 10$ [30].

5. NONTHERMAL PAIR PRODUCTION

Four main results emerge from detailed studies [31]–[40] on nonthermal pair production:

(i) When the acceleration process puts most of the power into the highest energy particles, then, in general, most of the 'primary' synchrotron and Compton radiation will be at the highest photon energies. If this energy exceeds $m_e c^2$ *and* the compactness is greater than unity, photon-photon interactions will initiate a pair cascade that redistributes the gamma-ray luminosity into the UV-X-ray range thereby changing the spectral shape.

(ii) Characteristic spectral breaks are introduced with the main one occurring around $m_e c^2$.

(iii) The pair yield, PY, i.e. the fraction of the injected power that ends up as pair rest mass, can reach values of 10 % [35],[39].

(iv) The Thomson depth of cooled nonrelativistic pairs can reach values of $\tau_{pair} \approx (PY \cdot \ell)^{1/2} \sim 20$ [31].

Another important result [41] for nonthermal plasmas is that a quasi-Maxwellian particle distribution forms at those Lorentz factors where self-absorbed synchrotron radiation is emitted. This requires synchrotron cooling to be the dominant radiation process and the cooling time scale at the relevant Lorentz factors to be shorter than typical dynamical time scales.

6. PAIR PRODUCTION IN ACCRETION FLOWS

Early considerations by Liang [42] in 1979 indicated that pairs may be important in accretion flows. A simple estimate [35] of the maximum number of nonthermal pairs per accreted proton onto a black hole is obtained as follows: With an accretion efficiency of 10% there is about $0.1 m_p c^2 \sim 100$ MeV per proton available to make radiation and ultimately at most 100 pairs. A fraction say 50%, may end up as nonthermal radiation and another fraction $PY < 10\%$ of this becomes pairs, i.e. 5 pairs or 10 particles per proton.

In spherical accretion flows the amount of pairs for sub-Eddington accretion rates is negligible (as then $PY \ll 1$). At Eddington rates the maximum pair-to-proton ratio derived above is reached, while at super-Eddington accretion rates the density is large enough for the pairs to annihilate on an inflow time scale reducing their numbers to negligible values [43].

In two-temperature accretion flows (ion-torus) the presence of pairs affects the onset of Coulomb-cooling by the ions and the consequent collapse of the torus [44]. Including the effects of pairs two-temperature accretion disks are found not to be steady for accretion rates that are marginally sub-Eddington or larger [45]–[46].

7. PAIR PLASMAS IN COMPACT OBJECTS

7.1 Active Galactic Nuclei (AGNs)

Great efforts [34],[36],[47]–[49] have been made to explain some or all of the following three properties of X-ray spectra from AGNs:

(i) the spectral index $\alpha_{2-10\mathrm{keV}}$ ranges from about 0.6 to 0.8,

(ii) the compactness $\ell_{2-10\mathrm{keV}} \sim 0.1 - 10$ (from observed luminosity and time variability),

(iii) the MeV-break required by observations of the cosmic X-ray background.

The best fit [36] is obtained by having most electron power injected at Lorentz factors of 200, allowing EUV photons from the UV-bump to be scattered no further then to about 1 MeV with little resulting pair production. Thus strong constraints are placed on the particle injection and the ultimate answer to why the X-ray spectra behave as they do may lie in the acceleration mechanism, not the radiation mechanism.

7.2 The Cosmic X-Ray Background

Leiter and Boldt [50] found that subtracting the contributions by AGNs with typical X-ray spectral index $\alpha_X \approx 0.7$ gives a flat residual background spectrum with $\alpha_X \approx 0$ and an exponential roll-over at 25 keV. This spectrum, containing most of the energy density of the background, has a spectral shape different from known possible sources. Leiter and Boldt [50]–[51] proposed the existence of a class of high-redshift ($z \approx 5$) unresolved precursor AGNs, whose spectra have a thermal character with $kT \approx 25(1+z)$ keV ≈ 150 keV. These objects would evolve from large to small compactnesses undergoing a change to the present-day nonthermal ($\alpha_X \approx 0.7$) spectrum when the compactness dropped below unity. Zdziarski [52] investigated spectra from thermal Comptonization of cyclotron photons in spherical accretion flows at large compactnesses and large redshifts finding spectral shapes and temperatures in agreement with the residual spectrum. Quite a different nonthermal scenario was suggested by Fabian, Done, and Ghisellini [53]. As more and more pairs are produced and share the available released power in a source, the individual particle energy after acceleration must decrease. This *pair loading* stops when the typical radiated (e.g. Compton scattered) photon has an energy of about 500 keV then being unable to pair produce. The objects (also precursor AGNs) must be at $z \approx 10 - 30$ in order for the MeV-break to redshift to 25 keV.

7.3 Gamma Ray Bursts

Risetimes of a few ms, observed fluxes, and distances greater than 20 pc imply compactnesses greater than unity and pairs are expected to be produced [54]. The flat X-ray spectra ($\alpha_X \approx 0$) and the steeper gamma ray spectra ($\alpha_X \approx 0.5 - 2$) above the MeV-break cannot be explained by thermal models. Zdziarski and Lamb [55] considered repeated Compton scatterings in a photon starved source (i.e. a source with few available soft photons) and got reasonable spectral fits.

8. SUMMARY

Three main points are to be made:

(i) Pairs are expected when the compactness ℓ is greater than unity.

(ii) Inclusion of pair effects allows a *large* variety of spectral shapes with little predictive power in the steady models considered so far.

(iii) Even so, it is hard to explain X- and γ-ray spectra from compact objects in detail.

That the success in explaining observed spectra has been limited so far may be due to too oversimplified assumptions. Yet to be considered and explored are dynamical and radiative transfer effects as well as timedependent nonsteady behaviour. It may also be that we are trying to hard to fit all data with one component models.

9. REFERENCES

1] Jelley, J. V., *Nature*, **211**, 472 (1966).
2] Herterich, K., *Nature*, **250**, 311 (1974).
3] Cavaliére, A., and Morrison, P., *Ap. J. (Letters)*, **238**, L63 (1980).
4] Bisnovatyi-Kogan, G. S., Zel'dovich, Ya. B., and Sunyaev, R. A., *Sov. Astr.*, **15**, 17 (1971).
5] Bonometto, S., and Rees, M., *M. N. R. A. S.*, **152**, 21 (1971).
6] Svensson, R., in *Radiation Hydrodynamics in Stars and Compact Objects*, IAU Coll. 89, eds. D. Mihalas and K-H. Winkler (New York: Springer Verlag), p. 325 (1986).
7] Zdziarski, A. A., and Lightman, A. P., in *Variability in Galactic and Extragalactic X-Ray*

Sources, ed. A. Treves, (Milano: Associazione per l'Avanzamento dell'Astronomia) (1987).
8] Blumenthal, G. R., and Gould, R. J., *Rev. Mod. Phys.*, **42**, 237 (1970).
9] Rybicki, G. R., and Lightman, A. P., *Radiative Processes in Astrophysics* (New York: McGraw-Hill) (1979).
10] Petrosian, V., *Ap. J.*, **251**, 727 (1981).
11] Svensson, R., *Ap. J.*, **258**, 321 (1982).
12] Svensson, R., *Ap. J.*, **270**, 300 (1983).
13] Stepney, S., *M. N. R. A. S.*, **202**, 67 (1983).
14] Stepney, S., and Guilbert, P.W., *M. N. R. A. S.*, **204**, 1264 (1983).
15] Dermer, C.D., *Ap. J.*, **280**, 328 (1984).
16] Dermer, C.D., *Ap. J.*, **295**, 28 (1985).
17] Dermer, C.D., *Ap. J.*, **307**, 47 (1986).
18] Baring, M., *M. N. R. A. S.*, **228**, 681 (1987).
19] Lightman, A. P., *Ap. J.*, **253**, 842 (1982).
20] Svensson, R., *Ap. J.*, **258**, 335 (1982).
21] Svensson, R., *M. N. R. A. S.*, **209**, 175 (1984).
22] Kusunose, M., and Takahara, F., *Progr. Theor. Phys.*, **69**, 1443 (1983).
23] Kusunose, M., and Takahara, F., *Progr. Theor. Phys.*, **73**, 41 (1985).
24] Zdziarski, A. A., *Ap. J.*, **283**, 842 (1984).
25] Zdziarski, A. A., *Ap. J.*, **289**, 514 (1985).
26] Zdziarski, A. A., *Ap. J.*, **303**, 94 (1986).
27] Guilbert, P. W., and Stepney, S., *M. N. R. A. S.*, **212**, 523 (1985).
28] Kusunose, M., *Ap. J.*, **321**, 186 (1987).
29] Sikora, M., and Zbyszewska, M., *Acta Astr.*, **36**, 255 (1986).
30] Moskalik, P., and Sikora, M., *Nature*, **319**, 649 (1986).
31] Guilbert, P. W., Fabian, A. C., and Rees, M., *M. N. R. A. S.*, **205**, 593 (1983).
32] Kazanas, D., *Ap. J.*, **287**, 112 (1984).
33] Zdziarski, A. A., and Lightman, A. P., *Ap. J. (Letters)*, **294**, L79 (1985).
34] Fabian, A. C., Blandford, R. D., Guilbert, P. W., Phinney, E. S., and Cuellar, L., *M. N. R. A. S.*, **221**, 931 (1986).
35] Svensson, R., *M. N. R. A. S.*, **227**, 403 (1987).
36] Lightman, A. P., and Zdziarski, A. A., *Ap. J.*, **319**, 643 (1987).
37] Zdziarski, A. A., *Ap. J.*, **335**, 786 (1988).
38] Aharonian, F.A., Kirillov-Ugryumov, V.G., and Vardanian, V.V., *Astr. Sp. Sci.*, **115**, 201 (1985).
39] Stern, B.E., *Sov. Astr.*, **29**, 306 (1985).
40] Ghisellini, G., *M. N. R. A. S.*, **224**, 1 (1987).
41] Ghisellini, G., Guilbert, P. W., and Svensson, R., *Ap. J. (Letters)*, **334**, L5 (1988).
42] Liang, E.P., *Ap. J.*, **234**, 1105 (1979).
43] Lightman, A. P., Zdziarski, A. A., and Rees, M., *Ap. J. (Letters)*, **315**, L113 (1987).
44] Begelman, M. C., Sikora, M., and Rees, M. J., *Ap. J.*, **313**, 689 (1987).
45] Kusunose, M., and Takahara, F., *Publ. Astr. Soc. Japan*, in press (1988).
46] White, T., and Lightman, A. P., *Ap. J.*, in press (1989).
47] Zdziarski, A. A., *Ap. J.*, **305**, 45 (1986).
48] Sikora, M., Kirk, J.G., Begelman, M.C., and Schneider, P., *Ap. J. (Letters)*, **320**, L81 (1987).
49] Dermer, C.D., *Ap. J. (Letters)*, **335**, L5 (1988).
50] Leiter, D., and Boldt, E., *Ap. J.*, **260**, 1 (1982).
51] Boldt. E., and Leiter, D., *Ap. J. (Letters)*, **322**, L1 (1987).
52] Zdziarski, A. A., *M. N. R. A. S.*, **233**, 739 (1988).
53] Fabian, A. C., Done, C., and Ghisellini, G., *M. N. R. A. S.*, **232**, 21p (1988).
54] Zdziarski, A. A., in *Thirteenth Texas Symposium on Relativistic Astrophysics*, ed. M. Ulmer (Singapore: World Scientific) p. 553 (1987).
55] Zdziarski, A. A., and Lamb, D. Q., *Ap. J. (Letters)*, **309**, L79 (1986).

COSMIC RAY SPECTRA FROM 1–100 TeV/AMU

*THE JACEE COLLABORATION**:

T. H. Burnett[i], S. Dake[b], J. Derrickson[h], W. Fountain[h], M. Fuki[d], J. C. Gregory[f], T. Hayashi[f], R. Holynski[j], J. Iwai[i], W. V. Jones[e,g], A. Jurak[j], J. J. Lord[i], O. Miyamura[c], H. Oda[b], T. Ogata[a], T. A. Parnell[h], E. Roberts[h], S. Strausz[i], T. Tabuki[h], Y. Takahashi[f], Y. Tominaga[g], J. W. Watts[h], J. P. Wefel[g], B. Wilczynska[j], H. Wilczynski[j], R. J. Wilkes[i], W. Wolter[j], B. Wosiek[j]

[a]ICR, Tokyo; [b]Kobe Univ.; [c]Osaka Univ.; [d]Matsusho Gakuen Junior College, Matsumoto; [e]NASA, Washington, DC; [f]Univ. of Alabama at Huntsville [g]Louisiana State Univ.; [h]NASA / Marshall Space Flt. Ctr.; [i]Univ. of Washington; [j]INP, Krakow.

JACEE uses balloon-borne emulsion chamber detectors to observe the spectra of cosmic ray protons and nuclei in the energy range 1-100A TeV[1]. For each interaction in the detector fiducial volume, the primary nucleus is identified in nuclear emulsion and/or CR39. Primary energy is estimated from the charged particle angular distribution and the total photon energy ΣE_γ observed in an electromagnetic calorimeter.

The data presented here are from short duration JACEE balloon flights prior to 1987, with total exposure approximately 120 m^2-sr-hr. The table below shows the number of events and exposure factors used for each category. Nearly the full exposure area is employed for the highest energies, while a more restricted fiducial volume has been fully analyzed for the lowest energy bins. The present data set does not include results from two recent long-duration southern hemisphere flights, which add approximately 160 m^2-sr-hr to the total exposure.

For protons and α particles, fitted power-law spectral slopes are consistent with previously published JACEE data[1] and with extrapolations of the Ryan *et al* [2] spectra, although a slightly higher intensity is obtained for the α spectrum.

For the heavier species, intensities are qualitatively consistent with extrapolations of the Spacelab-2 data[3]. These results are preliminary and may be sensitive to rebinning. Spectral slopes have not been estimated due to extremely limited statistics in the highest energy bins. The apparent enhancements in the CNO and Ne-S intensities above 10 TeV/amu should be treated cautiously until additional data, presently under analysis, are included.

component	protons	alphas	CNO	Ne-S	(Z>16)	"Fe"(Z>23)
Number of events	150	88	20	18	12	6
Exposure factor (10^5 m^2-sr-sec):						
highest energy bin	3.11	4.10	5.66	5.99	6.09	
lowest energy bin	0.03	0.27	1.04	0.41	0.80	

REFERENCES

1. Burnett, T., *et al,* 1983, Phys. Rev. Lett. 51 1010.
2. Ryan, M., *et al,* 1972, Phys. Rev. Lett. 28 985.
3. Grunsfeld, J., *et al,* 1988, Ap. J. 327 L31.
4. Hara, T. *et al,* 1983, *Proc. 18th ICRC* 9,198.

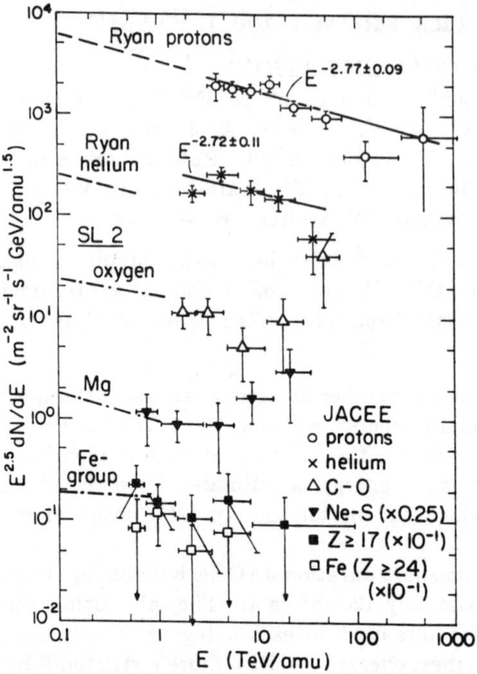

1. Differential spectra (in energy per nucleon) for protons, alphas, and nuclei, showing extrapolation of Ryan *et al* [2] and Grunsfeld *et al* [3] (SL-2) data.

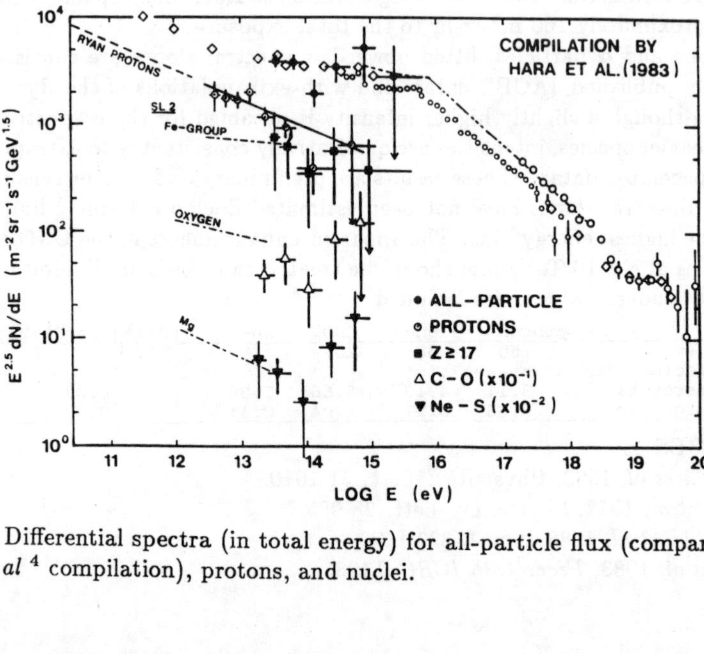

2. Differential spectra (in total energy) for all-particle flux (compared with Hara *et al* [4] compilation), protons, and nuclei.

The Muon Content of Extensive Air Showers and Limits on the Gamma Ray Component of Primary Cosmic Rays

J. Matthews
Harrison M. Randall Laboratory of Physics
University of Michigan, Ann Arbor, MI, 48109

Abstract

Results on the muon content of EAS above 100 TeV are presented from the Utah-Michigan array. Hadron rejection factors from 100 to 3000 are achieved, based on muon content. No anomalous excess of low-muon showers is observed. The ratio of gammas to hadrons in primary cosmic rays is found to be < 0.003 (90% CL) in showers with $N_e > 10^5$; similar limits are obtained for showers arriving from within 10° of the galactic plane.

Data and Analysis

The Utah-Michigan air shower experiment, located at Dugway, Utah, is the intial stage of the Utah-Michigan-Chicago Extensive Air Shower Experiment, more fully described elsewhere[1]. Data reported here were obtained from a 33 element surface scintillator array (radius ~100m) and 256 scintillators (632 m^2) buried under 3 m of earth. Muons with energies greater than 1 GeV can penetrate to the buried counters and there is negligible punch through of electrons at this depth.

Lateral distributions of particles are fit by least squares minimization: surface (electron) data are fit to an NKG function of fixed age while the buried (muon) counter hits are fit to a Greisen distribution. The total electron and muon size (N_e and N_μ) of the shower at the ground are obtained from the fit normalizations.

This is the simplest, fastest type of reconstruction technique but is subject to errors for hits near the core location; for this reason muon hits very near the core are excluded. This has the effect of giving a small subset of showers with $N_\mu = 0$. Such showers are included as candidates for primary gamma rays, giving the most conservative limits. Better fitting techniques are now developed which more accurately compute the muon size, but will not be discussed here.

Figure 1 exhibits N_μ and N_e for showers within 5.7° of the zenith; fitting yields a power law of the form $N_\mu \propto N_e^{.74}$. Deviations from the expected value of N_μ at a given N_e are analyzed to separate showers which are muon poor.

It is vital to account for the zenith angle dependence of both N_e and N_μ. The quantity $\log_{10}(N_\mu/N_e^{.74})$ is evaluated and the average value is computed versus $\cos(\theta)$. Applying this correction, the quantity $\log_{10}(N_\mu/\langle N_\mu\rangle)$ is computed for each shower, thus adjusting for zenith angle and electron size. Figure 2 shows the distribution obtained for showers with $N_e > 10^5$ and 10^6.

Discussion

Figure 2 shows no anomalous subset of air showers with low muon content (N_μ less than one-tenth the expected number or $\log_{10}(N_\mu/\langle N_\mu\rangle) < -1$). As mentioned,

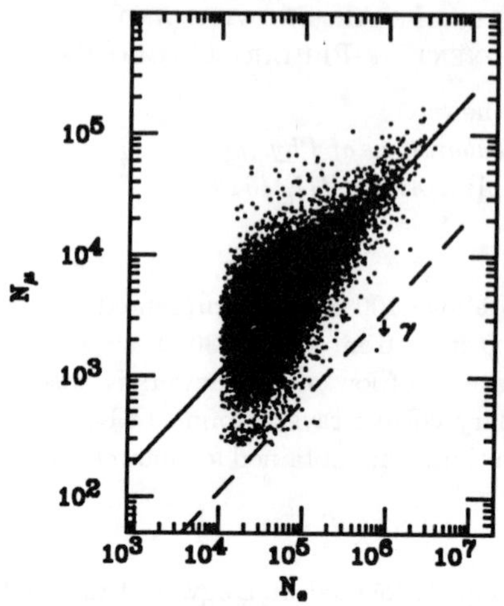

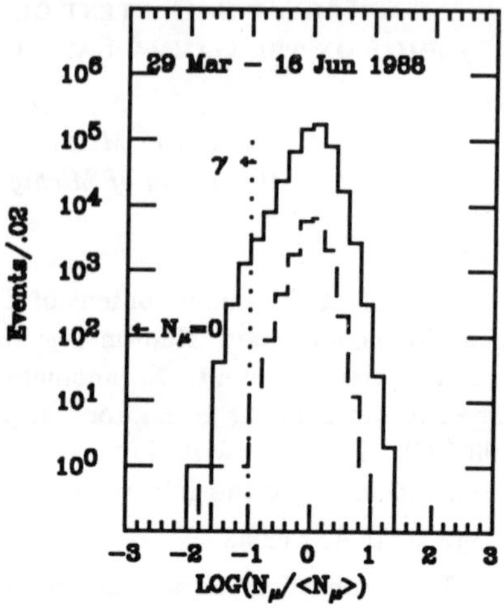

Figure 1. N_μ vs N_e for a 24 hr data run (10^5 events). Points below dashed line are γ candidates

Figure 2. Dispersion of N_μ from expected value. Solid: $N_e > 10^5$; Dashed: $N_e > 10^6$. γ showers should be left of dotted line.

those showers with $N_\mu = 0$ are believed to be fitting anomalies, but are included here to set the most conservative limits. Table 1 summarizes the results for showers of various sizes both over the whole sky and within 10° of the galactic plane. The "hadron rejection" factor is the fraction of showers observed with less than one-tenth the expected number of muons; The "γ fraction" is the 90% CL limit on the fraction of gamma induced showers in the total set, assuming all showers with $\log_{10}(N_\mu/\langle N_\mu\rangle) < -1$ are candidates.

Table 1

N_e	Hadron Rejection	γ fraction	γ fraction $\lvert b\rvert<10°$
$> 10^4$	1/150	.007	.007
$> 10^5$	1/300	.003	.003
$> 10^6$	1/4000	.0005	.0007

REFERENCE

1. K.G. Gibbs, et al., *Nucl. Inst. and Meth.* **A58**,67 (1988); D.F. Nitz, et al., *Proc. XXIV ICHEP*, Munich, 1988 (to be published)

REPORT ON THE WORKSHOP ON HIGH RESOLUTION GAMMA RAY COSMOLOGY UCLA, November 2-5, 1988

Ervin J. Fenyves
University of Texas at Dallas

The major goals of the Workshop on High Resolution Gamma Ray Cosmology were: (1) to identify possible sources of cosmic gamma rays, (2) to discuss exotic sources such as dark matter and cosmic strings, (3) to discuss new detector principles and a new generation of gamma ray telescopes, and (4) to discuss recent results in gamma ray astrophysics.

According to this the meeting consisted of eleven sessions:

1. Gamma Ray Spectra in Space: 3 papers presented by C. Fichtel, J. Trombka, and J. Audouze
2. Gamma Ray Bursters: 4 papers presented by K. Hurley, V. Petrosian, D. Hartmann and E. Fenimore
3. Point Sources : 1 paper presented by T. Weekes
4. Galactic Center: 4 papers presented by M. Morris, T. Prince, R. Lingenfelter and E. Chupp
5. Cosmic Strings: 4 papers presented by D. Spergel, A. Zdziarski, C. Thompson and D. Schramm
6. Dark Matter, Antimatter and WIMP Annihilation: 4 papers presented by F. Stecker, J. Primack, J. Silk and S. Rudaz
7. Compton and Pair Production Telescopes: 9 papers presented by S. White, J. Ryan, R. Hartmann, M. Atac, J. Park, T. Doke, K. Kasahara, T. Edberg and E. Fenyves
8. Gamma Background Produced by Reactors in Space: 3 papers presented by J. Primack, E. Chupp and J. Kurfess
9. New Detector Principles: 6 papers presented by D. Cline, G. Yodh, C. Akerlof, D. Koch, J. Matteson and R. Morse
10. Recent Results and New Projects in Gamma Ray Astrophysics: 7 papers presented by D. Nagle, T. Gaisser, R. Daly, K. Yu, F. Kajino, J. Grunsfeld and Y. Gao
11. Future Applications: Round table discussion moderated by E. Fenyves and D. Cline

The major conclusions of the workshop were that

(1) We have to develop a new generation of large area high angular and energy resolution gamma ray telescopes for studying the diffuse cosmic gamma rays, point sources, gamma ray bursters, the Galactic Center, gamma rays from antimatter, WIMP annihilation and decay, and from other exotic sources such as cosmic strings

(2) Applying new detector principles such as scintillating fiber detectors, liquid xenon calorimeters and ultrahigh energy gamma ray detectors

These detectors will permit to carry out extensive experiments in space, particularly in the Space Station, at mountain altitudes, sea level and underground, and open new windows into the universe. The most exciting new possibilities are:

(1) Looking back to High Z: The MeV Window, the study of the energy range 1-10 MeV which offers the best chance of observing discrete sources of very large redshifts, as discussed by D. Spergel, A. Zdziarski, C. Thompson and D. Schramm

(2) The Annihilation of Weakly Interacting, Neutral, Heavy Majorana Fermions into gammas, the GeV Window:

$$\chi\chi \rightarrow \gamma\gamma \tag{1}$$

giving rise to discrete gamma lines at GeV energies which may be observed experimentally, e.g. in a Space Station experiment, for a broad range of m_χ as it was shown by S. Rudaz (Fig. 1). Similar results supporting the findings of Rudaz have been obtained recently by D. Suson and L. Bergstrom.

The 3 1/2 day Workshop was attended by about 80 participants and the Proceedings will be published in Nuclear Physics B, Proceedings Supplement Section.

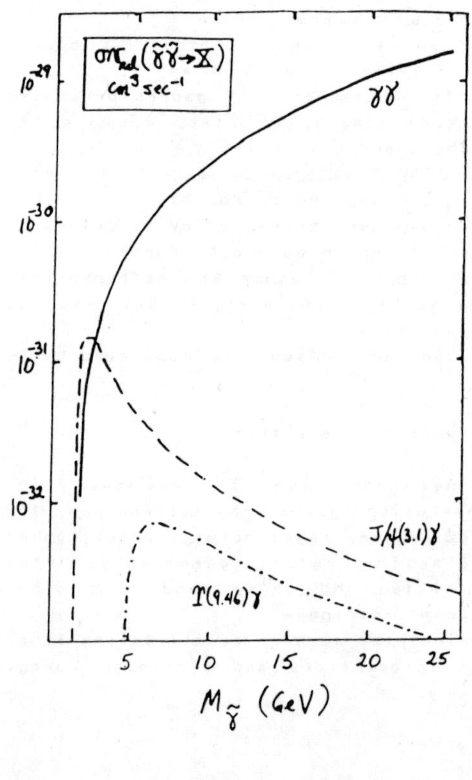

Fig. 1

Cross sections for photino annihilation vs. photino mass

NEUTRINO PHYSICS

ISSUES IN NEUTRINO PARTICLE ASTROPHYSICS

Lawrence M. Krauss[1]
Center for Theoretical Physics and Department of Astronomy
Yale University
New Haven CT 06511

Bitnet Address: Krauss@Yalehep

I. INTRODUCTION AND OVERVIEW

Attempting to summarize the relevant issues in neutrino particle astrophysics is a daunting task indeed. Neutrinos are the grand-daddies of the field. Almost every subfield discussed at this meeting is related, in one way or another, to some aspect of neutrino physics. If I try to think of the major developments which spurred the growing symbiotic relationship between particle physics and astrophysics, neutrinos were in at the ground level in almost every case. For example, it was the discovery in 1975 by Cowsik and McLelland that light neutrinos could dominate the mass density of the universe today which laid a precedent for the barrage of non-baryonic dark matter candidates on the one hand, and which instigated the impressive astrophysical simulations of large scale structure on the other. Similarly, the solar neutrino problem has over the years stimulated particle physicists to investigate such questions as the nature of quark and lepton families, mass matrices, and Grand Unification. At the same time, it has prompted astrophysicists to examine in detail the nature of the standard solar model.

The deep connection between particle physics and astrophysics through neutrino physics is made manifest if you examine the enshrined storehouse for our accumulated knowledge about particle physics-- the Particle Data tables. Alone among all the particles listed in the tables, one will find an entire section entitled "Neutrino Bounds from Astrophysics and Cosmology". As these tables are coincidentally compiled in part at the same location which which will now house a Science and Technology

---

[1] Also Visiting Scientist, Harvard Smithsonian Center for Astrophysics. Research supported in part by DOE and NSF Presidential Young Investigator Award.

Center in Particle Astrophysics, one cannot help but think that neutrinos even helped lay the groundwork leading to our hosts' good fortune.

Be that as it may, the difficulty of attempting an overview of neutrino particle astrophysics becomes apparent the minute one tries to think of ways of just outlining all of the neutrino physics /astrophysics connections. Any such attempt can quickly become hopelessly complex. For example, consider listing the important particle physics properties of neutrinos alongside the major astrophysics issues impacted by research in "particle astrophysics". If one attempts to connect items from the first column with those areas in which they play a role, one gets the following :

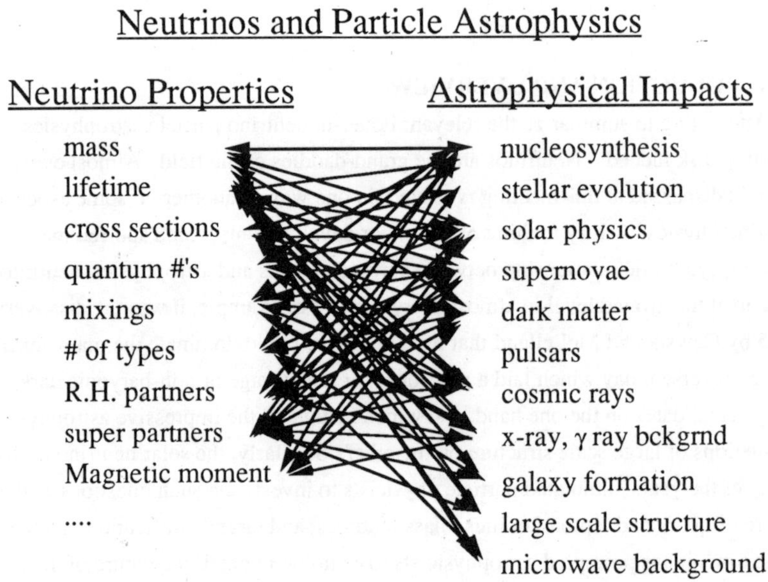

This looks like something the Maharishi might produce, not something which is likely to provide a good basis for an overview. Consider another approach however. The 1980's have seen the coming of age of "neutrino astronomy". Thus I am emboldened to propose the following figure, much as I might for electromagnetic astrophysics:

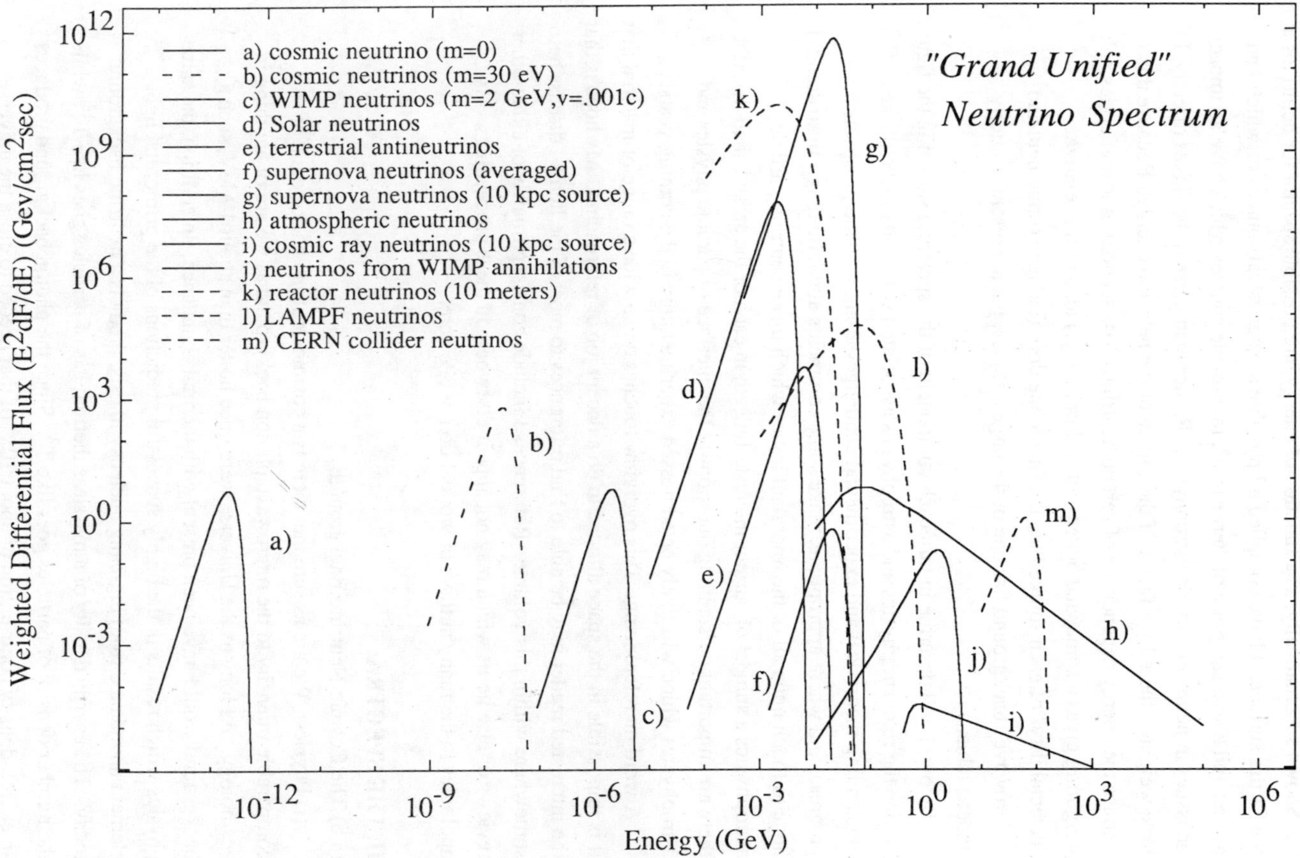

"Grand Unified"
Neutrino Spectrum
a) cosmic neutrino (m=0)
b) cosmic neutrinos (m=30 eV)
c) WIMP neutrinos (m=2 GeV,v=.001c)
d) Solar neutrinos
e) terrestrial antineutrinos
f) supernova neutrinos (averaged)
g) supernova neutrinos (10 kpc source)
h) atmospheric neutrinos
i) cosmic ray neutrinos (10 kpc source)
j) neutrinos from WIMP annihilations
k) reactor neutrinos (10 meters)
l) LAMPF neutrinos
m) CERN collider neutrinos
Weighted Differential Flux (E2dF/dE) (Gev/cm2sec)
Energy (GeV)

Shown schematically, as a function of energy, are possible spectra for neutrinos at the earth's surface. I have compiled all possible astrophysical sources of which I am aware, and also a sampling of "terrestrial", man-made sources which have an impact on issues in neutrino particle astrophysics. Rather than present the fluxes directly, I have weighted these by a factor of the square of the neutrino energy. Because this mimics the energy dependence of certain neutrino cross sections at low energies, this weighting gives a somewhat better (but still imperfect) idea of the relative "detectability" of each source than the flux alone does (i.e. the cosmic neutrino "microwave" background has one of the largest fluxes, but at present is utterly undetectable..)

Below, I will describe in turn each component of the spectrum shown in the figure, outline the relevant physics and astrophysics associated with it, describe the key outstanding theoretical and experimental challenges along with the prospects for progress, and where appropriate, make some comments about funding. In so doing, I hope to touch upon all of the important issues which make neutrino particle astrophysics a subject of current interest. Indeed, in each of the areas I shall describe there are important outstanding questions at the interface of particle physics and astrophysics which will likely be addressed experimentally in the coming years.

A word on referencing: This overview touches on so wide a range of material that it is impossible in the space allotted to do a decent job of referencing. My hope is that the interested reader may be able to find references to most of the items I discuss here somewhere within this volume. References I include consist primarily of either other reviews or else items which may not appear elsewhere in these proceedings. My apologies to the many authors of works I do not cite here.

II. THE SPECTRA:

(a,b) The Cosmic Neutrino Background:

(i) Physics: We are as sure that there is a cosmic neutrino background of comparable intensity to the microwave photon background as we are of anything in cosmology. As long as the Universe were once hotter than about 1 MeV, such a background would have been present with thermal abundance. In addition, the same physics which results in the highly successful predictions of the primordial light element abundance from big-bang nucleosynthesis, predicts a neutrino background today. The energy density of a massless thermal gas of particles goes like T^4, while the number density of particles goes like T^3. Since the photon background produces an energy density contribution of about 10^{-4} of that needed to close the universe today, the neutrino background could close the universe if its energy density were 10^4 times as large. How could this happen? If the neutrinos had a small mass M, then their energy density would not go as T^4, but rather as M times their number density, or as MT^3. Thus, the ratio of the energy density in the neutrino background to that in the photon background is $\approx M/T$. (actually since light or massless neutrinos are expected to be only about 1/10th as abundant as photons the ratio is $\approx M/10T$..). Setting $T \approx 10^{-4}$ eV, one finds that if $M \approx 30$ eV, light neutrinos would close the universe today.

(ii) Theoretical challenges: The realization that light, but massive neutrinos might form a significant fraction of the matter in the universe has stirred a great deal of activity among astrophysicists in the past 10 years. In particular, the possibility that such neutrinos might compose galactic halo dark matter has been studied in some detail. Numerical simulations of gravitational collapse in a neutrino dominated universe have been done by scores of individuals.[1)] The current wisdom is that the results are incompatible with what we actually observe. For example, the predicted clustering scale is too large, and there is too large a remnant anisotropy in the microwave background. The case is not closed however. Our knowledge of large scale structure in the universe is very sketchy, so surprises are likely. Also, recent inhabitants in the particle physics zoo, such as cosmic strings, may revitalize the neutrino picture.[2)] New measurements of large scale drift, as well as controversy over the cluster-cluster distribution function [3)] have also served to make things murky. There is plenty of room for more theoretical work before light neutrinos are laid to rest as dark matter candidates. It is clear, however, that light neutrinos cannot form *all* the dark matter. As Gunn and Tremaine discussed a long time ago [4)] phase space arguments restrict such neutrinos from contributing significantly to the mass of Dwarf Galaxies. The most recent direct application of this work to the Dwarf Galaxy DDO 154 has led to the claim that in order to account for the dark matter observed in this system the neutrino mass would have to exceed 90 eV.[5)] Of course, this is a somewhat lame result since no standard collapse scenario suggests that neutrinos would collapse in great number in such a system anyway.

Cosmic neutrinos, massless or massive, remain of great interest in primordial nucleosynthesis, where it has been claimed for some time that at best only one more light neutrino type is allowed[6)], thus possibly limiting the number of light quark and lepton families. Recently, potential loopholes in this bound have been raised by alternative scenarios for light element formation, including inhomogeneous nucleosynthesis[7)], and/or decays of exotic particles [8)]. All these scenarios predict different abundances of the isotopes of lithium than does the standard model.

However, controversies about lithium abundances, both actual and predicted, abound right now. With new theoretical work on lithium evolution in stars, we will hopefully soon be able to pin down the primordial lithium value and either prove these alternatives correct or false [9]. In addition, the neutron half-life has recently been re-measured and is now found to be shorter, easing somewhat the restriction on the number of neutrino families, even in the standard nucleosynthesis scenario. However, in order for 5 families to be allowed, the primordial Helium fraction by mass would have to exceed 25%, which is unlikely.[10]

Whether or not the neutrino has a mass, and what type of mass it may have, is of course closely tied up with our knowledge, or lack thereof, of mass generation for fermions, family structure, and possible Grand Unification. Developments in these areas on the particle physics front will be very important in finally settling the question of the mass of cosmic neutrinos.

Finally, nuclear and atomic theory play a very important role in the calculation of matrix elements which are relevant to experiments which attempt to limit the electron neutrino mass. Both the tritium end point measurements and double beta decay experiments are held hostage to the theoretical uncertainties associated with these calculations. Improvements in theory at this end could play a greater role than experimental improvements in tightening such laboratory constraints.

(iii) Experimental challenges: Of course, probably the greatest experimental challenge in all of particle astrophysics is to detect the cosmic neutrino background. Unlike other possible dark matter backgrounds, we know that a neutrino background must exist. Unfortunately, if neutrinos are light (i.e. 30 eV) or massless, no one has yet figured out a viable way to detect them. This is a pity, because detecting such a background would be even more significant than the discovery of the microwave photon background was, in that it would give us information about what the universe looked like at a time of 1 second, rather than 100,000 years.

On the other hand, experiments to directly detect a mass for the electron neutrino have multiplied in the last 10 years. These have focussed on either searching the shape at the endpoint of the Curie plot for the beta decay spectrum of tritium, or else on searching for peaks in double beta decay which would signal a neutrino-less mode (only possible if the neutrino has a Majorana mass). Unfortunately the results are still inconclusive. While SIN and Los Alamos based groups report upper limits of 18 and 27 eV respectively, the ITEP group still reports a non-zero measurement of 26±5 eV.[11] Supernova 1987a matched or exceeded the best terrestrial experiments in only 10 seconds of running, yielding an upper limit in the range of 15-20 eV. Finally,

Germanium double beta decay experiments put a limit (depending upon assumptions about nuclear matrix elements) of less than about 2 eV on a Majorana mass for a state which is primarily the electron neutrino.[12)]

How can these be improved? For a few Megadollars, the frozen tritium endpoint experiments might produce a limit in the range of 5-10 eV. Germanium double beta decay limits will not significantly improve for any amount of money, because they already have essentially no noise. However, it has been claimed to me that high pressure Xenon gas experiments might lower the Majorana mass limit to less than .1 eV, for about .5 Megadollars. Meanwhile, even if we are lucky for another supernova to go off right next door, say at a distance of 10 kpc, the neutrino mass limit will best improve by about a factor of three, to about 3-5 eV.

On the astrophysical front, observations of clustering on large scales will be very important before distinctions among different dark matter models can be made with any great confidence. This will require dedicated telescopes, and modern CCD devices, so the cost is in the Megadollar range. Observations of the microwave background anisotropy will provide probably the most powerful data limiting cosmological models, and I have been told at this meeting that all the interesting questions can be answered for 2.8 Megadollars. Finally, X-ray and Gamma ray observations are important for limiting the possibility that the cosmic neutrino background is unstable to decay.

Last, but not least, we are tantalizingly close to answering the question of the number of light neutrino species using direct probes. UA1 and UA2 have together, by examining the ratio of W decays into electrons versus Z decays into electrons, placed a 90% confidence upper limit of 5.9 neutrino types.[13)] This limit depends both on the fact that we know there are 3 neutrinos (actually we only assume the tau neutrino exists) and upon the fact that we don't know the top quark mass. If the top quark is discovered, this limit will improve. e^+e^- machines gives a slightly better limit based on the ratio of the processes, $e^+e^- \to \gamma$ +nothing vs. $e^+e^- \to \gamma + \mu^+\mu^-$. Of course perhaps the SLC, and most certainly LEP will tie down the Z width and settle the question once and for all within a year or two.

(c) WIMP Neutrinos:

(i) Physics: If neutrinos are heavy, they will remain in thermal equilibrium in the early universe until the temperature drops well below their mass. Thus, assuming no initial asymmetries, their abundance today compared to photons will be suppressed by a Boltzmann factor $\exp(-M/T_o)$ where T_o is the temperature at which their annihilations can no longer keep pace with the expansion rate. If Dirac neutrinos have a mass of around 2 GeV (5 Gev for Majorana neutrinos), their remnant density would be sufficient to close the universe today. If they are heavier, their remnant mass density decreases. Assuming that these particles form the galactic halo dark matter one can estimate a terrestrial flux by assuming they have a mean galactic virial velocity of about 10^{-3} c today, and by assuming a density characteristic of the inferred dark matter density in the solar neighborhood of about 0.3 GeV/cm^3. This yields a flux at the earth of about 10^7 (1 GeV/M) cm^{-2}sec^{-1}. The maximum energy they could transfer upon an elastic scattering with a nucleus would be about $2Mv^2$, or about 10^{-6} M, which would result in KeV range energy deposits for GeV mass particles.

(ii) Theoretical challenges: The interactions rates of heavy neutrinos are well established, assuming Z exchange in the standard model. Coherence is very important in leading to large event rates in materials, and must be carefully examined as both nuclei and WIMP candidates get heavier. (Of course, neutrinos cannot be too heavy if they are to be cosmologically relevant, unless one fudges things and adds asymmetries or something like that.) For Majorana neutrinos (or photinos), spin-dependence of the cross-sections becomes an issue, and recent EMC experiments have forced theorists to reconsider our notions of the contribution of sea quarks to the spin of the proton. Also important in this case is the calculation of the cross sections in different nuclei, not at all a straightforward exercise. As far as the flux of WIMPs is concerned, it is important for astrophysicists to figure out just how wide a range of density and velocities is allowed in the WIMP distribution. This will also be important for thinking of possible signatures for WIMPs in detectors. In fact, I think this is one area where theorists can give useful contributions. If experimentalists are lucky enough to see extra noise in their detectors, how can they know if it is due to WIMPs or not? Possibilities which have been suggested include: an annual variation in the energy dependence of the signal due to the earth's motion around the sun [14)], the obvious Z (or spin) dependence of the signal due to the scattering cross section [15)], or finally an additional signal due to a trapped distribution of WIMPs inside the earth [16)]. Condensed matter physicists might want to think of possible effects which might lead to a directionality determination when a nucleus recoils in a crystal after a WIMP collision.

(iii) Experimental challenges: This is one of the most active areas in particle astrophysics today. Double beta decay devices have already yielded significant bounds[12)], although experiments on the ionization by slow moving nuclei have yet to be performed to confirm that these devices are in fact sensitive to recoil nuclei with a few KeV. Similar ionization devices using silicon are now being built, which should allow a lower mass range to be explored. Finally, the explosion of interest in cryogenic detectors would require a whole separate article to properly convey. Let me just say that such devices may soon provide us with direct evidence of dark matter, one of the most important potential discoveries in physics. In any case they are leading to a whole new generation of particle detectors [14]. The cost of these devices, including prototype development will probably be at least a few Megadollars.

(d) Solar Neutrinos:

(i) Physics: The sun shines, therefore it must emit neutrinos. The flux on earth is almost 10^{11} $cm^{-2}sec^{-1}$, in the range up to about 14 MeV. The low energy pp neutrinos, which have not yet been directly measured, are closely tied to the major energy producing processes in the sun. The higher energy Boron neutrinos, which presumably are the source of events observed by Davis and collaborators over the past 20 years, are not so directly coupled, but nevertheless are very sensitive to the central core temperature of the sun. While the predicted flux of such neutrinos is about 7×10^6 $cm^{-2}sec^{-1}$, Davis infers a flux of less than 1.5×10^6 $cm^{-2}sec^{-1}$ coming out as electron neutrinos. Most recently, Davis has claimed a rise in this flux by at least a factor of 2-3 [17] At the same time Kamiokanda has reported a signal at a level of about 3×10^6 $cm^{-2}sec^{-1}$. Is this a fluctuation, or has the solar neutrino problem disappeared?

(ii) Theoretical challenges: The solar neutrino problem can be explained by one of three different alternatives: (a) the data is wrong, (b) the standard solar model is wrong, (c) standard particle physics is wrong. Many experimentalists doubt the first. Many astrophysicists doubt the second. Many particle physicists doubt the third. Happily, not everyone can be right. The first possibility has been eschewed in recent years, but the new Davis results have created some new interest in this. The second possibility seems difficult to reconcile with all the apparent successes of stellar modelling. Nevertheless, the inside of the sun is probably a pretty complex place, and there is room for tweaking. It may even be possible to affect neutrino production without altering many of the other successful features of the solar model. New results on the solar age, and on solar oscillations are not only providing new insights, but new challenges to attempts to revise the solar model. As far as particle physics is concerned, sufficiently exotic possibilities can remedy the problem, but usually at a price. The solution of choice in recent years has been the MSW mechanism for adiabatic neutrino mixing in the sun.[18] This is a very pretty mechanism, but requires neutrinos to have small masses. One way to tell the difference between the different possible solutions will be the new generation of experiments now being designed or coming on line to probe the solar neutrino spectrum, as I discuss below.

(iii) Experimental Challenges: To settle the issue we must perform new experiments to look not just at the Boron neutrinos, but also at the pp and Be neutrinos. If we obtain enough measurements we may eventually be able to unambiguously distinguish both the nature of the problem and the possible solutions. Right now, the first real time measurement of Boron neutrinos has been the famous Kamiokanda detection, down to about 6 MeV. The Sudbury Neutrino Observatory promises to push this threshold down to about 3 MeV, and maybe (although no one has demonstrated its feasibility to date) will be able to probe neutral current interactions of the solar neutrinos. As these are insensitive to possible mixing they might lay to rest the question whether the actual total flux is reduced, or just the component coming out as electron neutrinos. The Gallium experiments, soon to come on line, will be sensitive to low energy neutrinos from the sun, and can thus see whether the inferred deficiency is universal, or energy dependent. Finally a new generation of detectors, currently in the prototype stage, or else existing purely as paper proposals, will hopefully be able to detect in real time the entire solar neutrino spectrum. I bet on bolometry, but I am biased. I might also add that if we ever observe another supernova we may be able to test the MSW mechanism independently, although probably not unambiguously.

(e) Terrestrial antineutrinos:

(i) Physics: Radioactive decays of potassium and uranium make the earth a rich source of antineutrinos, with a flux at the surface in excess of the solar neutrino flux in certain energy ranges.[19)]

(ii) Theoretical considerations: Detecting this flux could provide a wealth of information on the interior processes governing the earth's dynamics. We do not even know if the earth is heating up or cooling down. The exact value of the flux is therefore uncertain. Also, no easy technique has yet been invented to look for them.

(iii) Experimental considerations: As of yet there are none...The techniques proposed are similar to the Davis chlorine experiment, but about 10 times as hard.

(f,g) Supernova Neutrinos:

(i) Physics: Essentially all of the energy emitted in the cataclysmic gravitational collapse of a star comes out in neutrinos, except for the 1% or so which creates the light show. The potential energy lost from the collapsing matter is about 10^{53} ergs, and thermal neutrinos are expected to be emitted with temperatures in the few MeV range over a time interval of 1-10 seconds.

(ii) Theoretical challenges: Remarkably, the gross features of the previously untested standard model of supernova dynamics were spectacularly confirmed by the observations of neutrinos from Supernova 1987a. However 19 measly events did not allow detailed tests of the model. Nevertheless, from the four observables from this supernova, and all future supernovae--- Neutrino Luminosity, Spectrum, Timing, Associated Signals----theorists have attempted to constrain : supernova models, production of new exotic particles such as axions, neutrino magnetic moments, neutrino masses, neutrino mixings, neutrino interactions, right handed currents, neutrino decays, neutrino gravitational couplings. Never has so little done so much for so many... What might we expect to derive from the next supernova? As far as the astrophysics is concerned, the details of stellar collapse could finally be probed. However as far as the particle physics is concerned the improvements possible are perhaps less dramatic than one might expect. The electron neutrino mass limit could be improved by about a factor of 3-5 at best. The muon neutrino mass limit might be reduced significantly from its present 250 keV value, as might the tau neutrino limit. If the millisecond scale deleptonization burst is observed directly, then one might hope to check such things as the MSW mechanism, and the weak equivalence principle for particles and antiparticles. Other than these new tests, the other limits derived from SN 1987a would merely be extended, but I do not think by very much. Before a quantum leap in what one can do with supernova neutrinos can take place we must observe enough supernovae to pin down the model parameters sufficiently well so that any variations away from these, even if subtle, can be observed. Don't hold your breath. I should mention that there is also a uniform background of neutrinos due to the time averaged signal from all supernovae [19]. If this is detected one might get information about the frequency of supernovae in our galaxy.

(iii) Experimental challenges: Ideally one would wish to be able to probe all the reactions of supernova neutrinos in detectors, i.e the neutral current scattering off both electrons and nuclei as well as the charge current reaction of antineutrinos off nuclei. Also, millisecond time resolution is essential, as is a big enough volume to detect 50-1000 events from the next supernova in our galaxy. To do all of this a new generation of detectors is required, but some new ones are imminent: Super Kamiokanda, MACRO, LVD, SNO. Even in their absence however, the old standbys: IMB and Kamiokanda will be able to do a lot. Unfortunately all these detectors may have to deal with a rep. rate which is rather small by particle physics standards, i.e. 10^{-9} Hz. To get around this Dave Cline [20] has proposed instrumenting a small mountain with neutron detectors to look for those neutrons ejected from nuclei by energetic neutrinos coming from supernovae as far away as the Virgo cluster, with a much higher frequency. Unfortunately, the mountain will have to be pretty clean. Finally, no techniques yet exist to measure the time averaged supernova signal, or to separate it from other backgrounds.

(h) Atmospheric Neutrinos:

(i) Physics: Cosmic ray protons hitting the atmosphere produce pions which decay and produce neutrinos. These provide a pesky background for proton decay detectors, but are now providing an interesting signal in their own right.

(ii) Theoretical Challenge: Kamiokanda sees almost 50% fewer muon neutrinos than expected, a claimed 3σ deviation [21] Is this due to muon-tau neutrino oscillations? Something else? Is it wrong? Theoretical estimates should, and are being re-checked.

(iii) Experimental Challenge: Repeat the experiments...

(j) WIMP annihilation neutrinos:

(i) Physics: WIMPs, whether they are neutrinos or not, may be captured by the sun[22] and earth[23] under favorable conditions . In general WIMPs and anti-WIMPS will annihilate in the cores of these objects [22], producing among other things light neutrinos[24]. Such neutrinos with energies in excess of 1 GeV will escape from these objects and may provide an observable signal amidst the atmospheric neutrino background described above.

(ii)Theoretical Challenges: The capture rate of WIMPs depends upon their scattering cross sections. These in turn depend on sometimes poorly known physics. In particular, as I mentioned above, the recent EMC experiments on deep inelastic scattering of muons off protons[25] suggests that spin dependent matrix elements may be smaller in the proton than previously expected. It is important to sort this out. It is also important to consider in some detail exactly what the expected branching ratio into neutrinos is for WIMP annihilations and what the resulting spectrum of neutrinos should be. Also, recent work which suggests that even particles whose interactions were thought to be primarily spin-dependent may have large spin-independent components in their interaction cross sections off matter could be very important in altering the predictions of capture and annihilation [26] Finally, also important in determining what the annihilation signal will be are considerations of the halo distribution in velocity space and in density.

(iii) Experimental Challenges: Further deep inelastic scattering experiments, neutrino based as well as muon and electron based, will be necessary to pin down all the SU(3) diagonal currents which enter into spin-dependent matrix elements. Collider experiments which search for supersymmetric particles continue to empirically determine the minimum possible mass scale at which supersymmetry may be broken. As such, they are slowly pushing up the mass at which the lightest supersymmetric particle might be expected to exist. Finally, the proton decay experiments which directly search for the annihilation neutrino signal can continue to try and beat down backgrounds and improve directionality in order to place more stringent limits on this possibility.

(i) Cosmic Ray High Energy Neutrinos:

(i) Physics: Cosmic rays of charged particles are observed up to energies of 10^{19} eV, while cosmic ray photons are observed up to 10^{16} eV. Clearly some very efficient cosmic accelerators must exist in order to produce these curiosities. Current theory suggests that perhaps at the site of supernovae charged particles are accelerated, which then may smash into neighboring companions, and like a vast beam dump experiment, produce secondary photons and neutrinos which we then should see among the cosmic rays bombarding the earth [27] Observations of such neutrinos could therefore elucidate the nature and sources of all high energy cosmic rays.

(ii) Theoretical Challenges: A great deal of work has been devoted to developing mechanisms by which cosmic ray acceleration could take place with the necessary energies and fluxes. Clearly, more data will be necessary before definitive statements will be needed. Nevertheless there remains lots of theoretical work to do in this area. In the process we may develop a whole new subfield of particle astrophysics. One of the great puzzles associated with present observations is the apparently high muon content originating from high energy cosmic ray photon showers. Is this real? Could it be due to QCD, as Halzen and collaborators have suggested, or is it some signal of new physics? Also, what exotic new physics at center of mass energies in excess of 10 TeV might we expect to deduce from observations of high energy neutrinos? Finally, neutrino scattering cross sections at ultra high energies are themselves of some interest. Their logarithmic growth is dependent upon quark structure functions at low x, and therefore may be useful to examine, if the primary neutrino spectrum can itself ever be deduced.

(iii) Experimental challenges: This is a field which is just beginning in earnest. Fly's Eye and other cosmic ray shower arrays are being supplemented by muon counters, and other neutrino detectors such as MACRO are coming on line. A future generation of detectors such as DUMAND and GRANDE may then allow concurrent observations of neutrinos and photons. It will be very important to check the muon content of photon and hadronic showers to settle the present puzzle. In addition, directionality will be important if we are ever to associate these cosmic rays with specific point sources. Detectors must also be very big, in order to obtain the rates necessary. In any such new field surprises are likely, and so detectors must be able to evolve to meet the needs brought on by the analysis of new data.

III. CONCLUSIONS:

I hope this terse and somewhat scattered review has at least help demonstrate both the wide scope of neutrino-related particle astrophysics, and the areas which could involve exciting new breakthroughs in the near future. If history is any guide, one thing is certain: whatever directions particle astrophysics evolves in, neutrino physics will be likely to play a central role. If the 1980's were the decade of the birth of neutrino astronomy, the 1990's may be the decade in which at least some of these outstanding issues associated with neutrino physics may finally be settled.

References:

1. see for example: *Inner Space, Outer Space,* ed. by E. Kolb et al., University of Chicago Press (Chicago 1986);
2. i.e see: *Cosmic Strings: The Current Status,* ed. by F. Accetta and L.M. Krauss, World Scientific (Singapore, 1988)
3. see G. Blumenthal et al, University of Santa Cruz preprint 1988
4. S. Tremaine and J. Gunn, Phys. Rev. Lett. 42, 407 (1979)
5. D.S. Spergel, D. H. Weinberg, J. R. Gott, Phys. Rev. D38, 2014 (1988)
6. J. Yang et al., Astrophys. J. 282, 493 (1984)
7. i.e. see G. Fuller, G. Matthews, C. Alcock.Phys. Rev. D37, 1380 (1988)
8. S. Dimopoulos et al, Astrophysics J. 330, 545 (1988)
9. see for example, C. Deliyannis et al, Yale Physics Preprint YCTP-P13-88
10. see L.M. Krauss and P. Romanelli, YCTP-P1-89
11. see for example *Proceedings Moriond January 1988 Meeting,* Editions Frontieres (Gif Sur Yvette, 1988)
12. D. Caldwell, UCSB preprint UCSB-HEP-87-10.
13. i.e. see C. Albajar, Phys. Lett.B198, 271 (1987)
14. see for example B. Sadoulet in *Proceedings Moriond March 1988 Meeting,*

Editions Frontieres (Gif Sur Yvette, 1988)
15. i.e. see M. Goodman, E. Witten, Phys. Rev. D31, 3259 (1985)
16. i.e. see L.M. Krauss, P. Romanelli, Phys. Rev. D39, Feb 15th 1989 issue
17. see Physics World, 1, 15 (1988)
18. see for example L. Wolfenstein, this meeting, and CMU preprint HEP88-14.
19. i.e. see L. M. Krauss, S. Glashow, D. Schramm, Nature, 310, 191 (1984)
20. see D. Cline, this meeting
21. K.S. Hirata et al, Phys. Lett. B205, 416 (1988)
22. i.e. see L. Krauss, HUTP-85/A008a, (unpublished); L.M. Krauss et al., Astrophys. J. 299, 1001 (1985)
23. see for example A. Gould. Astrophys. J. 321, 571 (1987)
24. i.e.see M. Srednicki, K.A. Olive, J. Silk , Nucl. Phys. B279, 804 (1987)
25. J. Ashman et al., Phys. Lett B206, 364 (1988); V. Hughes et al, Phys. Lett B212 511 (1988)
26. see for example J. Ellis, R. Flores, S. Ritz, Phys. Lett. B198, 393 (1987)
27. T. Gaisser, Bartol preprint BA-88-64

WHAT CAN WE HOPE TO LEARN VIA NEUTRINOS FROM A SUPERNOVA IN OUR GALAXY?

Adam Burrows
Departments of Physics and Astronomy
University of Arizona, Tucson, AZ 85721

ABSTRACT: Plenty.

1. CONCLUSIONS

The detection[1,2] of neutrinos from SN1987A is probably a once-in-a-lifetime scientific event. If it is not, and we are to be treated once again to such neutrino data before we are dust, what could we hope to learn from a galactic supernova burst, with its bonanza of events, about neutrinos, supernovae, neutron stars, and black holes that SN1987A did not teach us?

SN1987A verified the basics of collapse supernova and protoneutron star theory: a) neutrinos are associated with supernovae, b) $\bar{\nu}_e$'s are an important component of the burst, c) $\langle \epsilon_{\bar{\nu}_e} \rangle \sim 15$ MeV, d) $E_{\bar{\nu}_e} \sim 5\times10^{52}$ ergs ("6"$\times 5\times10^{52}$ ergs $= 3\times10^{53}$ ergs $\cong$ binding energy of neutron star), e) the duration of burst is seconds[3], not milliseconds[4,5], and f) that indeed we now have the required detectors. Furthermore, SN1987A illustrated Mark Twain's dictum about "fascinating" science: "one gets such a wholesale return of conjecture from such a trifling investment of fact." Amen. With more than ten papers per event (real or coincidental), a future of hundreds of events seems daunting. Nevertheless, data are required to legitimize a scientific field. However now legitimized in the eyes of even the most skeptical, collapse theory needs more constraints. Even with the sparse IMB and Kamioka data, neutrino mass[6], charge[7], magnetic moment, and decay rate upper limits have been derived. Furthermore, useful information concerning axions[8], majorons, and the weak principle of equivalence has been garnered for free. A bufferful of events from a nearby, galactic collapse should provide the blow-by-blow diagnostics of collapse, bounce, shock formation and breakout, accretion, and explosion that will finally nail the beast at the core of the dead star and allow it to be carefully dissected.

The total event yields in some detectors, existing or planned[9], for a galactic center burst (D = 8.5 kpc) are given in Table 1, taken from Burrows 1989[10]. Note in particular the large "ν_μ" and total yields expected in the SNO heavy water detector and the good prompt ν_e burst yields expected in SNO and ICARUS (if 3.6 ktonnes). One thousand events should, when digested, answer many of the outstanding questions that remain after SN1987A. Among these are: a) What is the mechanism of collapse supernovae (prompt, delayed, other)? b) What is the role of convection in protoneutron stars and supernovae? c) What is the mix of emitted neutrino species? d) What are the neutrino <u>spectra</u> and how do they evolve? e) Are there any plateaus or pulsations in the neutrino "light" curve? f) What remains: a neutron star or a black hole? g) What is the nature of the prompt breakout burst? h) How quickly can the ν_μ and $\bar{\nu}_e$ light curves turn on? i) What is the collapse rate in the galaxy? j) What is the neutron star equation-of-state? k) Is there a neutrino signature for other supernova types? l) What is the reason for the skewness in the angular distribution of the IMB events? m) What were the Mont Blanc events?

However, a major question that remains is when and where will the next nearby supernova show itself?

If I knew the answer to this question, I surmise that I would know the answers to the others. The one truly diabolical thing about SN1987A is that it has whet our appetites for an encore. Do we have the patience?

REFERENCES

1) Bionta, R. M. et al., Phys. Rev. Lett. 58, 1494 (1987) (IMB).
2) Hirata, K. et al., Phys. Rev. Lett. 58, 1490 (1987) (Kamiokande, KII).
3) Burrows, A., Ap. J. 334, 891 (1988); Burrows, A. and Lattimer, J. M., Ap. J. 307, 178 (1986).
4) Colgate, S. A. and White, R. H., Ap. J. 143, 626 (1966).
5) Arnett, W. D., Canadian J. Phys. 45, 1621 (1967).
6) Burrows, A., Ap. J. (Letters) 328, L51 (1988).
7) Bahcall, J. N. (1989), in press.
8) Burrows, A., Turner, M., and Brinkmann, R. F., Phys. Rev. D (1989), in press.

9) Mak, H. B., in the proceedings of the VIIth Recontre de Moriond, Les Arcs, France, Jan. 21-28, 1988.

10) Burrows, A., in "Supernovae," ed. A. G. Petschek (Springer-Verlag, 1989).

Table 1. Sample future detector event totals from the galactic center (8.5 kpc) (A. Burrows '89).

	Total #	Prompt ν_e's x($E_{\nu_e}/10^{51}$ ergs)	Infall ν_e's x($E_{\nu_e}/10^{51}$ ergs)	"ν_μ"'s
KII (H_2O)	548	0.5	0.5	20
SNO* (D_2O+H_2O)	1179	4.1*	4.1	500*
ICARUS (^{40}Ar)	169	3.6*	3.6	25
Homestake (^{39}Cℓ)	5.7	0.15	0.15	--

*Nota bene.

What particle physics can we learn from a Milky Way Supernova ? (That we don't already know.)

David Seckel
Bartol Research Institute
University of Delaware
Newark DE 19716

January 4, 1989

In my talk at Berkeley I gave a list of 9 items that we might understand better depending upon the data and its quality. After my talk, Dr. Cline suggested that such lists were not very useful in determining what were the crucial experiments. I must concur, and so these two pages will read a little differently than my talk.

To gain insight into what might be the crucial experiments, we need only look at how we draw inferences from the neutrino data of SN87A. Almost all dependable conclusions are drawn solely from the fact that neutrinos were observed at all. Other conclusions are possible based on gross features of the neutrino data. On the other hand, inferences based upon details of the signal generally require a more detailed model for the supernova than is available. As an example, mass limits for ν_e based primarily on the ten second time scale are fairly rigorous, but those based upon correlations in a subset of the observed events are not. From this experience, the crucial experiments are those that confirm general features expected of the neutrino flux. Conclusions based upon a low number of events or 'peculiar' features in the data will not be solid.

With this perspective, the crucial experiments to improve our knowledge are those that unambiguously confirm expected features of the neutrino flux. It is generally accepted that neutrino emission from a supernova will consist of five phases[1].

1. ν_e's from deleptonization during infall. This phase lasts a few 10's of milliseconds.
2. 'prompt' ν_e's when the shock breaks out of the neutrino sphere. This phase lasts for a few milliseconds.
3. $\sim$ 1 sec diffusive cooling as the shocked mantle below the neutrino sphere cools off.
4. $\sim$ 20 sec diffusive cooling as the core deleptonizes
5. long term core cooling by volume emission of neutrinos. This phase is dominated by ν_e and $\bar{\nu}_e$ emission with a temperature that decreases quickly as the core cools.

Phases 1 and 2 may account for $\sim$ 1% of the total energy budget. In phases 3 and 4 the luminosity is roughly flavor independent, although the mean energy of the ν_μ's and ν_τ's is expected to be about twice that for ν_e's. The $\bar{\nu}_e$'s are expected to be slightly cooler than the ν_e's.

The observations of SN87A confirm that $\bar{\nu}_e$ emission during phases 3 and 4 occurred generally as predicted. A galactic supernova at 10 kpc would have a flux $\sim$ 25 times that of SN87A. That, coupled with the development of new detectors, should allow other features to be identified.

From a particle physics point of view the most important experiment is to observe neutral current interactions and confirm the emission of ν_μ and/or ν_τ. If the expected fluxes are observed, limits on (or even a measurement of) the masses of ν_μ and ν_τ may be obtained. A difficulty is that neutral currents do not distinguish between ν_μ and ν_τ. We may find that one neutrino has a mass of, say 100 eV, but not know which it is. Another key experiment would show that the spectrum of μ and τ neutrinos is more energetic than for ν_e. Depending upon the data, inferences may be derived concerning neutrino decays and oscillations; however, without good spectral information uncertainties in the supernova model might cloud the picture.

The next item would be to observe the 'prompt' burst of ν_e's. Millisecond time resolution is desirable for any experiment. In principle, the short duration of this burst would allow improved limits on the mass of ν_e. In practice, uncertainty in the event rate predicted for the prompt pulse is an order of magnitude and confusion with the infall and cooling phases may obscure the sharpness of the prompt signal, so that a finite mass for ν_e may be difficult to infer. Similar remarks may be made concerning the use of the burst as a diagnostic for neutrino oscillations.

Although difficult because of the low luminosity, observations of the 5th phase of neutrino emission would provide information directly about the core of the collapsed object. Information about the nature of nuclear matter might be inferred or, perhaps, a quark star or strange star (instead of a neutron star) might be indicated.

The detection of neutral current events and the prompt pulse are the two most important experiments for particle physics implications. In addition, there is a host of particle physics issues that may be discussed if more detailed information is available and the supernova mechanism is well understood. In addition to neutrino masses, decays, and oscillations; one could, for example, study neutral current interactions of hadrons and neutrino-neutrino scattering. Both these phenomena affect the surface of last scattering for the neutrinos and therefore the details of their emission spectra.

Although this session concentrated on neutrino detection, we should not overlook the need for an improved gamma-ray facility. A 1 m^2 γ-ray telecope observing a galactic supernova would allow a 10^5 increase in sensitivity over SN87A when used as a probe of radiative neutrino decays.

Finally, it is unfortunate, but limits on direct core cooling by exotic processes are not likely to improve dramatically. Based upon SN87A, we know that exotic processes do not dominate cooling of the proto-neutron star. Quantitative improvements will depend upon details of the supernova mechanism. Perhaps a better diagnostic would be a search for signals in the energy range 100 – 200 MeV, albeit with cross sections much smaller than weak.

1. For a short review, see A. Burrows this proceedings.

Detecting Neutrinos from Future Supernovae

Eugene W. Beier
Department of Physics
University of Pennsylvania
Philadelphia, PA 19104

The success of the Kamiokande II and IBM light water detectors in observing the neutrinos from supernova SN1987a[1] has made the community aware of the need to maintain active detectors for neutrinos from future supernovae. In this brief review of detector potential we comment on the detector requirements and capabilities for astrophysics and particle physics, with reference to particular detectors or classes of detectors.

An essential design parameter of new detectors is the number of events which can be handled during the short burst of neutrinos from a supernova. To illustrate this point, Table 1 displays the event rates for supernovae at various distances scaled from the number of events which were observed by Kamiokande II from SN1987a using a fiducial mass of 2140 tons of light water. From Table 1 it is clear that this type of detector has no chance of observing a supernova from Andromeda, the nearest galaxy to our own galaxy. Undaunted by this paucity of events, David Cline presented[2] a design study for a detector which would have sufficient mass to be sensitive to supernovae in the Virgo cluster of galaxies.

A second observation from Table 1 is that detectors designed to observe supernovae in our galaxy should be capable of recording up to 10^4 events in the several seconds of the neutrino burst. Although this is not an enormous data rate, it does require that the experimenters plan ahead. If a potential core collapse star as near as Betelgeuse is to be detected, the detector should be able

KII Supernova Event Rates

SOURCE	DISTANCE (Kpc)	KII EVENTS
Andromeda	700	.05
LMC	50	10
Galactic Center	10	250
Crab Pulsar	2	6000
Betelgeuse	0.1	2.5×10^6

Table 1: Numbers of events expected in the Kamiokande II light water detector for supernova sources at various distaances.

to go into a sampling mode, rather than become dead for a significant portion of the collapse event.

The principal classes of neutrino reactions which can be used to detect neutrinos of the energies expected from supernovae are listed in Table 2. With a sufficient variety of detectors, or sufficiently clever individual detectors, much can be learned about supernova dynamics and about particle physics. While the Kamiokande II and IMB observations of SN1987a informed us of the time dependence of the energy distribution of $\bar{\nu}_e$, future supernovae at closer range can tell us about the fluxes of both $\bar{\nu}_e$ *and* ν_e, and with neutral current detection capability can tell us about the ν_μ and ν_τ fluxes. In particular, if the neutral current reactions occur at the same time as the ν_e and $\bar{\nu}_e$ for a galactic supernova, as can be determined by the characteristic energy deposit in the detector, the masses of ν_μ and ν_τ can be constrained to be less than about 50 eV, much less than present kinematic limits on the masses of those neutrinos. The masses of these neutrinos could furthermore be measured through the time delay with respect to the ν_e if the mass lies in the range between 50 eV/c^2 to 200 keV/c^2[3].

Both Čerenkov detectors and scintillator detectors have neutral current ca-

Neutrino Reaction Types

Reaction	Comments
I. $\nu_e + Z \rightarrow (Z+1) + e^-$	E_e determines E_ν (if measured)
II. $\nu_x + e^- \rightarrow \nu_x + e^-$	All flavors; $\sigma(\nu_e) \approx 6 \times \sigma(\nu_\mu, \nu_\tau)$ Electron direction $\approx$ Neutrino direction
III. $\nu_x + Z \rightarrow \nu_x + Z^*$	Need signature for NC excitation
IV. $\bar{\nu}_e + Z \rightarrow (Z-1) + e^+$	Similar to I, determines E_ν
V. $\bar{\nu}_x + e^- \rightarrow \bar{\nu}_x + e^-$	Similar to II, smaller σ

Table 2: Neutrino reaction types useful for detecting supernova neutrinos.

pability. The light water Čerenkov detectors are sensitive to the elastic scattering of ν_μ and ν_τ by electrons, a reaction in which the direction of the recoil electron is strongly correlated with the direction of the incident neutrino. In a light water detector, massive ν_μ or ν_τ reactions would be delayed relative to the ν_e and $\bar{\nu}_e$ reactions, and would point to the direction of the source. In a heavy water detector, this process is augmented by the neutral current dissociation of the deuteron with the produced neutron thermalizing and capturing on a deuterium nucleus or some other nucleus added to the heavy water to enhance the neutral current signal. The gamma ray emitted following neutron capture is detected, and its characteristic energy and isotropic angular distribution allow separation of this signal from other processes induced by neutrinos. The cross section for neutral current dissociation of the deuteron is much larger than the cross section for neutrino electron elastic scattering. Scintillator is sensitive to neutral current interactions through neutral current excitation of the carbon

Supernova Detector Types

Detector Type	Reaction	Measured Quantity			
		Number of Events	Time	Neutrino Energy	Neutrino Angle
Radiochemical	$\nu_e Z$	X			
Scintillator	$\bar{\nu}_e p$	X	X	X	
	$\nu_x e^-$	X	X		
	$\nu_x C$	X	X		
H_2O Čerenkov	$\bar{\nu}_e p$	X	X	X	
	$\nu_x e^-$	X	X		X
D_2O Čerenkov	$\bar{\nu}_e d$	X	X	X	
	$\nu_e d$	X	X	X	
	$\nu_x e^-$	X	X		X
	$\nu_x d$	X	X		
	$\bar{\nu}_e p$	X	X	X	

Table 3: Information which can be extracted from different types of detectors which are sensitive to supernova neutrinos

nuclei and the characteristic de-excitation gamma ray.

Notice also that the inverse beta decay reactions in Table 2 can be used to determine the energy of the incident neutrinos, while the elastic scattering from electrons can be used to determine the direction of the source. The presumed initial burst of ν_e from electron capture should have both a unique time structure and a characteristic angular correlation with the source.

The generic capabilities of four different types of neutrino detectors are displayed in Table 3. Since most radiochemical detectors are designed to be sensitive to the ν_e from the sun, they typically give only the integrated number of ν_e from the supernova. The other detectors all operate in real time, and yield valuable data on the collapse and explosion details of the supernova. Scintillation detectors are able to measure the spectrum of the $\bar{\nu}_e$, provided the gammas from the neutral current excitation of the carbon and the recoil electrons from neutrino electron elastic scattering can be fit simultaneously to the measured energy spectrum. The light water Čerenkov detectors can distinguish ν_e and $\bar{\nu}_e$ on a statistical basis, but measure the energy spectrum of only the $\bar{\nu}_e$. The heavy water Čerenkov counter[3] is surrounded by light water, and in principle can measure the spectrum of the $\bar{\nu}_e$ in that light water, and subtract it from the summed $\bar{\nu}_e$ and ν_e energy spectrum observed in the heavy water, thereby yielding measurements of both the ν_e and $\bar{\nu}_e$ spectra.

Finally, it is important in comparing data sets from the various detectors that all the detectors have accurate absolute time reference. It is also useful to maintain contact with the gravity wave detectors in the event that they should detect the collapse of a stellar core. Communication between the operators of detectors sensitive to neutrinos from supernovae should be maintained to assure that some detectors are operating all the time. It would be a sad day for science if neutrinos from the next galactic supernova were not detected.

This work was supported in part by the United States Department of Energy and the University of Pennsylvania Research Fund.

References

[1] K. Hirata, et al., Phys. Rev. Lett.**58** , 1490, (1987); R.M. Bionta, et al., Phys. Rev. Lett.**58** , 1494, (1987).

[2] D. Cline, these proceedings.

[3] A.B. McDonald, these proceedings.

THE SUDBURY NEUTRINO OBSERVATORY (SNO)

A.B. McDonald, R.T. Kouzes, M.M. Lowry, R.M. Key, *Princeton University*, H.C. Evans, G.T. Ewan, H.W. Lee, J.R. Leslie, J.D. MacArthur, H.B. Mak, W. McLatchie, B.C. Robertson, P. Skensved, *Queen's University*, M. Bercovitch, W.F. Davidson, C.K. Hargrove, H. Mes, *National Research Council of Canada*, E.D. Earle, G.M. Milton, E.T.H. Clifford, *Chalk River Nuclear Laboratories*, R.C. Allen, G. Bühler, P.J. Doe, F. Reines, *University of California, Irvine*, E.W. Beier, W. Frati, M. Newcomer, R. Van Berg, *University of Pennsylvania*, T.J. Bowles, B.T. Cleveland, S.R. Elliott, M.M. Fowler, R.G.H. Robertson, D.J. Vieira, D.L. Wark, J.B. Wilhelmy, J.F. Wilkerson, J.M. Wouters, *Los Alamos National Laboratory*, P. Jagam, J.J. Simpson, *University of Guelph*, E.D. Hallman, R. Haq, *Laurentian University*, B. Hollebone, D. Kessler, *Carleton University*, C. Waltham, *University of British Columbia*, D. Sinclair, N.W. Tanner, *Oxford University*.

The proposed SNO experiment will measure in real-time the flux and spectrum of electron neutrinos emitted by the ^{8}B and *hep* processes in the Sun, together with the total flux of neutrinos above 2.2 MeV, independent of neutrino type. In the event of a stellar collapse, it will be very sensitive to muon and tau neutrino fluxes, enabling sensitive mass measurements as well as providing details of the physics of stellar collapse. The experiment will consist of a 1000 tonne heavy water Čerenkov detector, surrounded by 5000 tonnes of light water shielding and located at the 6800 foot level of the Creighton mine near Sudbury, Ontario, Canada.

Solar Neutrino Detection

With independent measurements of electron neutrino and total neutrino fluxes, it will be possible to carry the search for neutrino oscillations and mass far beyond the levels achievable with terrestrial sources, and at the same time to delineate solutions to the Solar Neutrino Problem (SNP). The measurement of total solar neutrino flux, irrespective of neutrino type, will test solar models, independent of the possibility of oscillations. Comparison of the total flux above 2.2 MeV with the flux and spectrum of electron neutrinos with energies above 7 MeV then provides information about oscillations that is largely independent of solar models.

The SNO experiment will also be sensitive directly to the Mikheyev-Smirnov-Wolfenstein effect by measuring the energy spectrum of electron neutrinos reaching the earth. The energy dependence of oscillations introduced by this so-called MSW effect means that the gallium experiments in progress, which are sensitive to the low energy *pp* neutrinos, can no longer by themselves give an unambiguous answer to the SNP.

The proposed SNO detector makes use of three complementary neutrino reactions. The charged current (CC) reaction, $\nu_e + d \rightarrow p + p + e^-$, of the electron neutrino on the deuteron is unique to the SNO detector. It offers excellent spectral information, thereby providing a sensitivity to the MSW effect. The large cross

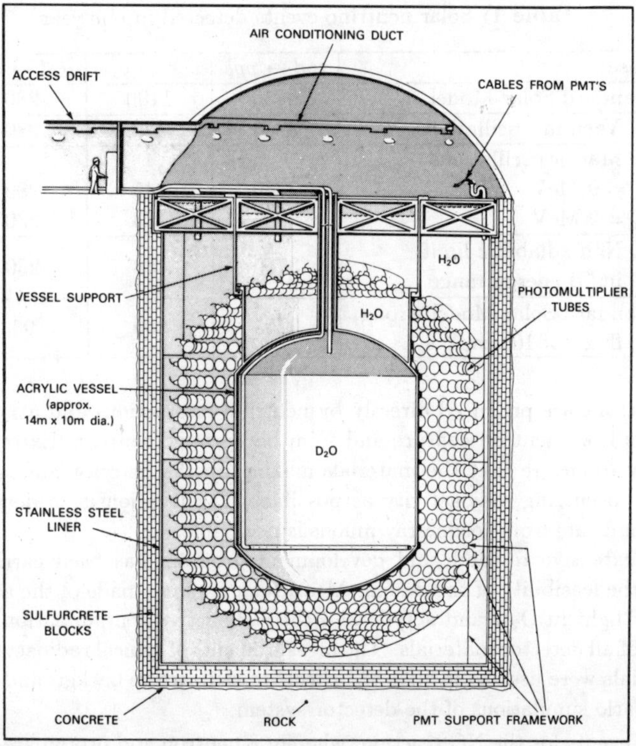

Figure 1: SNO detector. The diameter of the cavity is 20 m.

section will result in more than ten events per day. The neutrino-electron elastic scattering (ES) reaction, $\nu_x + e^- \rightarrow \nu_x + e^-$, is sensitive to all neutrino types, but is dominated by the electron neutrino. This reaction offers excellent directional information, but provides little information about the energy spectrum of electron neutrinos.The third reaction occurring in the SNO detector is the neutral current (NC) disintegration of the deuteron, $\nu + d \rightarrow \nu + p + n$, observed by the detection of the gamma rays resulting from the subsequent neutron capture. This reaction is equally sensitive to all neutrinos and would be used to measure the total flux of neutrinos with energies above the threshold of about 2.2 MeV.

The proposed SNO detector is illustrated in Figure 1. It contains 1000 tonnes of D_2O (isotopic purity > 99.85%), contained in a transparent acrylic vessel. This vessel, constructed from 5 cm thick panels, is surrounded by 4 m of high-purity H_2O and 1 m of low-activity concrete. Mounted uniformly around the vessel, at a distance of 2.5 m, are 1955 50-cm diameter photomultiplier tubes (PMTs), covering about 40% of the available area. The PMT array is sensitive to Čerenkov light produced by electrons and muons in the central regions of the detector.

Table 1: Solar neutrino events detected in one year.

Case	$\nu_e d \to ppe$	$\nu_x e \to \nu_x e$	$\nu_x d \to \nu_x pn$
Standard Solar Model	9750	1100	2800
A. Vacuum oscillations	3260	480	2800
B. Matter oscillations			
$E_c = 9$ MeV	3750	645	2800
$E_c = 2$ MeV	98	173	2800
C. Non adiabatic limit in ^{8}B energy range	3750	488	2800
Standard Solar Model wrong, flux 2×10^6 cm^{-2} s^{-1}	3255	360	933

The electrons are produced directly by neutrino interactions and also indirectly by γ-rays from neutron capture and from background sources. Extreme care is necessary to ensure that the materials making up the detector are as free from naturally occurring radioactivity as possible. The location is so deep that the background rate from cosmic-ray muons is negligible.

An extensive research and development program has been carried out to confirm the feasibility of the design. Measurements were made of the attenuation length of light in D_2O and acrylic and the radioactive contamination in typical samples of all detector materials. The measurements of typical radioactive content of materials were used to include the effects of radioactive background in the full Monte Carlo simulations of the detector system.

The events for the NC reaction, wherein a neutron and proton are generated by the inelastic scattering of all neutrino types from deuterium, will be observable when about 2.5 tonnes of NaCℓ is added to the heavy water. Then 83% of the neutrons will be captured by ^{37}Cℓ, producing γ-rays with energies typically around 8 MeV, producing a counting rate of 8 events per day for a total neutrino flux corresponding to the Standard Solar Model (SSM). An additional background must be considered for this reaction, corresponding to neutrons produced by the photodisintegration of deuterium by γ-rays with energies greater than 2.2 MeV. Considerable work has been done by the SNO collaboration to demonstrate that this background rate will be less than about 3 events per day. This background rate will arise predominantly from radioactive decay of members of the ^{232}Th and ^{238}U decay chains contained in the D_2O. This radioactive contamination can be sampled on-line to determine the background rate accurately. The techniques to be used for removal of radioactive impurities and monitoring of the residue involve the use of manganese-dioxide-coated acrylic beads which have been used successfully for many years in the detection of Ra and Th isotopes in oceanographic studies.

The expected data-taking sequence for the experiment would involve an initial filling of the acrylic tank with H_2O for a counting period of about 6 months. This would be followed by a filling with pure D_2O for a counting period of about 1 year, and finally the addition of NaCℓ for subsequent operation. The H_2O fill would enable events from the ES reaction to be observed, as well as providing an initial confirmation of the detector operation. The D_2O fill would provide a clean energy spectrum from the CC reaction, to be inspected for evidence of distortions from enhanced neutrino oscillations in the Sun. Finally, the addition of NaCℓ will provide a measure of the counting rate from the NC reaction, by comparing the observed spectrum with that obtained with pure D_2O.

Table 1 indicates the expected number of counts in one year for the three reactions under various scenarios. The rates are calculated assuming a ^{8}B flux corresponding to the Standard Solar Model (6×10^6 cm^{-2} s^{-1}), except for the last row, as indicated. The two entries under matter oscillations indicate different possible values of the effective cutoff energy.

Supernova Neutrinos

In the event of a supernova event within our Galaxy (predicted to occur every 10 to 50 years), the SNO detector would provide detailed information about the fluxes of all neutrino types. In a stellar collapse, nearly equal numbers of all neutrino and antineutrino types are expected to be produced by pair production during the cooling phase, lasting up to 10 seconds after an initial short burst of electron neutrinos. For an event at the middle of our Galaxy, SNO would observe about 17 events from the initial burst, and about 900 events from the cooling phase, primarily from the NC reaction. In addition, events from electron antineutrinos would be observed from interactions with both deuterium and hydrogen in the detector as indicated in Table 2. Since the events from the NC reaction would occur at about 8 MeV or lower, they would be separable from the majority of the other events expected at typical energies of 15 MeV or higher. The time distribution of events from the NC reaction could be used to define the mass of muon and tau neutrinos within the range 50 to 100 keV.

Table 2: Predicted number of events from a stellar collapse at 10 kpc at 100% detection efficiency

	Reaction	Target medium	Events in ν_e burst kt^{-1}	Events in cooling phase kt^{-1}
I	$\nu_e + d \to p + p + e^-$	D_2O	10	33
II	$\nu_x + e \to \nu_x + e$	D_2O/H_2O	1	16
III	$\nu_x + d \to \nu_x + p + n$	D_2O	6	760
IV	$\overline{\nu}_e + d \to n + n + e^+$	D_2O	0	20
V	$\overline{\nu}_e + p \to n + e^+$	H_2O	0	120

Status of SUPER-KAMIOKANDE

Kasuke TAKAHASHI

KEK, National Laboratory for High Energy Physics
1-1 Oho, Tsukuba-shi, Ibaraki-ken, 305, Japan

§1. Introduction

With a dramatic observation of the neutrino burst from the supernova explosion of SN1987A on February 23, 1987[1], and a detection of solar neutrino flux by the KAMIOKANDE-II[2], we are now even more confident that an excellent and very large underground facility such as the one, planned as "SUPER-KAMIOKANDE" will make an essential contribution to not only fundamental particle physics but also cosmology or astrophysics. In this brief talk, I would like to give a very brief description on how capable the proposed SUPER-KAMIOKANDE will be and a short note on status of its R&D work[3].

§2. General Layout of SUPER-KAMIOKANDE

SUPER-KAMIOKANDE, a very massive, imaging water Cherenkov detector, is a natural extension of the present, excellent KAMIOKANDE-II detector with various improvement and up-grade. Table I gives principal parameters of the detector in comparison of its size with that of its predecessor. A sketch of the proposed detector is shown in Fig. 1.

TABLE I Principal Parameters of SUPERKAMIOKANDE

Principal Parameters	SUPERKAMIOKANDE
Vessel	40 mh × 38 mϕ
Total mass	32000 t
Fiducial mass	22000 t
Number of PMTs	11000
% photosensitive coverage	40%
Photoelectron yield/MeV	5.8
Resolution for EM shower	$0.03/\sqrt{E}$ (E in GeV)
Anti-counter layer	depth > 2m
PMT time resolution at 1 photoele'n	3 ns
$\mu^+ \to e^+$ decay detection eff.	> 95%
Accuray in vertex determination[(a)]	30 cm
M/S separation[(b)]	up to ~80% overlapping rings
π/μ separation	for p ≲ 400 MeV/c
e/γ separation	partly yes

(a) for 100 MeV/c electrons
(b) M··π or μ, S··γ, e or π°.

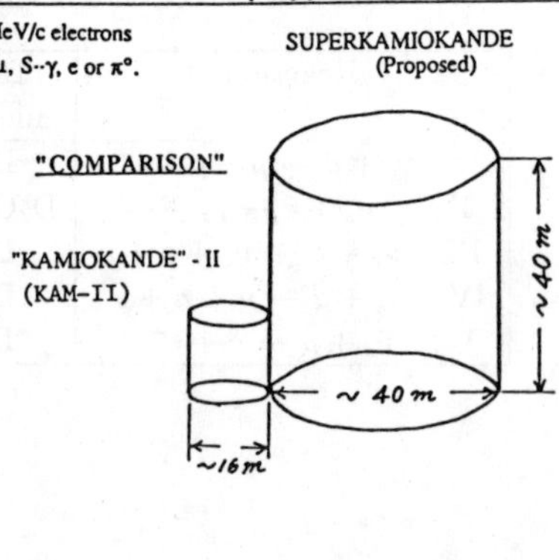

Physics motivation of this huge detector is of wide variety and of great significance from the point of view of Particle and Astrophysics. The detector is optimized to have an excellent performance for possible phenomena related to Grand Unified Theories (such as nucleon decay, n-n̄ oscillation and magnetic monopole), Neutrino Physics and Neutrino Astronomy (solar, supernova and atmospheric ν's), Dark Matter Problems and so on. In particular, the detector will be one of the most powerful detector to probe solar-ν problem, to look for ν-burst from various supernovae and to study atmospheric and cosmic ν's, covering most of grand unified ν-spectrum in the universe. Figure 2 shows expected number of events from ν-bursts from supernova as a function of distance. For solar-ν, the detection efficiency will be improved by an order of magnitude compared with the KAM-II (more than ∿2500 ev/yr with $E_e \gtrsim 7$ MeV). Of course, the SUPER-KAMIOKANDE will extend its sensitivity of nucleon decay to $10^{33} \sim 10^{34}$ years.

Fig. 1 SUPER-KAMIOKANDE: A Sketch of the Detector

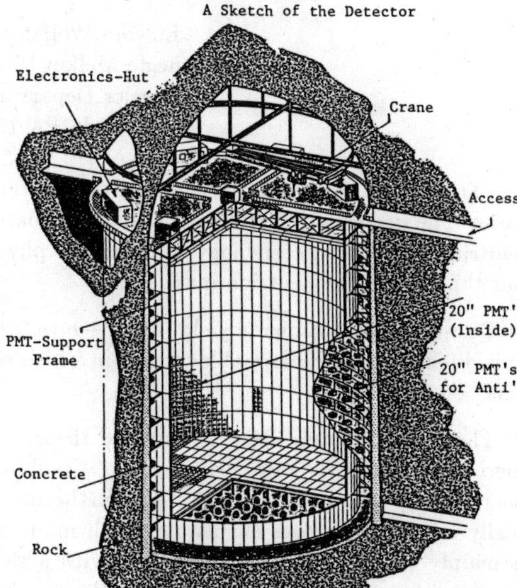

§3. Status

Since the SUPER-KAMIOKANDE will be equipped with two times more 20" PMTs than the KAM-II detector, and have light collection efficiency of about 40%, a factor of 2 improvement, energy and event-pattern resolutions will be much improved. This upgrade will be even enhanced by the new 20" PMT upgraded under the present R&D. The results of this work show a better single photoelectron peak and a smaller transit-time jitter of factor 2 ($\lesssim$3 ns). Water proof technique for PMT due to 40 m depth of the water tank is also under R&D. Water purification to get rid of radioactivities has been extensively explored in connection with solar-ν detection at the KAM-II. Upgrading the electronic system to get better charge and timing measurement is almost completed with semi-custom analogue integrated circuit[4].

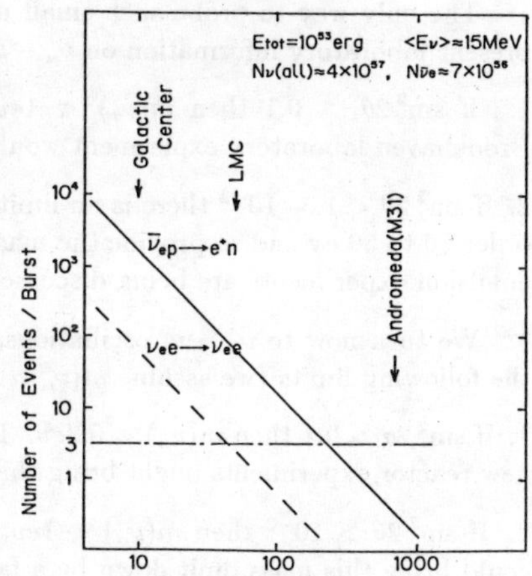

Fig. 2 No. of neutrino events from supernova burst vs distance.

References

1) K. Hirata et al.; Phys. Rev. Lett. 58 (1987) 1490.
2) M. Nakahata et al.; to be published.
3) Y. Totsuka; Proc. of 7-th Workshop on Grand Unification/ICOBAN'86 (ed. J. Arafune) Toyama, April 1986, P-78, and A. Suzuki; Proc. of the Workshop on Elementary Particle Picture of Universe, KEK, Feb. 1987, P-136.
4) T. Tanimori et al.; KEK Preprint 88-87 (1988).

NEUTRINO OSCILLATIONS

Lincoln Wolfenstein
Carnegie Mellon University
Physics Department
Pittsburgh, PA 15213

There is no compelling theoretical or experimental evidence that neutrinos have mass. The possibility that neutrinos do have mass can have profound implications for neutrino astronomy, cosmology, and particle physics. Theoretical ideas which govern our thinking about neutrino mass are:

1. Cosmological arguments yield $m(\nu_\tau) + m(\nu_\mu) + m(\nu_e) < 80ev$ for stable neutrinos. It is difficult to find reasonable models in which neutrinos decay fast enough to evade this bound.

2. Theoretical models like grand unified theories suggest that neutrinos have a mass hierarchy $m(\nu_\tau) >> m(\nu_\mu) >> m(\nu_e)$. Combining this with (1) we have $m(\nu_\tau) < 80ev$; $m(\nu_\mu) < 1ev$; $m(\nu_e) \leq 10^{-2}ev$. Furthermore the neutrino weak eigenstates are really mixtures of mass eigenstates with moderately small mixing angles. Thus ν_e is mainly the lightest mass eigenstate with a small mixture of ν_μ and a very small mixture of ν_τ. In analogy with quarks the $\nu_\mu - \nu_e$ mixing angle θ might be given by $\sin^2 2\theta = 0.01$ to 0.2. A similar angle might describe $\nu_\mu - \nu_\tau$ mixing.

The only way to probe such small masses is through neutrino oscillations. Our present laboratory information on $\nu_\mu - \nu_\tau$ oscillations may be summarized:

1. If $\sin^2 2\theta > 0.1$ then $m(\nu_\tau) < 1ev$ provided $m(\nu_\tau) >> m(\nu_\mu)$. A proposed Brookhaven laboratory experiment would bring this down to 0.1 ev.

2. If $\sin^2 2\theta < 1.3 \cdot 10^{-2}$ there is no limit. The possibility that $m(\nu_\tau)$ could be of the order 20 to 50 ev and so provide the mass needed for $\Omega = 1$ is so important that new emulsion experiments are being discussed to go down to $\sin^2 2\theta$ of a few times 10^{-4}.

We turn now to $\nu_\mu - \nu_e$ oscillations. Reactor and accelerator experiments yield the following limits (we assume $m(\nu_\mu >> m(\nu_e))$:

1. If $\sin^2 2\theta > 0.1$ then $m(\nu_\mu) < 0.3ev$. This limit also holds for $m(\nu_\tau)$. Proposals for new reactor experiments might bring this mass limit down by an order of magnitude.

2. If $\sin^2 2\theta > 10^{-2}$ then $m(\nu_\mu) < 1ev$. Proposals for new accelerator experiments could bring this mass limit down by a factor of 4.

To proceed to lower masses one must go to larger distances and this leads us to the use of neutrinos from cosmic rays, the sun, or a supernova.

Solar neutrinos from ^{8}B decay (Davis chlorine experiment) together with pp neutrinos (Gallium experiment) allow the exploration of $m(\nu_\mu)$ between $3 \cdot 10^{-4}$ and 10^{-2} if $\sin^2 2\theta > 10^{-2}$ and down to 10^{-4} if $\sin^2 2\theta > 0.1$. Indeed this also holds for $m(\nu_\tau)$ where θ is then the $\nu_e - \nu_\tau$ mixing angle. It is very important to realize that *solar neutrino experiments can explore a larger range of neutrino masses* suggested by theoretical ideas *than all laboratory experiments can.* This depends critically on the assumption $m(\nu_e) < m(\nu_\mu)$ because the analysis uses the MSW (Mikhaeyev-Smirnov-Wolfenstein) effect in which the effects of solar matter enhance the oscillation even for quite small values of θ.

To clearly identify possible oscillations for solar neutrinos a new generation of experiments will be needed. The goal of these experiments will be to look for energy dependence and possible time dependence of the neutrino flux and to look for ν_μ or ν_τ into which the ν_e may have oscillated. Among the proposed experiments are the Sudbury D_2O Neutrino Observatory (^{8}B neutrino spectrum including ν_μ, ν_τ detection), the Icarus liquid argon detector (^{8}B neutrino spectrum), the detection of ^{7}Be neutrinos using ^{115}In, and in the longer term the use of cryogenic detectors to see the spectrum of the pp neutrinos.

The mass range between 10^{-2} and 0.3 ev (Δm^2 between 10^{-4} and 10^{-1}) is difficult to explore for $\sin^2 2\theta \leq 0.1$. Because of the large density range outside the neutrinosphere the MSW effect for supernova neutrinos would be effective for values of Δm^2 up to 10^4 ev^2 and so could in principle explore this range. This really requires detection of the neutronization burst of ν_e and being able to check whether ν_μ or ν_τ appear in the same burst. The other way to explore this region is via cosmic ray neutrinos in proposed experiments like Superkamiokande, LVD, or Grande. To go to small mixing angles one would hope to exploit the MSW effect for neutrinos going through the earth. For values of Δm^2 near 10^{-4} some limits already exist from the IMB experiment using cosmic ray neutrinos around 250 Mev. More thought needs to be given to the use of cosmic rays to explore the entire range.

This research was supported in part by the U.S. Department of Energy under contract number DE-AC02-76ER03066.

THE LARGE VOLUME DETECTOR (LVD) AND ITS PHYSICS POTENTIALS

Presented by **Kwong H. Lau**
Department of Physics
University of Houston
Houston, Texas 77204-5504

THE LVD COLLABORATION

C. BARI, M. BASILE, G. BRUNI, G. CARA ROMEO, A. CASTELVETRI, L. CIFARELLI, A. CONTIN, C. DEL PAPA, P. GIUSTI, G. IACOBUCCI, G. MACCARRONE, T. MASSAM, R. NANIA, V. O'SHEA, F. PALMONARI, E. PEROTTO, G. PRISCO, G. SARTORELLI, and M. WILLUTZKY (*University of Bologna and INFN, Bologna, Italy*); J.A. CHINCELLATO, C. DOBRIGKEIT CHINCELLATO, A. C. FAUTH, and A. TURTELLI (*Department of Rajos Cosmicos, Campinos, Brazil*); M. ARYAL, K. DE, A. M. SHAPIRO, and M. WIDGOFF (*Brown University, Providence, Rhode Island 02912, USA*); F. ROHRBACH and A. ZICHICHI (*CERN, 1211 Geneve 23, Switzerland*); P. CAPUTI and G. L. SUSINO (*Colabrie University and INFN/INF, Cozenza, Italy*); A. BIZZETI, A. COMPAGNELI, G. CONFRONTO, G.LANDI, B. MONTELEONI, and P. PELFER (*INFN and University of Firenze, Firenze, Italy*); G. ANZIVINO, S. BIANCO, R. CASACCIA, F. CINDOLO, M. DEFELICE, M. ENORINI, F. FABBRI, C. JING, I. LAAKSO, S. QIAN, Z. SHI, A. SPALLONE, Y. SUN, L. VOTANO, and A. ZALLO (*Frascati INFN/LFN, Frascati, Italy*); K. LAU, J. LIU, B. MAYES, G. H. MO, L. PINSKY, J. PYRLIK, D. SANDERS, and R. WEINSTEIN (*University of Houston, Houston, TX 77204-5504, USA*); Y. DAI, L. DIN, C. JING, G. JING, Z. LU, P. SHEN, and Q. ZHU (*IHEP, Beijing, China*); D. Alyea (*Indiana University, Bloomington, IN 47401, USA*); T. KITAMURA and Y. MINORIKAWA (*Kinko University, Japan*); R. SCRIMAGLIO (*University of L'Aquila, L'Aquila, Italy*); P. ROTELLI (*University of Lecce, Lecce, Italy*); G. E. KOCHAROV (*Ioffe Physical Technical Institute, Leningrad, USSR*); M. DEUTSCH, E. S. HAFEN, P. HARIDAS, B. JECKELMANN, G. JI, H. H. KUANG, A. PITAS, I. A. PLESS, Y. YUAN, and C.Z. ZHAO (*Massachusetts Institute of Technology, Cambridge, MA 02139, USA*); V. S. BEREZINSKI, V. L. DADYKIN, F. F. KHALCHUKOV, E. V. KOROLKOVA, P. V. KORTCHAGUIN, V. B. KORTCHAGUIN, V. A. KUDRYAVTSEV, A. S. MALGUIN, M. A. MARKOV, V. G. RYASSNY, O. G. RYAZHSKAYA, V. P. TALOCHKIN, V. F. YAKUSHEV, and G.T. ZATSEPIN (*Institute for Nuclear Research, Moscow, USSR*); J. MOROMISATO, E. SALETAN, and E. VON GOELER (*Northeastern University, Boston, MA 02115, USA*); T. Wada (*Okayama University, Japan*); N. TAKAHASHI, and I. YAMAMOTO (*Okayama University of Science, Japan*); G. D'ALI and DE PASQUALE (*University of Palermo and INFN, Bologna, Italy*); B. ALPAT, F. ARTEMI, C. CAPPELLETTI, P. DIODATI, and P. SALVADORI (*University of Perugia and INFN, Perugia , Italy*); A. MISAKI, and INOUE (Saitma University, Japan); T. HARA (*Institute for Cosmic Ray Research, University of Tokyo, Japan*); C. AGLIETTA, G. BADINO, L. BERGAMASCO, C. CASTAGNOLI, A. CASTELLINA, G. CINI, M. DARDO, W. FULGIONE, P. GALEOTTI, C. MORELLO, G. NAVARRA, L. PANARO, L. PERIALE, P. PICCHI, O. SAAVEDRA, G. C. TRINCHERO, P. VALLANIA, and S. VERNETTO (*Institute of Cosmo-Geophysics CNR, University of Torino, and INFN, Torino, Italy*); F. GRIANTI and F. VETRANO (*University of Urbino, Italy*).

ABSTRACT

LVD is a large underground neutrino and muon detector located at the Gran Sasso Laboratory in Italy. The experiment is a collaboration of physicists from Italy, Russia, the United States, China, Brazil, and Japan. The detector features modular scintillation detectors and a finely segmented tracking system. The scintillation system consists of 1.8 ktons of liquid scintillator operated at a low energy threshold, providing detection of low energy neutrinos. The tracking system comprises 20,000 streamer chambers with good spatial resolution and high detection efficiency. The total mass of the detector is 3.8 ktons. The detector was designed for observation of supernova neutrinos, cosmic ray physics, searches for neutrino mixing and oscillations, dark matter, point sources, and proton decay.

1. THE LVD DETECTOR

The LVD detector has been described in great detail elsewhere.[1] Only the salient features of the liquid scintillation and tracking systems are described here.

1.1 THE LIQUID SCINTILLATION SYSTEM The 1.8 ktons of liquid scintillator is distributed in 190 identical modules, each holding eight tanks. Each tank weights about 1.2 ton and measures 1.5 m x 1.0 m x 1.0 m. The modules are arranged in a multilayer steel structure organized in 5 towers. Each tank is viewed by three 5" photomultiplier tubes. Timing resolution is about 3 ns, providing over 10σ separation for upgoing and downgoing muons. The light output of the liquid scintillator averages 5 photoelectrons per MeV of deposited energy with an energy resolution better than 20 %. The low radioactivity in Gran Sasso and careful selection of materials for the detector components allow the system to be operated at an energy threshold of 5-7 MeV. Neutrinos are detected by elastic scatterings off the electrons:

$$e^- + \nu \rightarrow e^- + \nu. \qquad (1)$$

The cross sections are well-known and the energy spectrum of the incident neutrinos can be obtained from the energy of the recoil electrons. Antielectron neutrinos are detectable via the inverse-beta reaction:

$$\bar{\nu_e} + p \rightarrow n + e^+. \qquad (2)$$

The energy of the positron is practically the same as that of the neutrino and hence provides a measure of the incident neutrino energy on an event-by-event basis. The neutron released in (2) is thermalized in about 170 μs before being captured and emitting a 2.2 MeV gamma ray which is also detected. The coincidence detection of the neutrons is achieved with over 50 % efficiency. Throughgoing muons deposit on the average 160 MeV in each scintillation tank. Muons that stopped in the tanks have in addition a delayed (about 2 μs) pulse from the decay. Electrons with an energy above a few hundred MeV exhibit characteristic electromagnetic showers and can be readily indentified when combined with information from the tracking system. More details of the scintillation system are given in Table I.

1.2 THE TRACKING SYSTEM Each scintillation module is surrounded by PVC streamer chambers, totaling 20,000 for the whole LVD. Each 6.3 m x 8 cm x 1 cm chamber has eight cells (each with a cross section of about 9 mm x 9 mm) sharing common gas volume and high voltage. Three sides of the cell walls are coated with a carbon suspension (DAG) in MIBK. The surface resistance is controlled to lie between 0.2 and 3 MΩ per square. The fourth side is the nonconducting outer sheath. The chambers are operated in the limited streamer mode in a gas mixture of 25 % argon and 75 % isobutane at a high voltage of 4.8 kV. The wire signals are picked up by the cathode strips outside the chamber in both the x and y directions. The width of the strips is 4 cm. The staggered double-layer chamber and strip planes translate to a spatial resolution of .58 cm (σ) in each coordinate. There are 9 horizontal and 6 vertical layers of tracking chambers, covering a geometric acceptance of 7700 m^{2}sr. Neglecting multiple scatterings, the angular resolution for throughgoing muons is less than a degree. More details of the tracking system are also given in TABLE I.

2. PHYSICS GOALS

2.1 STELLAR COLLAPSE The standard model of stellar evolution predicts a star with more than a few solar masses undergoes rapid collapse at the final stages of its evolution, resulting in a neutron star or possibly a black hole.[2] The finale is spectacularly accented by the ejection of its mantle and the release of typically 10^{58} neutrinos with an average energy of about 15 MeV. The model was recently confirmed in several respects by the neutrino bursts associated with the supernova SN 1987a.[3] In spite of the paucity of data, due to the large distance (50 kpc), a lot has been learned from SN1987a, ranging from obtaining the most stringent mass limit for the neutrinos ($<$ 20 eV), to limits on the equivalence principle violation.[4] A supernova in our galactic center is expected to yield close to a thousand events in the LVD detector, mostly $\bar{\nu_e}$ events detected by reaction (2). In addition to separately identifying ν and $\bar{\nu_e}$, LVD has almost 100 % efficiency for observing neutrino interactions down to 5 MeV. Combined with accurate arrival time and energy measurements of the neutrinos, this will provide invaluable information on the collapsing star and the hitherto unveiled physics of the neutrinos.[5]

2.2 NEUTRINO OSCILLATIONS Neutrino species with unequal masses are expected to oscillate from one kind to another.[6] Such oscillations have not been seen in experiments carried out so far. LVD will continue the search for evidence for oscillations by using atmospheric neutrinos produced by cosmic rays. The ratio of upgoing to downgoing neutrino fluxes, after correcting for geomagnetic effects, is sensitive to oscillation effects. For example, up to 10% deviation is expected for neutrino mass differences in the range of 10^{-4} to 10^{-3} eV^2. Over 1000 atmospheric neutrino induced contained muon events are expected in a few years of running in LVD. A statistically significant comparision can be made between observations and expectations without oscillations. It was pointed out that matter effects need to be taken into consideration.[7] Calculations for LVD including matter effects for the upgoing neutrinos have been done for the case of oscillation between ν_e and ν_μ.[8] The results are shown in Fig 1. Plotted in Fig. 1 are the 3% (solid line) and 5% (dashed line) effect contours relating $\Delta m^2 = m_1^2 - m_2^2$ and $\sin^2 2\alpha$ such that the integrated (60 degrees from the zenith) updown ratio of contained muon events differs from expectation (without oscillation) by 3 and 5 % respectively. The calculations have been done for models in which the density of the Earth is (i)constant(5.5 g/cc) and (ii) higher in the core (16 g/cc) and 3.2 g/cc elsewhere. The two models give essentially the same results. The search region for Δm^2 is from 10^{-3} to 10^{-4} eV^2 and $(\sin 2\alpha)^2 > .2$. It is noteworthy that this search region compliments searches performed in accelerators ($> 10^{-2}$ eV^2) and proposed searches using solar neutrinos (10^{-4} to 10^{-7} eV^2).[9]

TABLE I. The main characteristics of LVD

Area	2332	m^2
Geometric acceptance	7700	m^2sr
Length, height, and width	40,13,12	m
Volume of scintillator	2280	m^3
Mass of sicintillator	1800	ton
Mass of steel	1800	ton
Number of tracking channels	90000	
Number of streamer tubes	20000	
Number of phototubes	6400	
Number of analog channels	6400	
Tracking spatial resolution	4	mrad
Energy resolution	20	%
Energy threshold	5-7	MeV

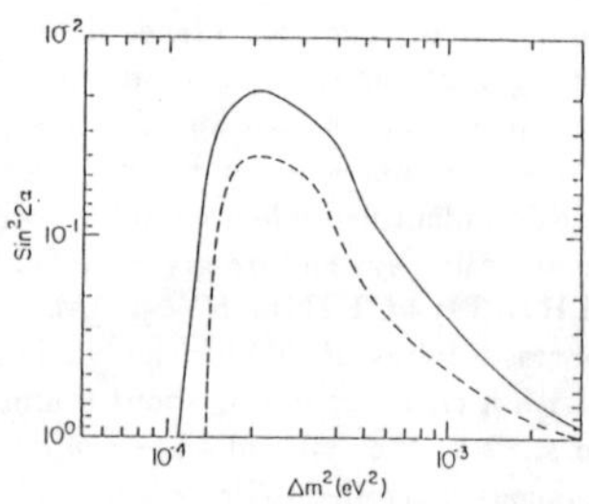

Fig. 1 Search regions for Δm^2 and $\sin^2 2\alpha$. See text for details.

2.3 DARK MATTER SEARCHES Some dark matter candidates postulated to occupy our galactic halo have cross sections with solar matter in the neighborhood of 10^{-36} cm^2, causing them to be trapped in the gravitational field of the sun as a result of collisions with the sun's matter.[10] A plausible candidate for dark matter is photino, the sypersymmetric partner of the photon.[11] The annihilations of photinos collected in the sun's core over its lifetime produce a detectable neutrino signal in LVD. Monte Carlo calculations, using the computed photino flux [12] predict a few events per year in LVD.[13] Light photinos annihilate to low energy neutrinos creating mostly contained-events in LVD whereas heavy photinos produce significant number of muons by neutrino interactions in the rock outside the detector. Background rejection is achieved by selecting only muons within an acceptance cone in the direction of the sun. A typical signal to background ratio is 1. It is worth pointing out that the rate of muons coming from interactions in the rock stays fairly constant for a photino mass > 20 GeV, up to a few TeV. This detection method will ultimately surpass searches conducted in accelerators which are limited by the available energy.

2.4 Solar neutrinos Solar neutrino detection depends on the background conditions at Gran Sasso. Given the experience of the LSD experiment running in the Mont Blanc tunnel which employed similar scintillation tanks,[14] and the measured lower radioactivity background at Gran

Sasso, it appears feasible to observe ^{8}B neutrinos from the sun; even though a higher energy cut may be necessary in the final analysis.

2.5 Proton decay Current searches for proton decay in the conventional decay modes are limited by atmospheric neutrino backgrounds. The present limit is $> 10^{31}$ years. A more sensitive search for

$$p^+ \to K^+ + \bar{\nu}_e, \tag{3}$$

a mode preferred by supersymmetry can be performed by LVD.[15] The combination of the primary kaon signal followed by the kaon decay pulse 12 ns later is readily distinguishable from the atmospheric neutrino background events. The subsequent decay of the muon or pion provides additional background rejection. Over 30 % of protons decaying via (3) involve 2 or more tanks.

2.6 Point sources Point sources of high energy gamma rays and neutrinos remain an interesting topic. However, Cygnus and Hercules, with the present luminosity, each yields less than 1 muon event per year.

3. SUMMARY

The LVD experiment is presently under construction at the Gran sasso laboratory. Its operation in the near future will add to our understanding of neutrino mixing and oscillations. Sensitive searches for some dark matter candidates, point gamma and neutrino sources, and exotic particles (e.g. monopoles) can be carried out. If a local supernova takes place after the detector begins data-taking, as many as 1000 events will be collected.

REFERENCES

1. C. Alberini *et al.*, Nuovo Cim. 9C, 237 (1986) and C. Bari *et al.*, NIM A264, 6(1988).
2. S. A. Colgate and R. H. White, Astrophys. J. 143, 626 (1966); H. A. Bethe *et al.*, Nucl. Phys. A324, 487(1979); S. Bludman *et al.*, Astrophys. J. 261, 611 (1982); R. L. Bowers and J. R. Wilson, Astrophys. J. suppl. ser. 50, 115 (1982); W. D. Arnett, Astrophys. J. 263, L55 (1983); J. R. Wilson, Numerical Astrophysics (eds. J. Centrella *et al.,*) 422 (Jones and Barlett, Boston, 1986); A. Burrows and J. M. Lattimer, Astrophys. J. 299, L19 (1985); and J. R. Wilson *et al.*, Ann N. Y. Acad. Sci. 470, 267(1986).
3. K. Hirata *et al.*, Phys Rev. Lett 58, 1490 (1987); R. Bionta *et al.*, Phys Rev. Lett. 58, 1494 (1987) ; M. Aglietta *et al.*, Europhys. Lett. 3(12) 1315(1987); and M. Aglietta *et al.*, Europhys. Lett. 3(12) 1321 (1987).
4. See, e.g., J. Bachall and S. Glashow, Nature 326, 476 (1987); W. Arnett and J. Rosner, Phys. Rev. Lett. 58, 1906 (1987); D. Dar and S. Dado, Phys. Rev. Lett. 59, 2368 (1987); J. Arafune and M. Fukujita, Phys. Rev. Lett. 59, 367 (1987); M. Longo, Phys. Rev. Lett. 60, 173 (1988); I. Goldman *et al.*, Phys. Rev. Lett. 60, 1789 (1988); G. Raffelt and D. Seckel, Phys. Rev. Lett. 60, 1793 (1988); and M. Turner, Phys. Rev. Lett. 60, 1797 (1988).
5. M. Turner, this proceedings.
6. Z. Maki *et al.*, Prog. Theor. Phys. 28, 870 (1962); and B. Pontecorvo, Soviet Physics, JETP 26, 984(1968).
7. L. Wolfenstein, Phys. Rev. D17, 2369(1978); and S. P. Mikheyev and A. Smirnov, Nuovo Cimento 9C, 17(1986).
8. K Lau, UH-LVD-038, January 1989.(Unpublished.)
9. L. Wolfenstein, this proceedings.
10. W. H. Press and D. N. Spergel, Astrophys. J. 296, 697(1985).
11. See, e.g., J. Ellis, D. V. Nanopoulos, and S. Rudas, Nucl. Phys. B202, 43(1982).
12. J. S. Hagelin *et al.*, UMN-TH-566/86 and MIU-THP-86/12, July 1986.
13. K. Lau, UH-LVD-005, January 1988.(Unpublished.)
14. O. G. Ryazhskaya, Proceedings of the 12th International Conference on Neutrino Physics and Astrophysics , p.286 (1986).
15. V. S. Berezinsky *et al.*, Nucl. Phys. B262, 383 (1985).

CORRELATION ANALYSIS OF THE DATA RECORDED BY THE BAKSAN, KAMIOKA AND MONT BLANC DETECTORS DURING SN 1987A

M.Aglietta[a], G.Badino[a], G.Bologna[a], C.Castagnoli[a], A.Castellina[a], V.L.Dadykin[b], W.Fulgione[a], P.Galeotti[a], F.F.Kalchukov[b], V.B.Kortchaguin[b], P.V.Kortchaguin[b], A.S.Malguin[b], V.G.Ryassny[b], O.G.Ryazskaya[b], O.Saavedra[a], V.P.Talockin[b], G.Trinchero[a], S.Vernetto[a], G.T.Zatsepin[b], V.F.Yakushev[b]

(a) Istituto di Cosmogeofisica del CNR, Torino, Italy, and Istituto di Fisica Generale dell'Università di Torino, Italy

(b) Institute of Nuclear Research of the Academy of Sciences of USSR, Moscow, USSR

During the explosion of the LMC supernova, on feb. 23, 1987, four underground experiments were running and reported evidence for detection of supernova neutrinos, within the time uncertainty of the instant of the collapse as given by the optical data. Very soon after, we suggested[1] to perform a correlation analysis among all the available data. Unfortunately, the IMB data are difficult to compare with the other ones because of the much higher energy threshold of this detector. On the contrary, the other detectors: Baksan (BST), Kamioka (K2) and Mont Blanc (LSD) have similar energy thresholds, being the 50% efficiency at about 10, 7.5 and 5 MeV respectively.

In this paper, we report preliminary results of the correlation analysis of the data recorded by these three detectors during the SN 1987a explosion. The overlapping time of the data is 36 hours for K2 and LSD (from feb.22, 10:00 UT to feb.23, 22:00 UT), and 4.5 hours for BST and LSD (feb.23, from 0:30 UT to 5:00 UT). The event rate in this period is 2.0, 1.4 and 0.7 per minute for BST, K2 and LSD respectively.

Since the accuracy in the absolute time of BST and K2 is rather poor (but their relative timing is fairly good), and assuming these detectors recorded a burst in time coincidence with IMB at 7:35:35 UT, their universal times must be shifted by +7.7 sec for K2, and by -30.4 sec for BST. In addition, we allow for a ± 2 sec uncertainty in the coincidence time of these two detectors with IMB.

Since a correlation has already been found[2] between LSD and the Maryland and Rome gravitational wave detectors in a period of time of about 2 hours, centered at 2:45 UT, we firstly searched for coincidences between BST, K2 and LSD, in a time interval from 1:45

to 3:45 UT of feb.23. In this 2 hours period, 240 events have been recorded in BST, 191 in K2, and 91 in LSD.

Thus, adopting a coincidence time $\Delta t = \pm\ 0.5$ sec, the expected number of random coincidences during this 7200 sec time interval is:

$$C_{(K2\text{-}LSD)} = 2.41, \quad C_{(BST\text{-}LSD)} = 3.03, \quad C_{3\text{-}fold} = 0.08$$

From the present analysis, the maximum number of coincidences:

$$C_{(K2\text{-}LSD)} = 9, \quad C_{(BST\text{-}LSD)} = 10, \quad C_{3\text{-}fold} = 1$$

is obtained by shifting the BST and K2 times (we used steps of 0.1 sec) when the shift is in the interval from +6.9 to +7.0 sec for K2, and in three intervals from -29.1 to -29.8 sec for BST. Both time shifts are well inside the allowed ranges required for a coincidence with IMB. The other time shifts (in total we made 601 shifts for BST from -55 to +5 sec, and 1201 shifts for K2 from -60 to +60 sec in order to take into account all the possibile shifts allowed from the published uncertainty in the UT of these detectors) give a number of coincidences in a very good agreement with the expected ones. Taking into account all the possibilities on the time shifts, the probabilities that the above coincidences are due to random fluctuations are of the order of the percent.

A search for coincidences over a longer period of time is possible only for K2 and LSD, for which we have 17 time-intervals of 2 hours each. During this period we have 1462 events in LSD and 2890 events in K2; hence, the expected number of coincidences is 2.03 in a 2 hours time-interval. With the K2 time shifted by +6.9 to +7.0 sec, the only period with a significant excess of coincidences is the period previously considered, i.e. from 1:45 to 3:45 UT, during which 9 coincidences were observed.

The other 16 time intervals have been further analysed in the same way as discussed previously: i.e. by shifting the K2 times in steps of 0.1 sec, from -60 to +60 seconds, for a total of 19216 shifts. In all cases, the number of observed coincidences agrees very well with the number expected from a Poisson distribution. Further analysis is in progress with a more extended set of data from BST.

References

1. M.Aglietta et al., Proc. ESO Workshop on SN 1987a, ESO, Garching bei Munchen, 1987, I.J.Danziger Ed., p.207
2. M.Aglietta et al., Proc. 2nd Rencontres de Physique de la Vallee d'Aoste, La Thuille, 1988, M.Greco Ed., p.107

Interpretation of the Kinematical Properties of SN1987A Neutrino Data by an Admixture of $\nu_{\mu|\tau}$

Danuta Kiełczewska

Phydics Dept., University of California, Irvine, Ca. 92717
and
Institute of Experimental Physics, Warsaw University
Hoza 69, Warsaw, Poland

Abstract

Statistical tests suitable for small data samples were used to compare the observed energies and angles of the SN1987A neurino events with distributions expected for a range of assumed temperatures and relative luminosities of different neutrino species. The highest confidence levels were obtained for the simulated samples with an admixture of $\nu_{\mu|\tau}$ to the $\overline{\nu_e}$ interactions.

The detection of interactions caused by neutrinos from the SN1987A explosion resulted in numerous studies providing far reaching conclusions about supernova explosions and neutrino properties. Most of the studies are based on the assumptions that the events are due to the interactions of $\overline{\nu_e}$, with maybe a small admixture of ν_e .

However the angular distributions in both underground water Cherenkov experiments [1,2] have the same trend to be more forward peaked that would fit to nearly isotropic distribution appropriate for the interaction of $\overline{\nu_e}$ on protons [3]. Although the data sample is too small to exclude statistical fluctuation and invoke new phenomena, the anomaly suggests that a critical check should be made to what extent the assumption about the type of neutrinos is justified.

It is widely accepted that μ and τ neutrinos emitted as a result of a supernova collapse have higher energies than electron neutrinos due to greater opacity of the infalling matter for the latter species. In addition trigger efficiencies in both experiments strongly favor higher energies. The cross sections rising with E_ν suggest then that $\nu_{\mu|\tau}$ may become a competition for $\overline{\nu_e}$.

For this analysis samples of neutrino interactions were simulated to get for both detectors and their trigger efficiencies the expected rates as well as the energy and angular distributions of electrons. Neutrino energies were distributed according to the Fermi - Dirac function, assuming a variety of temperatures for different neutrino species. A free of binning test of Smirnov - Cramer - Von Mises (SCVM) was used to quantify the goodness of fit of expected distributions to the data.

The best solutions for different assumptions about maximum luminosity and relative fluxes of different species, tend always occur for low temperatures $T_{\overline{\nu_e}}$ and $T_{\nu_{\mu|\tau}}$ $\geq$10 MeV. As an example the confidence levels obtained using the SCVM test for $T_{\overline{\nu_e}}$=2.5 MeV, $T_{\nu_{\mu|\tau}}$=10 MeV and T_{ν_e}=4 MeV are given in Table 1. For comparison the results of the SCVM test applied to samples generated assuming the neutrino spectra and luminosities obtained by Mayle et al.[4] are also shown. The energy distributions were tested independently in both experiments, because they cover distinctly different energy range.

Table 1
Confidence levels (%) - SCVM test

	this work	Mayle et al.[5]	
		12M⊙	25M⊙
IMB and Kamioka angles (19 events)	6.3	0.8	0.6
IMB energies(8 events)	33.0	13.0	26.0
Kamioka energies (11 events)	93.0	27.0	13.0

The results presented in Table 1 were obtained assuming the same luminosities for different neutrino species. For an assumed total luminosity of 6×10^{53} *ergs* the expected number of events is 14.8 and Poisson probability to detect 19 or more events is then 15% .

The expected content of both samples normalized to the observed 8 and 11 events is given below.

	$\overline{\nu_e}$	$\nu_{\mu\|\tau} + \overline{\nu_{\mu\|\tau}}$	ν_e
IMB	4.02	3.58	0.41
Kamioka	9.31	1.11	0.58

It is seen that the IMB and Kamioka detectors are sensitive to different kind of neutrinos. While it looks that the latter detected mainly $\overline{\nu_e}$'s, it is possible that the IMB collaboration recorded a few $\mu|\tau$ neutrinos. The IMB sample appears to be a mixture of different kind of interactions, including CC absorptions of μ neutrinos on protons or oxygen nuclei, which in the water Cherenkov detectors produce subthreshold muons and the observed μ decay electrons.

The data (especially the Kamioka sample) display a characteristic correlation: forward peaking comes from the events with higher energies. Such a correlation cannot be accounted for by ν_e scattering, because the recoil electrons take per average only half the neutrino energy and therefore there is no reason why they should be more energetic than the positrons from the inverse beta decay. On the other hand significantly higher energies of the $\mu|\tau$ neutrinos justify this correlation.

Acknowledgement

I gratefully thank A.Burrows, A.Dar, M.Goldhaber, J.Schultz and L.Sulak for discussing the material in this report. This research was supported by the U.S. Department of Energy.

References

1. Bionta,R., et al., *Phys. Rev. Lett.* **58**,*1494(1987)*
2. Hirata,K., et al., *Phys. Rev. Lett.* **58**,*1490(1987)*
3. Bratton,C.B., et al., *Phys. Rev.* **D37**,*3361(1988)*
4. Mayle,R., Wilson,J.R. and Schramm,D.N, *Astrophys. J.* **318**,*288(1987)*

NEUTRINO WORKSHOP SUMMARY

Frederick Reines and LeRoy Price

Dept. of Physics
University of Calif., Irvine 92717

We are witnessing the emergence of an exciting field of particle astrophysics in which yesterday's dream is becoming today's reality. The topics on which we would like to comment in this brief summary are: the solar neutrino situation, neutrino oscillations, supernovae, and point sources of high-energy neutrinos. Details can be found in the other contributions to this workshop.

Solar Neutrinos

The historic experiment of Ray Davis gave great stimulus to astrophysics and to questions regarding the properties of neutrinos. How does one explain his Cl^{37} radiochemical results which gave numbers markedly different, perhaps by a factor of three, from the numbers predicted by the so-called standard solar model. Long ago these results called for an explanation and yet for many years there was only the one experiment. Additional information has recently come from the elegant work of the Japanese/American collaboration called Kamiokande. Their detector is a large, water-Čerenkov detector, and we will mention it in several connections because there are various aspects to the information one can get from such a device. This detector is of the order of a few thousand tonnes sensitive volume. It is extraordinarily well populated by photomultiplier tubes, and consequently the threshold energy can go low enough to encompass the B^8 neutrino signal. Interestingly, recent results from the radiochemical measurements of Davis, as well as those from the water-Čerenkov detector, are not the same as the classic results previously obtained by Davis. Although they are not equal to the prediction of the standard model, they are in better agreement with it. All are eagerly waiting to see what Davis' longer-term results will show, because it's not yet clear whether his current results are really significantly different from his earlier results; something quite interesting may be going on. Davis, as all of us who know him can attest, is very thorough and reliable; however, we are grateful that there are now two experiments because it is the nature of science to require experimental verification and checks.

The previous paucity of detectors is finally being remedied as new detectors come on line. Two gallium experiments are pretty far along in preparation. One is called SAGE (Soviet-American); the other GALEX (a European group). Gallium, as you know, is sensitive to neutrinos from the p-p reaction, the most plentiful source of solar energy. The detection techniques have been worked out, and the detectors are now being constructed. Within a year, or so, they will be giving results.

In addition there is the intriguing idea of using a large, heavy-water Čerenkov detector called the Sudbury Neutrino Observatory (SNO); it is under active consideration by an international collaboration. The idea is to make use of the charged- and neutral-current interactions of solar neutrinos with deuterons.

There is another detector that has been talked about, but not yet funded. This is the scaled up Kamiokande detector, called Super Kamiokande. It may be as large as 20,000 tons; indeed an enormous device.

Neutrino Oscillations

The idea of neutrino oscillations was first suggested by Pontecorvo to explain the solar neutrino puzzle. It's a tribute to the credibility of the experimentalist (Davis) that there has been a veritable blizzard of paper and innumerable discussions in connection with such oscillations. These studies have become a very popular and wide spread activity which promises to continue to engage dedicated searches.

There are many different ways to look into the question. Many qualitatively different sources such as reactors, accelerators and cosmic rays have been used to map out limits on oscillations. But so far, there is no convincing experimental evidence for oscillations.

One of the techniques uses large water-Čerenkov detectors to observe atmospheric neutrinos having different path lengths through the earth. Some of the results are suggestive but not persuasive.

The supernova (SN 1987A) can also be used to say something about oscillations, but the source uncertainties are large, and the number of events small. It is difficult to draw meaningful conclusions at this juncture.

Supernovae

The most remarkable aspect of the detection of SN1987A via neutrinos is that despite the small number of events (19) there is no reasonable doubt that a signal was observed which is attributable to SN1987A. It is evident from a consideration of the background rate in the IMB detector (~1 in 2 days) and the observed signal rate (8 in 6 seconds) that the probability of the observation being "accidental" is minuscule. In addition the result cannot be due to systematics because there were two detectors on opposite sides of the earth and they both saw the same thing. The signal has all the earmarks of being due to neutrinos and is taken to be such. However an interesting, if irreverent, question was raised during one of the workshop sessions, which Mike Turner ran: "How do you really know that the signal is due to neutrinos? Prove it, give us an argument." Well, that is a tall order. At present all we can do is bow in the direction of Sir William of Ockham.

A partial list of the particle physics results already deduced from the data includes:

- limits on neutrino mass, not red hot, but free
- limits on the neutrino magnetic moment
- limits on the neutrino lifetime
- and one very pleasant to contemplate: weak-interaction gravitational-interaction equivalence. Space looks the same for both these fields with really quite good accuracy and that is a rather profound and interesting conclusion.

The astrophysical (i.e. supernova modeling) results seem to fit rather well. They aren't perfect — the "nominal" model doesn't seem to want to explode — but details, details.

A lot of interesting and important questions have been raised by SN1987A. How do we obtain enough data to answer them? A supernova in our galaxy would yield about 100 times as many events in the existing detectors as did SN1987A. We expect there should be neutrino bursts from optically-visible supernovae in our galaxy every 10 to 50 years. In addition there might also be neutrino bursts from supernovae which give very little visible light and from those whose visible light is obscured by interstellar material.

It's important, if you want to obtain maximal information, to be sensitive to a tremendous data rate, perhaps as great as 1000 in a tenth of a second, and to have very precise timing. Unfortunately, the timing of the SN1987A data (± 50 ms IMB, ± 1 min. Kamiokande) was not sufficiently precise; thus one has another parameter to fiddle around with. It would also be really nice to have a very sharp time which could be used as a detector trigger for the gravitational wave pulses which might be produced in a supernova blast.

These are not altogether fanciful conjectures; the difficult part, it would appear, is the long wait between supernovae. We just have to decide how long we are willing to wait. What is our attitude going to be towards these things? Should we not have a supernova watch which involves at least the existing detectors? Should we try to make a detector which is so big ($\sim 10^6$ tons) that we can see out to the Virgo cluster? The point is, one has to think somewhat differently about doing this kind of physics. You have to decide whether it's worth while to have a supernova watch. If we have one, hopefully it will be coordinated — you don't want all the detectors turned off at the same time because it's Christmas or something.

What are the present detectors? There are IMB, Kamiokande, Baksan, and Mount Blanc. Coming on line are Macro and LVD, which are elegant detectors, not as large — but getting there. And then, as mentioned, there are the proposed SNO and Super Kamiokande.

Point Sources of High Energy Neutrinos

So far no point sources of high-energy neutrinos have been identified. But we are optimistic, since if one can see stellar production of high-energy gammas, then almost certainly pions are being produced. This implies simultaneous neutrino production. Such production models have been looked into, to some extent, and their results are most encouraging. The present generation of detectors can't see these high-energy sources, they're just not large enough. However, if we had something about 100 times bigger than the existing detectors, it looks as though one might well detect several such sources. The question is how do you make detectors that big? What are they like? Two have been around in discussion periods and workshops over the years. DUMAND has been under study for more than a decade. GRANDE has been under consideration for several years. The DUMAND approach uses a deep-ocean, water-Čerenkov detector (~20,000 m^2). GRANDE uses a shallow-lake, water-Čerenkov detector (~65,000 m^2).

The details of GRANDE have been described in this workshop. The angular resolution is excellent ($\sim 1^\circ$). A key question for any detector at the surface of the earth is: Can it distinguish between the neutrinos that come up through the earth (the desired signal) and the background due to all the downward-going cosmic-ray muons? The GRANDE group has shown in a very convincing way that you need not go deep under the ground or under the water in order to distinguish the signal from the background.

Because the high-energy neutrinos come from hadronic interactions, they provide a unique window on these interactions in the stellar sources. They also provide information on hadronic interactions at extraordinarily high energies — far above the energies one can get at accelerators — up to the PeV region.

We started some 25 years ago with the observation of atmospheric neutrinos; the reason was simply to see them. There wasn't any detailed quantitative reason, it was just our policy — our hunch — that they would be interesting to look for. Attitudes changed with the emergence of minimal SU_5, which has to do with the unification of the four fundamental forces which made predictions concerning proton decay. Driven by those predictions, very large detectors were built (IMB and Kamiokande). Decaying protons have not seen. But these large detectors were appropriately modified to be sensitive to supernova neutrinos and, lo and behold, to everyone's delight, neutrino astronomy began. The late Luis Alvarez was quoted as having said "the best gift was the supernova and that came from looking for proton decay." But how else would we ever have had such large detectors?

Our field of neutrino astrophysics has made a notable, serendipitous beginning. *Now is the time to be bold and move strongly forward.*

WORKSHOP SUMMARY

THE EMERGENCE OF PARTICLE ASTROPHYSICS CONCLUDING REMARKS

Bernard Sadoulet
Center for Particle Astrophysics
and
Department of Physics
University of California, Berkeley

Abstract

I am presenting a few personal remarks about the workshop, its general atmosphere, the science discussed and some institutional issues raised by the emergence of Particle Astrophysics as a new field.

I. THE EMERGENCE OF A NEW FIELD

The most striking experience for most of us at this workshop has been, I think, the realization of the true emergence of a new field, Particle Astrophysics, at the border (Fig. 1) between Particle Physics, Astrophysics and Cosmology. In the organization of this workshop, we may have forgotten Gravity which is coming back from a period of splendid isolationism to the center of the Physics arena. The quantum gravity theory that we hope to have with superstrings is a clear link to particle physics, the magnificent General Relativity tests with the binary pulsar and the suspicion that black holes are the central engines of quasars are examples of the involvement in astrophysics and, of course, Gravity is central to cosmology. Already the meeting at Rochester[1] on non-accelerator physics gathered two hundred physicists and discussed mostly high energy astrophysics (Extensive Air Showers in particular). This time, we were more than three-hundred and fifty and probably it was one of the first times in this country that particle physics and astronomers were speaking about the same physics. The 1986

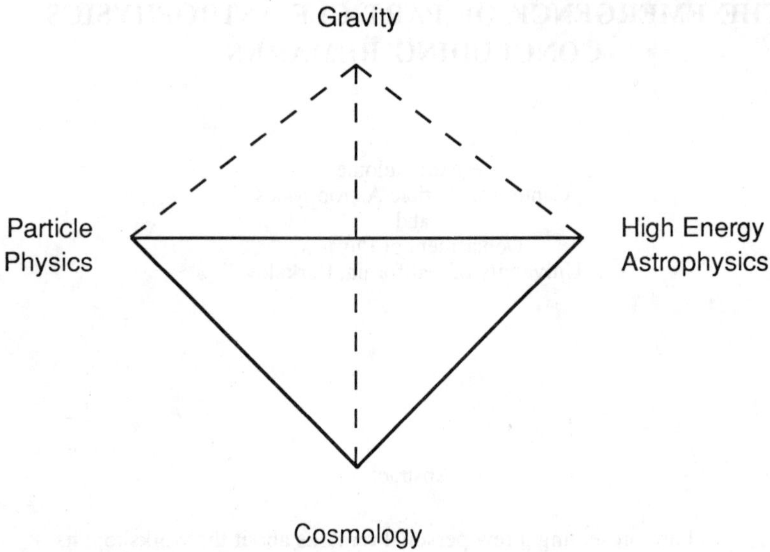

Particle Astrophysics

(Fig. 1)

"Texas" Symposium on Relativistic Astrophysics which took place in Chicago[2] and the ESO-CERN symposiums in Europe[3] are probably the most similar previous occurrences. And here in Berkeley, all of us experienced a new kind of vibrant enthusiasm.

This enthusiasm arose first from the fascinating science that we discussed and that I will review in the next section. It came also from the feeling that the field is beginning to have a constituency of his own, since an attendance of three-hundred and fifty may have to be multiplied by say four to have an idea of the number of physicists involved: 1400 represents a fair proportion of the community. Two additional factors certainly played a role. The presence of DOE and NSF representatives (Drs. J. O'Fallon and R. Nicholson) was very much appreciated and certainly gave the funding agencies a direct impression of where the field is going. Moreover, by coincidence, the National Science Foundation had just announced its decision to establish a Center for Particle Astrophysics in Berkeley in the framework of the Science and Technology Centers. The competition (800 letters of intent, 323 proposals) had spanned all fields of science and the review process has been unprecedented (extensive mail reviews, forty-eight site visits, ranking by a national committee, decision by the National Science Board to fund eleven centers). In those conditions, the fact that Particle Astrophysics was deemed worthy to be supported is, of course, highly significant, and provides an official recognition of the scientific importance of the field.

II. THE SCIENCE HIGHLIGHTS

Let me now attempt to outline the major questions discussed at this workshop. Such a sketch can only be extremely personal and biased. I will regroup my remarks around the three subfields that we considered, Cosmology, High Energy Astrophysics and Particle Physics.

II.a Cosmology

The problem of **dark matter** emerged as the central theme. The fact that at least ninety percent of the mass in the universe does not shine and that its nature is uncertain, is a major scientific puzzle which involves all facets of the field. I will isolate

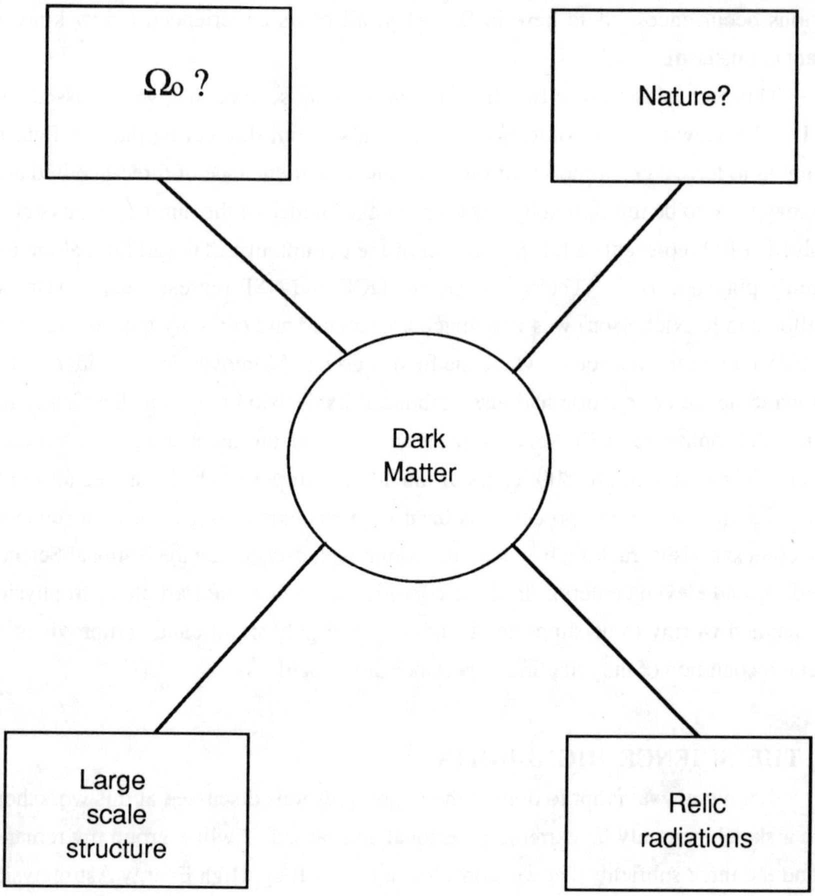

The four questions about Dark Matter

Fig. 2

four questions (Fig. 2), and comment on the new experimental tools which may provide the answers.

• <u>How much Dark Matter is there?</u> The values of the **average density** ρ of the universe or of its ratio $\Omega = \rho/\rho_c$ to the critical density is still uncertain within a factor ten. While the virial determinations of Ω provide a value of 0.2 ± 0.1 at the cluster scales, the comparison[4] of our peculiar velocity with that inferred from the distribution of galaxies of the IRAS survey leads to rather large values of Ω between 0.3 and 1.2 depending on the exact treatment (in particular the correction for the absence of ellipticals in the sample). However, the first methods sample only density fluctuations and are not sensitive to a uniform component and are expected to give underestimates of Ω. The second method may be similarly biased if galaxies are formed preferentially in regions of very high density fluctuations or anti-biased if the correction made for the ellipticals is insufficient. Therefore these methods are at the moment, unable to prove or disprove that Ω is equal to unity.

We need better probes of the global density. It is likely that we will have to resurrect the traditional cosmological tests which measure directly the geometry. Loh and Spillar[5] have recently attempted to trace the volume evolution as a function of the redshift. They have been criticized on the details of their treatment but it is clear that this type of method is very promising. As discussed in the supernova parallel session[6], supernova at redshift of 0.3 or more may be used to measure the deceleration parameter q_0. Type I_a may be sufficiently homogeneous to be used as standard candles and it is in principle possible to use the Baade method based on the expanding shell to determine absolutely the distance of type II supernova.

• <u>What is the nature of Dark Matter?</u> This question was, of course, much discussed with the usual two camps arguing respectively that it can be baryonic, and that this is very difficult.

As remarked by D. Schramm in his overview[7], **some dark matter has to be baryonic,** at least if we take seriously the standard primordial nucleo-synthesis. The most significant recent advance has been made with the measurement of ^{7}Li abundances at the surface of old (population II) stars. Although there is a heated debate around the question of whether it is primordial, if we take the lithium 7 abundances from population II stars, the agreement between the various determinations of primordial abundances is spectacular. It leads[8] to a baryon density Ω_b (in units of the critical density)

$$\Omega_b h^2 = (1.6 {}^{+0.16}_{-0.22}) \, 10^{-2} \tag{1}$$

where, as usual, h is the Hubble constant H in units of 100km/s/Mpc. If H is as low as 50 km/s/Mpc, we need $\Omega_b \sim 6\%$ much more than we expect from stars and gas in clusters (< 1%). Where can these baryons be hidden and how can we detect them? This is an important question that would require much more experimental data to be answered. At least two types of measurements have to be done to check whether these baryons are in the intergalactic medium or in the form of failed small stars, the so called "Jupiters":

- A better simultaneous mapping of the spatial distribution and temperature of the gas flowing into clusters[9], would give not only the depth of the potential well but also the amount and the origin of the gas flowing in. This critical measurements may be done by ROSAT to be launched in 1990 and most probably by AXAF which will have the required instrumentation but unfortunately has not been approved yet.

- It is also crucial to try to detect condensed objects in the galaxy halos by microlensing[10].

The most fascinating possibility is of course that most of dark matter is made of non-baryonic particles (especially if $\Omega = 1$). Two cases have to be distinguished depending whether these particles have been or not in the thermodynamic equilibrium with the rest of matter in the early universe.

If there has not been any equilibrium, the experiment has to rely heavily on the specific properties predicted by the theory to be tested. This is the case for axions which have been reviewed in this conference: the sensitivity level of the two present experiments is too low by a factor 100 to really test conventional axion models[11] and there appears no simple way to gain this factor.

If these particles have been in thermodynamic equilibrium in the early universe, their properties are to a large extent determined by their density today. In particular, if they are massive enough to have been non-relativistic at the time at which they dropped out of equilibrium, their annihilation rate then is inversely proportional to their density today[12]. What is surprising is that this argument gives a weak interaction cross section while no where in the argument, has this fact been used. This may be a profound hint that dark matter is due to some new physics occurring at the weak scale (e.g.

supersymmetry) or a numerical coincidence. In any case, this fixes the scale of the interaction allowing the experimenter to search simultaneously for a wide class of particles, the Weakly Interacting Massive Particles (WIMP's).

In particular, the above argument can be extended, by crossing , to show that these particles concentrated in the halo of our galaxy should **scatter** elastically on matter in the laboratory. The expected event falls between 0.1 and 1000 events per kilogram of target and per day. The best search strategy appear to rely in the short run on the use of ionization detectors while developing simultaneously cryogenic detectors. These may, in the long run, be necessary to give the background rejections and redundant signatures demanded by some of the particle candidates. The first type of techniques are already producing results while new and very interesting ideas and experimental tests have been presented at this workshop[11] on both types.

A final possibility is that Dark Matter is made of light massive neutrinos. There are no good ideas to detect such light neutrinos directly. However constraints may be imposed on their presence in halos of dwarf galaxies. The application DD0 154 by Spergel et al.[13] of the Gunn and Tremaine phase space argument, to limits the mass of a dark matter neutrino to be above 90eV. Even though it may be argued that some of the assumptions made in that paper are rather optimistic, the systematic study of dwarf galaxies may eventually allow to exclude the neutrino as the main component of Dark Matter.

• <u>Large scale structure.</u> Another central question of cosmology is the origin of the large scale structure of the universe. What has to be explained are the large sheets and filamentary structures enclosing voids that the recent surveys are displaying[14] and the more controversial large scale flows. These flows seem to arise when comparing the measured radial velocity of galaxies and the Hubble velocity expected from their distances as obtained with distance indicators. In particular, the data obtained by the 7 Samourais[15] for elliptical galaxies show a global movement of a very large neighborhood (> 30 $h^{-1}M_{pc}$) at 600 km/s with respect to the cosmic microwave background which could be interpreted as the effect of a "great attractor." Note that this massive object carefully hides itself behind the plane of the galaxy! Although new data collected beyond this massive object by Dressler and Faber[16] indicate a decrease of the average velocity and therefore support that interpretation, these finding seems to be in contradiction with spiral galaxy IRAS data[17] and no clear conclusion can yet be made.

A third piece of data was very much discussed in parallel session, the correlation functions of galaxy clusters. The basic selection criteria of Abel clusters may produce artificially a high value. A new computer based survey by Efstathiou and collaborators will probably clarify the question.

Up to now, the galaxy formation and large scale structure was reasonably understood within a "standard model" based on a homogeneous and isotropic universe on which random "white noise" fluctuation of the gravitational potential has been superimposed, for instance by inflation. The large scale structure grows by collapse of the density fluctuations, with the Dark Matter playing an essential role to allow this growth during the time when photons damp the fluctuation of the baryonic component. With a yet poorly understood "biasing" preventing bright galaxies to form in regions of low density, the model is able to give a rather good description of the overall features. However, if indeed large scale flow exist, they will be difficult to explain within this model, where the seed fluctuations does not have enough power at large scale. Numerous discussion occurred in this workshop on what aspect should we give up. Probably the most extreme proposition was that of Turner at this workshop where he proposed to give up homogeneity with a "tilted" universe that some inflation scenarios could apparently produce.

I am not sure that we have to go to this extreme and it may be better to gather a more complete set of data before jumping off the cliff. It would be extremely interesting to extend the present distance survey to $100h^{-1}M_{pc}$ which I am told is possible. The CFA group is also proposing a dedicated 4m class telescope which would allow a complete red shift survey up to 10^6 galaxies. Meanwhile, they are intending to extend gradually their statistics of 10^4 galaxies. Selection of clusters by their X-ray emission with ROSAT will probably give an unbiased measurement of deep gravitational potential cluster distribution and be more reliable that on optical selection. It is also probably fair to say that in spite of the efforts of numerous groups, the Cold Dark Matter model has not yet really reached the solidity to be expected of a "standard model." For instance, the inclusion of hydrodynamics currently being attempted is required if we want to understand the bases for "biasing."

• Radiative diffuse backgrounds from the early universe. The understanding of the dark matter problem and the large scale structure is deeply linked to that of the diffuse background radiations which give us information about the early universe.

The best understood relic so far, is, of course, the cosmic microwave background. The corresponding parallel session discussed, in particular, the possible explanations and implications of the submillimeter excess reported by the Berkeley - Nagoya[18] group. New data from a second flight and from COBE are expected in the coming year. Another focus of the session was the anisotropy. The positive result of the Tenerife group[19] obtained at one wave length is still regarded as tentative. And it is clear that a systematic effort to make maps at various scales with $\Delta T/T \sim 10^{-6}$ becomes a priority of this field. If no fluctuation is yet seen at that level, the fundamental ideas of present cosmology may have to be revised. If fluctuations are seen, their statistical properties will tell us about the relevance of the gaussian model of density fluctuations or the presence of topological defects (cosmic strings). The discussion on experimental methods was also quite informative with, for instance, the presentation of innovative design of small interferometers which are insensitive to the fluctuation of the atmosphere, or the incredible expected sensitivity of bolometers at 100 mK.

The X-ray and γ-ray diffuse extragalactic background is another potential important source of information on the early universe, once the contribution from active galactic nuclei has been subtracted. However before more progress can be made, we need more sophisticated instrumentation, especially in the hard X-ray and soft γ-ray region.

A final form of relic radiation background was not actually discussed in the workshop. It is possible that a gravitational wave background exist as a relic of violent events in the early universe. A particularly interesting case is that of the gravitational waves expected from the oscillation and collapse of cosmic strings. It so happens that the timing of an array of millisecond pulsars may soon allow us to put a significant limit[20] on strings of tension suitable for galaxy formation. In order to do so, the timing accuracy has to increase by a factor 10, which seems within reach both with improved hardware and greater observation time span.

II.b High Energy Astrophysics

Let us now turn to the field of high energy astrophysics, in which many particle physicists are getting increasingly interested. Note that the motivations are **very different** than in the 1950's. The main goal is not so much to use cosmic sources as accelerators to perform particle physics measurements, but to use instrumentation from high energy physics in order to understand compact celestial objects which seem to accelerate particles at least up to 10^{18} eV! But surprises and feedback on fundamental particle physics cannot be excluded.

We will divide our remarks between two topics.

• High energy γ-rays. It is fair to say that the experimental situation of the field of extensive air showers (10^{12} - 10^{19} eV) in the last few years have been one of extreme confusion. Statistically marginal detections of compact binary objects have been reported by numerous groups but because of the small collecting area of the instruments and the erratic behavior of the sources, the picture is clearly dominated by fluctuations and is far from being consistent. Given this state of affairs, the wild speculations which have been made on the existence of new particles seems somewhat premature. It appears now obvious that only very large instruments will clarify the situation. This is the approach taken by the Chicago - Michigan group, which is building, an instrument in Utah which will have a 500 x 500 m^2 area and a sensitivity more than 100 times that of small air shower arrays. This should bring clear confirmation or rejection of the present hints. Even larger instruments[21] are being considered, for instance GRANDE which is designed to observe all the shower components simultaneously (e, γ, hadrons and μ) with 100% coverage, down to 5 TeV or the Heidelberg proposal (Heintze et al.) for a large proportional chamber array to be installed at high altitude (3000 m) whose sensitivity should go down to 10^{13} eV. At the other energy extreme, the Utah team is proposing an ambitious upgrade of the Fly's Eye to be able to measure the cosmic ray (not necessarily γ rays) spectrum around 10^{20}eV and determine, in that way, if these cosmic rays have a cosmological origin.

In showers seemingly coming from known objects, there have been indications of muon excess with respect to what is expected for incident photons. There also, we should get unambiguous confirmation within two years. Explanations in terms of neutron and charged particles are precluded because of the distance of the objects and the galactic magnetic field. F. Halzen[22] has suggested that the muon excess may be due to QCD effects making the γ's to appear hadronic. An interesting complementarity with Particle Physics at accelerators may emerge there, since such effect could be tested with virtual photons at HERA.

• Neutrino astrophysics. With the detection of the supernova 1987A by Kamiokande and IMB, neutrino astrophysics was born. And very much has thus been learned both on the supernova itself and the properties of particles. More recently the Kamiokande has detected unambiguously ^8B neutrinos from the sun, and confirmed the deficit seen by Ray Davis and his team. This will be discussed in the next section.

Where do we go from here? There is an obvious interest in trying to observe high energy sources in neutrinos. The relative ratio of the neutrino and γs will give an interesting hint on the geometry of the source. However because of the low ν cross section, the size (and cost) is increasing rapidly. If we assume that 10 times more ν are emitted than γs by a source, 10^4 m^2 of detectors of upgoing muons emerging from ν interactions in the rock are equivalent to approximately 0.1 m^2 and 10^2 m^2 of γ detectors above the atmosphere at 1 TeV and 100 TeV neutrino energy respectively. The DUMAND team has been exploring for some time the possibility of using Cerenkov techniques in the deep ocean. For the first time from their recent test and the progress of oceanographic methods allowing *in situ* connection of cables, it appears that such deep-sea methods may be feasible. The GRANDE group proposes to have a similar detector in an artificial pool (280m x 280m x 70m). In the latter case, dowgoing muons have to be rejected at a high efficiency level. In both cases, the effective area is of the order of a few $10^4 m^2$.

II.c Particle Physics

Let us finally look at this new field from the point of view of Particle Physics and the link to Cosmology. The particle physicists expect hints on the physics at ultra high energy from the impact it may have on the universe since the Big Bang. Conversely, the cosmologists are turning themselves toward particle physicists to request some help with the major puzzles of their field: Dark Matter, the matter-antimatter asymmetry, and, more technically, the flatness and horizon problems (solved by inflation).

• Supersymmetry. As an example of this complementarity let us take the searches for supersymmetry. By combining the absence of events in $e^+e^- \Rightarrow \gamma +$ weakly interacting particles (ASP, MAC, CELLO) with the cosmological constraints on the abundance of stable particles, it can be deduced in a rather model independent way that the least massive supersymmetric particle has to have a mass of at least a few GeV[23]. In more specific models (e.g. minimum supergravity), the absence of events observed in both the collider experiments and the germanium detection searches for dark matter particles prevents the Higgs particle to be too light[24]. More generally, we witness a friendly competition between accelerators and the direct detection groups. If dark matter particles were to be observed first underground, it is likely that part of the community would not believe the evidence before such particles have been produced at accelerators. Reversely, if supersymmetry or some similar phenomenon are discovered at accelerators, the next most urgent task would be to check whether these particles could explain dark matter.

There are also close interactions between Particle Physics and High Energy Astrophysics as the cosmos may provide energies, masses or distances unavailable in the laboratory.

• Neutrinos. The most obvious example discussed in this workshop has been the properties of the neutrinos. The detection of neutrinos from SN 1987A gives by time of flight, limits on the mass of the electron neutrino of 20 eV which is, by coincidence, similar to present laboratory limits. Much better limits could be obtained from the detection of supernovae in the Virgo Cluster. In addition over there, there is one supernova exploding per two months. The technical problems of making a large enough detector are formidable. Among the far-out ideas being explored are a dense packing of phototubes in DUMAND or the use of neutron detectors in a small mountain as proposed by D. Cline at this workshop. It has to be demonstrated that the necessary background rejection and redundancy can be achieved at reasonable costs! But the observation of neutral current events will even put a limit on the ν_μ and ν_τ mass!

Closer to home and feasibility is the use of astronomical distances to study the mixing between neutrinos species. The solar neutrino problem has been, of course, very much discussed in this workshop. In particular, the disturbing new result of the chlorine experiment have been interpreted by Ray Davis as potentially related to the solar cycle. Another highlight of this year has been the direct detection by KAMIOKANDE of ^{8}B solar neutrinos and the confirmation of Ray Davis' basic claim for a deficit. A large number of experiments sensitive to the neutrinos emitted by the pp reactions, are being constructed (in particular the Baksan and GALLEX gallium experiments) and being planned, including some involving cryogenic detectors (Stanford, Brown University). The possibility of atmospheric neutrinos oscillations was also considered with the still controversial claim of KAMIOKANDE of the deficit of ν_μ neutrinos.

We are therefore in the presence of a very active field with numerous questions, very few answers, yet many prospects for an experimental verification in the foreseeable future!

III INSTITUTIONAL ISSUES

The problem arises then of how we organize ourselves to allow this field to bloom! There are significant potential problems that I would like to comment upon.

This field is by definition **interdisciplinary**. That is, most of our proposals fall between the cracks. Funding agencies are organized naturally along historical fields and some assignments of sub-fields to particular division may not be to the best interest of the physics. Typically the jurisdiction of Nuclear Science at DOE over solar neutrinos do not correspond anymore to the basic physics motivation behind the field. Therefore, while recognizing the effort of most DOE and NSF program officers to transcend bureaucratic rigidity and without being unduly Cartesian, I would claim that a certain amount of evolution may be necessary in the organization of the funding agencies, to correspond more closely to the scientific reality of our field.

Even if this evolution occurred, a large number of difficulties would still be ahead. In a period of limited resources, it is better to choose to do a few things well with **a critical mass** rather than spread out resources too thinly. This means that after small scale prototypes we have to dare to do things right, at the required scale. The early history of the ISR in Particle Physics or that of the extensive air showers arrays are clear examples of the risks of not being ambitious enough. At the same time, we have to maintain an **overall balance** of the program since a great deal of disparity exists between the needed funding levels. There are essential developments to be done at less than half a million dollar a year (e.g. cryogenic detectors or new calorimetry). There may be critical experiments to be done with a few millions dollars (e.g. cosmic microwave background anisotropy). The need may arise to build very expensive detectors (in the 30 million dollars range, e.g. for high energy gamma or neutrino astrophysics). There are long term maintenance needs of facilities (e.g. for a Supernova watch). If we allow one size to kill the other (by monopolization of resources or fear of starting an expensive experiment), our scientific output may not be optimal. A minimum of coordination and long term policy outlook is therefore necessary both nationally and internationally, without preventing a stimulating competition to exist and new ideas to reshuffle the cards.

These remarks brings me to comment on the evaluation procedures. As an outsider, I have the feeling that the review process occurs too much case by case without a global sense of priority. The traditional argument that funding has to be driven by proposals, may be partially flawed in the case of Particle Astrophysics. In accelerator based experiments, the large national laboratories are by definition involved and, before appearing in the funding agencies, the proposals are filtered by the laboratory program advisory committee, which partially enforce a medium range vision. In the case of Particle Astrophysics, this is not the case and the funding agencies are bombarded by uncorrelated proposals. Some **continuity** is needed in the advisory structure so as to help the funding agencies to define a long term policy. I am aware of the practical and legal obstacles to a standing committee on Particle Astrophysics linked to HEPAP and NSAC but even if this form is not adopted, creative thinking, assuring for instance some continuity in the membership of ad hoc committee, may provide the same result. The ranking of the physics objectives cannot be avoided and some comparison across fields (including comparisons between accelerator and non-accelerator experiments) will become absolutely necessary, although they are very difficult. Too large discrepancies between the real physics priorities and the absolute funding levels would be harmful in the long run.

A final comment may be appropriate. The number of participants in this workshop suggests that the field has begun to have a constituency of its own. Gradually we are convincing our colleagues who are solely involved in particle physics at accelerators and our funding agencies, that not only the questions we tackle are fun, but they are fundamental, exciting and complementary to conventional Particle Physics or Astrophysics. We are gradually changing the point of view that this is just play, on which you can afford to loose a few percent of your budget. It is, on the contrary, a fundamental scientific endeavor falling in the **mid-stream** of the mission of the funding agencies. At the same time, we are competing for a limited pool of resources, and the existing structure of accelerator labs give them an advantage over our rather diverse community. Our only hope to establish ourselves is to present, in spite of our different needs, a **united front**. The Center for Particle Astrophysics established in Berkeley will try to work in this direction. Its existence is already a major step forward in the recognition of the importance of the field. It will attempt as much as possible within its own resources to work as "a center without walls" helping to establish the field, and to resist the natural tendency to consider that it is *the* answer to the Schramm-Wilkinson committee request for a major NSF cosmology initiative.

IV. CONCLUSION

Coming back to science, we are witnessing a closer and closer connection between the infinitely small and the infinitely large. For instance, dark matter which dominates the dynamics of the universe may be made of elementary non-baryonic particles. If this can be proven experimentally to be the case, this will be the ultimate Copernican revolution. Not only are we not at the center of the universe, but we would not even be made of the stuff of which the universe is made. Another fascinating suggestion is that the large scale structure with the millions of galaxies each of them with billions of stars originate in fluctuations at the quantum level.

Our amazement comes back to that of Blaise Pascal, a French mathematician and philosopher of the seventeenth century, who wrote immortal pages on the two infinitudes, the infinitely small and the infinitely large: "One depends on the other and one leads to the other. These extremities are touching and being so distant, they are getting united [25]". He could not be more correct. His intuition of the scale involved was, of course, much more limited because of the instrumentation available at his time. Even more, the progress of experimental methods is such that we are now speaking of testing **experimentally** our speculations.

Acknowledgements

It is a pleasure for me as the last speaker to express thanks for the excellent job of the local organizing committee: C. Pennypacker, E. Norman, R. Johnson. W. O'Connor, and the LBL conference staff (in particular Peggy Little and Pam McNey) who made this conference a success. We are all very grateful to the LBL management for the financial support which made everything possible.

As far as my foregoing remarks are concerned, it is a pleasure for me to acknowledge the many conversations I had with my colleagues in the Center for Particle Astrophysics and else where. Many experts contributed patiently to my education in this field which is new for me and I would like to thank them warmly. The remaining misunderstandings and naiveties are, of course, mine.

---

[1] A. C. Melissinos and B. E. Moskowitz, Nucl. Instr. and Meth. **A264,** (1988).

[2] M. P. Ulmer, Edit., Proceedings of 13th Texas Symposium on Relativistic Astrophysics, Chicago, Dec 14-19,1986 (Singapore, World Scientific, 1987)

[3] G. Setti and L. Van Hove, Proceedings of the First ESO-CERN Symposium (1984).

[4] M.A. Strauss, "IRAS Galaxies and the Value of Ω", in this volume.
O. Lahav,M. Rowan-Robinson,and D. Lynden-Bell,MNRAS,**234**,677(1988).

[5] E. D. Loh and E. J. Spillar, Ap. J. **303,** 154 (1986).

6 R.Kirshner, "Supernovae, Summary of the parallel session", in this volume.
7 D.N.Schramm,"Dark Matter, an Overview", in this volume.
8 D. Denegri, B. Sadoulet and M. Spiro, Submitted to Rev. of Modern Physics LBL -88-26014 (1989).
9 See e.g. G. Fabbiano, "Gaseous Hot Halos and Dark Matter: An Observational Viewpoint", in this volume.
10 B. Paczynski, Ap. J. **301,** 503 (1986).
B. Paczynski, Nature **319,** 567 (1986).
11 B.Cabrera, "Status of Dark Matter Searches", in this volume.
12 B. W. Lee and S. Weinberg, Phys. Rev. Lett. **39,** 165 (1977).
13 D.S. Spergel, D.H. Weinberg, J.R. Gott, Phys. Rev **D 38**, 2014 (1988).
14 M.J. Geller, "Experimental Issues in Research on the Large Scale Structure of the Universe", in this volume
15 A. Dressler, D. Lynden-Bell and D. Burstein et al., Ap. J. **313,** 42 (1987).
16 Presented by A.Dressler at this workshop.
17 K. Gorski, M.Davis, M.Strauss, S.D.M. White and A. Yahil , preprint (1989).
18 T. Matsumoto et al.,Ap.J.,**329**,567 (1988).
19 R.D. Davies et al., Nature, **326**,467 (1987)
J.E. Beckman et al., "The Tenerife CMB Fluctuation Experiment: Proogress to Date",
in this volume.
20 M.M. Davis et al., Nature, **315**,547 (1985).
21 See for instance the Proceedings of the Snowmass workshop 1988.
22 F.Halzen, "Gamma Ray Astronomy: The Importance of Muons", in this volume.
23 K. Griest and B. Sadoulet, "Model Independence of Constraints on Particle Dark Matter," in Proceedings of The Second Particle Astrophysics School on Dark Matter,
Erice (Italy), March 1988, Edit. by D. Schramm and P. Galeotti (In press - Fermilab Preprint FERMILAB-Conf-89/57A, 1989)
24 R. Barbieri, M. Frigeri, and G.F. Giudice, Pisa preprint, IFUP-TH 8/88 (1988).
G. Gelmini, P. Gondolo, E. Roulet, "Bounds on the lightest Neutralino as Dark Matter", in this volume.
25 B. Pascal, "Pensées" #352.

LIST of PARTICIPANTS

LIST OF PARTICIPANTS

NAME	INSTITUTION
AHLEN, STEVEN P.	BOSTON UNIVERSITY
AKERLOF, CARL WILLIAM	UNIVERSITY OF MICHIGAN
ALESSANDRELLO, ANGELO A.	INFN, MILANO
ALEXANDREAS, DIMITRIS E.	LOS ALAMOS NATIONAL LABORATORY
ALLEN, KENNETH W.	LAWRENCE BERKELEY LABORATORY
ANDERSON, GREG W.	LAWRENCE BERKELEY LABORATORY
ARONS, JONATHAN	UNIV. OF CALIFORNIA, BERKELEY
AURIEMMA, GIULIO	INFN, ROME
AVIGNONE, FRANK T.	UNIV. OF SOUTH CAROLINA
BACKER, DONALD C.	UNIV. OF CALIFORNIA, BERKELEY
BALATTI, LANA	CALIFORNIA STATE COLL., STANISLAUS
BAND, DAVID L.	UNIV. OF CALIFORNIA, BERKELEY
BARISH, BARRY C.	CALIFORNIA INSTITUTE OF TECHNOLOGY
BARON, EDWARD	SUNY, STONY BROOK
BARTLETT, JAMES G.	UNIV. OF CALIFORNIA, BERKELEY
BARWICK, STEVEN W.	UNIV. OF CALIFORNIA, BERKELEY
BEACOM III, JOHN F.	UNVERSITY OF KANSAS
BEATTY, JAMES J.	BOSTON UNIVERSITY
BECKMAN, JOHN	INSTITUTODE ASTROFISICA DE CANARIAS
BEIER, EUGENE W.	UNIVERSITY OF PENNSYLVANIA
BENNETT, DAVID P.	PRINCETON UNIVERSITY
BENSADOUN, MARC J.	LAWRENCE BERKELEY LABORATORY
BERGER, MICHAEL	LAWRENCE BERKELEY LABORATORY
BERGSTROM, LARS	UNIVERSITY OF STOCKHOLM
BERNSTEIN, GARY M.	UNIV. OF CALIFORNIA, BERKELEY
BERTSCHINGER, EDMUND	MASSACHUSETTS INST. OF TECHNOLOGY
BIRGE, ROBERT W.	LAWRENCE BERKELEY LABORATORY
BLAKE, BERN	THE AEROSPACE CORPORATION
BLANCHARD, ALAIN	OBSERVATOIRE DE PARIS
BLUDMAN, SIDNEY A.	UNIVERSITY OF PENNSYLVANIA
BOCK, JAMES J.	UNIV. OF CALIFORNIA, BERKELEY
BOLDT, ELIHU	NASA
BOND, J. RICHARD	UNIVERSITY OF TORONTO
BOOTH, N. E.	UNIVERSITY OF OXFORD
BRAHM, DAVID E.	LAWRENCE BERKELEY LABORATORY
BRANDENBERGER, ROBERT H.	BROWN UNIVERSITY
BREEDON, RICHARD	UNIV. OF CALIFORNIA, DAVIS
BURMAN, ROBERT	LOS ALAMOS NATIONAL LABORATORY
BURNS, SHANE	HARVEY MUDD COLLEGE
BURROWS, ADAM S.	UNIVERSITY OF ARIZONA
BURTON, JOSHUA W.	LAWRENCE BERKELEY LABORATORY
CABRERA, BLAS	STANFORD UNIVERSITY
CALDWELL, DAVID	UNIV. OF CALIFORNIA, SANTA BARBARA
CARLSON, ERIC D.	LAWRENCE BERKELEY LABORATORY
CARLSON, PER	RESEARCH INSTITUTE OF PHYSICS
CASE, TOM	UNIV. OF CALIFORNIA, BERKELEY
CHANG, CHUNG Y.	UNIVERSITY OF MARYLAND
CHARTOCK, MICHAEL	LAWRENCE BERKELEY LABORATORY
CHENG, ED	MASSACHUSETTS INST. OF TECHNOLOGY
CHERRY, MICHAEL	LOUISIANA STATE UNIVERSITY
CLINE, DAVID B.	UNIV. OF CALIFORNIA, LOS ANGELES
COAN, THOMAS E.	BOSTON UNIVERSITY
COLGATE, STIRLING A.	LOS ALAMOS NATIONAL LABORATORY

NAME	INSTITUTION
COOPER, SUSAN	MASSACHUSETTS INST. OF TECHNOLOGY
COTTINGHAM, DAVID	MASSACHUSETTS INST. OF TECHNOLOGY
COUCH, WARRICK J.	ANGLO-AUSTRALIAN OBSERVATORY
CRANE, PHILIPPE	EUROPEAN SOUTHERN OBSERVATORY
CRAWFORD, HENRY J.	LAWRENCE BERKELEY LABORATORY
CUMMINGS, ALAN K.	UNIV. OF CALIFORNIA, BERKELEY
DALY, RUTH A.	PRINCETON UNIVERSITY
DAS, ANADI J.	SIMON FRASER UNIVERSITY
DASILVA, ANGELA J.	UNIVERSITY OF BRITISH COLUMBIA
DASILVA, LUIZ	UNIVERSITY OF BRITISH COLUMBIA
DAVIS, MARC	UNIV. OF CALIFORNIA, BERKELEY
DAWESE, L.	VICOLO OSSERVATORIOS
DEAMICI, GIOVANNI	LAWRENCE BERKELEY LABORATORY
DEBERNARDIS, PAOLO	INFN, ROME
DECARVALHO, REINALDO RAMOS	CALIFORNIA INSTITUTE OF TECHNOLOGY
DERMER, CHARLES	LAWRENCE LIVERMORE NATIONAL LAB.
DICKINSON, MARK	UNIV. OF CALIFORNIA, BERKELEY
DION, GERARD M.	UNIV. OF CALIFORNIA, IRVINE
DJORGOVSKI, S. G.	CALIFORNIA INSTITUTE OF TECHNOLOGY
DONNELLY, RALPH	UNIV. OF CALIFORNIA, SANTA CRUZ
DOUGHERTY, BRIAN L.	LAWRENCE BERKELEY LABORATORY
DRESSLER, ALAN	HALE OBSERVATORIES
DRUKIER, A. K.	APPLIED RESEARCH CORPORATION
DUBANN, DIONEL R.	UNIV. OF CALIFORNIA, BERKELEY
DZIARSKI, ANDRZEJ Z.	SPACE TELESCOPE SCIENCE INSTITUTE
EASTMAN, RONALD G.	SMITHSONIAN ASTROPHYSICAL OBSERVATORY
EDBERG, TIMOTHY	LAWRENCE BERKELEY LABORATORY
ENGELAGE, JON	LAWRENCE LIVERMORE NATIONAL LAB.
ENSMAN, LISA M.	UNIV. OF CALIFORNIA, SANTA CRUZ
EVRARD, AUGUST E.	UNIV. OF CALIFORNIA, BERKELEY
FABBIANO, GIUSEPPINA	SMITHSONIAN ASTROPHYSICAL OBSERVATORY
FELDMAN, HUME A.	SUNY, STONY BROOK
FENYVES, ERVIN J.	UNIV. OF TEXAS AT DALLAS
FILIPPENKO, ALEXEI V.	UNIV. OF CALIFORNIA, BERKELEY
FIREMAN, EDWARD L.	SMITHSONIAN ASTROPHYSICAL OBSERVATORY
FISCHER, MARC L.	UNIV. OF CALIFORNIA, BERKELEY
FISHER, KARL	UNIV. OF CALIFORNIA, BERKELEY
FLATTE, MICHAEL E.	UNIV. OF CALIFORNIA, SANTA BARBARA
FLORES, RICARDO A.	UNIVERSITY OF MINNESOTA
FOSTER, ROGER S.	UNIV. OF CALIFORNIA, BERKELEY
FRANSSON, CLAES	STOCKHOLM OBSERVATORY
FRAZER, WILLIAM	UNIV. OF CALIFORNIA, BERKELEY
FREESE, KATHERINE	MASSACHUSETTS INST. OF TECHNOLOGY
FRENK, CARLOS S.	UNIVERSITY OF DURHAM
FREUDENREICH, HENRY	ST SYSTEMS CORPORATION
FRIEMAN, JOSH	FERMI NATIONAL ACCELERATOR LABORATORY
FULLER, GEORGE M.	UNIV. OF CALIFORNIA, SAN DIEGO
GAISSER, THOMAS K.	UNIVERSITY OF DELAWARE
GALEOTTI, PIERO	UNIVERSITY OF TORINO
GELLER, MARGARET J.	SMITHSONIAN ASTROPHYSICAL OBSERVATORY
GELMINI, GRACIELA	SCUOLA INTERNAZIONALE SUPERIORE
GILL, MANDEEP	NASA
GIURICIN, GIULIANO	INFN - SEZIONE DI TRIESTE
GLENDENNING, NORMAN K.	LAWRENCE BERKELEY LABORATORY
GOLDHABER, GERSON	LAWRENCE BERKELEY LABORATORY

NAME	INSTITUTION
GOLDSMITH, DONALD W.	
GOULDING, FRED	LAWRENCE BERKELEY LABORATORY
GRANNAN, SABRINA M.	UNIV. OF CALIFORNIA, BERKELEY
GRAZIANI, FRANK R.	UNIVERSITY OF MINNESOTA
GREGORY, RUTH A. W.	FERMI NATIONAL ACCELERATOR LABORATORY
GRESS, JOSEPH	UNIVERSITY OF NOTRE DAME
GRIEST, KIM	UNIVERSITY OF CHICAGO
GRINDLAY, JOSH	HARVARD UNIVERSITY
GROOM, DONALD E.	SSC CENTRAL DESIGN GROUP
GUSH, HERBERT P.	UNIVERSITY OF BRITISH COLUMBIA
HAGMANN, CHRIS A.	UNIVERSITY OF FLORIDA
HAILEY, CHARLES J.	LAWRENCE LIVERMORE NATIONAL LAB.
HALL, LAWRENCE J.	LAWRENCE BERKELEY LABORATORY
HALPERN, MARK	UNIVERSITY OF BRITISH COLUMBIA
HALZEN, FRANCIS L.	UNIVERSITY OF WISCONSIN
HANNA, DAVID S.	MCGILL UNIVERSITY
HARKNESS, ROBERT P.	UNIVERSITY OF TEXAS AT AUSTIN
HARRISON, FIONA A.	UNIV. OF CALIFORNIA, BERKELEY
HARTMANN, DIETER H.	UNIV. OF CALIFORNIA, SANTA CRUZ
HAUBOLD, HANS JOACHIM	UNITED NATIONS
HODGES, HARDY M.	UNIV. OF CALIFORNIA, SANTA CRUZ
HODGES, MARK	CALIFORNIA INSTITUTE OF TECHNOLOGY
HOFLICH, PETER	MAX PLANCK INSTITUT, ASTROPHYSIK
HOLLIMAN, JOHN	UNIV. OF CALIFORNIA, BERKELEY
HSU, STEPHEN D-H	LAWRENCE BERKELEY LABORATORY
HURLEY, K.	UNIV. OF CALIFORNIA, BERKELEY
JAIN, VIDYUT	LAWRENCE BERKELEY LABORATORY
JOHNSON, ROB	LAWRENCE BERKELEY LABORATORY
KAUFMANN, WILLIAM J.	UNIV. OF CALIFORNIA, SAN DIEGO
KERTH, LEROY	LAWRENCE BERKELEY LABORATORY
KIELCZEWSKA, DANUTA M.	UNIV. OF CALIFORNIA, IRVINE
KIM, D. Y.	UNIVERSITY OF REGINA
KIRSHNER, ROBERT	HARVARD UNIVERSITY
KOGUT, ALAN J.	UNIV. OF CALIFORNIA, BERKELEY
KOWITT, MATTHEW S.	LAWRENCE BERKELEY LABORATORY
KRAUSS, LAWRENCE M.	YALE UNIVERSITY
KREYSA, ERNST	MAX PLANCK INST. FUR RADIOASTRONOMIE
KROPP, WILLIAM R.	UNIV. OF CALIFORNIA, IRVINE
LABOV, SIMON E.	LAWRENCE LIVERMORE NATIONAL LAB.
LACEY, CEDRIC G.	UNIV. OF CALIFORNIA, BERKELEY
LAMB, RICHARD C.	IOWA STATE UNIVERSITY
LANGE, ANDREW E.	UNIV. OF CALIFORNIA, BERKELEY
LANOU, ROBERT E.	BROWN UNIVERSITY
LARIMER, RUTH-MARY	LAWRENCE BERKELEY LABORATORY
LAU, KWONG	UNIVERSITY OF HOUSTON
LAWRENCE, CHARLES R.	CALIFORNIA INSTITUTE OF TECHNOLOGY
LEAHY, DENIS A.	UNIVERSITY OF CALGARY
LEARNED, JOHN GREGORY	UNIVERSITY OF HAWAII AT MANOA
LEE, ADRIAN	STANFORD UNIVERSITY
LEGROS, MARK	UNIVERSITY OF BRITISH COLUMBIA
LEIBUNDGUT, BRUNO	UNIVERSITY OF BASLE
LESKO, KEVIN T.	LAWRENCE BERKELEY LABORATORY
LEVIN, JAY	LAWRENCE BERKELEY LABORATORY
LEVIN, STEVE M.	LAWRENCE BERKELEY LABORATORY
LIEDAHL, DUANE A.	UNIV. OF CALIFORNIA, BERKELEY

NAME	INSTITUTION
LIN, ROBERT P.	UNIV. OF CALIFORNIA, BERKELEY
LINDER, ERIC V.	MAX PLANCK INSTITUT, ASTROPHYSIK
LINDLEY, DAVID	NATURE
LIPARI, PAOLO	INFN, ROME
LOH, EDWIN D.	MICHIGAN STATE UNIVERSITY
LOH, EUGENE C.	UNIVERSITY OF UTAH
LOWDER, DOUGLAS M.	UNIV. OF CALIFORNIA, BERKELEY
MAGNUSSON, BENGT	UNIV. OF CALIFORNIA, SANTA BARBARA
MARSHAK, MARVIN L.	UNIVERSITY OF MINNESOTA
MARVIN, HEIDI J.	LAWRENCE BERKELEY LABORATORY
MASEK, GEORGE E.	UNIV. OF CALIFORNIA, SAN DIEGO
MASI, SILVIA	INFN, ROME
MAST, TERRY S.	UNIV. OF CALIFORNIA, BERKELEY
MATHEWS, GRANT J.	LAWRENCE LIVERMORE NATIONAL LAB.
MATTHEWS, JAMES M.	UNIVERSITY OF MICHIGAN
MAYLE, RONALD W.	LAWRENCE LIVERMORE NATIONAL LAB.
MAZURE, ALAIN	LABORATOIRE D ASTRONOMIE
MCCRAY, RICHARD	UNIVERSITY OF COLORADO
MCDONALD, ARTHUR B.	QUEEN'S UNIVERSITY
MCGRAW, JOHN T.	UNIVERSITY OF ARIZONA
MCKEE, CHRISTOPHER	UNIV. OF CALIFORNIA, BERKELEY
MELKONIAN, LEON G.	UNIV. OF CALIFORNIA, BERKELEY
MEYER, HINRICH	BERGISCHE UNIVERSITAET
MILLER, DONALD H.	NORTHWESTERN UNIVERSITY
MILLIKEN, BARRETT	CALIFORNIA INSTITUTE OF TECHNOLOGY
MORRIS, DONALD E.	LAWRENCE BERKELEY LABORATORY
MORSE, ROBERT M.	UNIVERSITY OF WISCONSIN
MOULD, JEREMY R.	CALIFORNIA INSTITUTE OF TECHNOLOGY
MULLER, DIETRICH	UNIVERSITY OF CHICAGO
MULLER, RICHARD A.	LAWRENCE BERKELEY LABORATORY
MYERS, ERIC A.	BOSTON UNIVERSITY
NAGLE, DARRAGH E.	LOS ALAMOS NATIONAL LABORATORY
NAKANO, GEORGE H.	LOCKHEED PALO ALTO RES. LAB.
NAZRARENO, MANDOLESI	ISTITUTO NAZ. FIZ. NUCLEARE
NELSON, JERRY E.	LAWRENCE BERKELEY LABORATORY
NEWPORT, BRIAN J.	UNIVERSITY OF CHICAGO
NG, KIN-WANG	UNIVERSITY OF MINNESOTA
NOETZOLD, DIRK	UNIVERSITY OF CHICAGO
NOMOTO, KEN'ICHI	UNIVERSITY OF TOKYO
NORMAN, ERIC B.	LAWRENCE BERKELEY LABORATORY
O'FALLON, JOHN R.	US DEPARTMENT OF ENERGY
ODDONE, PIERMARIA	LAWRENCE BERKELEY LABORATORY
OLINTO, ANGELA V.	FERMI NATIONAL ACCELERATOR LABORATORY
OLIVIER, SCOT	UNIV. OF CALIFORNIA, SANTA CRUZ
ONG, RENE A.	UNIV. OF CHICAGO, ENRICO FERMI INST.
ORMES, JONATHAN F.	NASA
PARSONS, ANN M.	UNIV. OF CALIFORNIA, BERKELEY
PARTRIDGE, BRUCE	HAVERFORD COLLEGE
PEARSON, TIMOTHY J.	CALIFORNIA INSTITUTE OF TECHNOLOGY
PEMMARAJU, AMMIRAJU	STEVENS INSTITUTE OF TECHNOLOGY
PENNYPACKER, CARL	LAWRENCE BERKELEY LABORATORY
PERLMUTTER, SAUL	LAWRENCE BERKELEY LABORATORY
PETERS, JAMES G.	SAN FRANCISCO STATE UNIV.
PFAFMAN, TIMOTHY E.	UNIV. OF CALIFORNIA, BERKELEY
PHILLIPS, JAMES D.	SMITHSONIAN ASTROPHYSICAL OBSERVATORY

NAME	INSTITUTION
PIANO-MORTARI, GIORANNI B.	UNIVERSITY OF L'AQUILA
PINTO, PHILIP A.	UNIV. OF CALIFORNIA, SANTA CRUZ
PLESS, IRWIN A.	MASSACHUSETTS INST. OF TECHNOLOGY
POTTER, MICHAEL	LOS ALAMOS NATIONAL LABORATORY
PRICE, LEROY R.	UNIV. OF CALIFORNIA, IRVINE
PRICE, P. BUFORD	UNIV. OF CALIFORNIA, BERKELEY
PRIMACK, JOEL R.	UNIV. OF CALIFORNIA, SANTA CRUZ
QUIGG, CHRIS	SSC CENTRAL DESIGN GROUP
RABY, STUART A.	LOS ALAMOS NATIONAL LABORATORY
RAFFELT, GEORG G.	PRINCETON INST. FOR ADVANCED STUDY
RALSTON, JOHN P.	UNVERSITY OF KANSAS
REINES, FRED	UNIV. OF CALIFORNIA, BERKELEY
REPHAELI, YOEL	TEL-AVIV UNIVERSITY
RICH, JAMES	CENTRE D'ETUDES NUCLEAIRES, SACLAY
RICHARDS, PAUL L.	UNIV. OF CALIFORNIA, BERKELEY
RITZ, STEVEN M.	COLUMBIA UNIVERSITY
ROGERS, JOSEPH T.	UNIVERSITY OF ROCHESTER
ROSENBERG, LESLIE J.	UNIV. OF CHICAGO, ENRICO FERMI INST.
ROSS, RONALD R.	LAWRENCE BERKELEY LABORATORY
RUSSELL JR., ROBERT S.	UNIV. OF COLORADO AT BOULDER
RYU, DONGSU	FERMI NATIONAL ACCELERATOR LABORATORY
SADOULET, BERNARD	UNIV. OF CALIFORNIA, BERKELEY
SALATI, PIERRE	UNIV. OF CALIFORNIA, BERKELEY
SASSEEN, TIMOTHY PAUL	LAWRENCE BERKELEY LABORATORY
SCHACHTER, JON	UNIV. OF CALIFORNIA, BERKELEY
SCHMIDT, BERNHARD W.	UNIVERSITAT HEIDELBERG
SCHRAMM, DAVID N.	UNIV. OF CHICAGO, ENRICO FERMI INST.
SCHULTZ, JONAS	UNIV. OF CALIFORNIA, IRVINE
SECKEL, DAVID	UNIVERSITY OF DELAWARE
SHAPIRO, GILBERT	LAWRENCE BERKELEY LABORATORY
SHUTT, THOMAS A.	UNIV. OF CALIFORNIA, BERKELEY
SIKIVIE, PIERRE	UNIVERSITY OF FLORIDA
SILK, JOSEPH	UNIV. OF CALIFORNIA, BERKELEY
SILVER, ERIC	LAWRENCE LIVERMORE NATIONAL LAB.
SIRONI, GIORGIO	INFN, MILANO
SMITH, CRAIG K.	UNIV. OF CALIFORNIA, BERKELEY
SMITH JR., FRANK D.	
SMITH, P. F.	RUTHERFORD APPLETON LABORATORY
SMOOT, GEORGE	LAWRENCE BERKELEY LABORATORY
SNOWDEN-IFET, DANIEL P.	UNIV. OF CALIFORNIA, BERKELEY
SOBEL, HENRY W.	UNIV. OF CALIFORNIA, IRVINE
SPERGEL, DAVID N.	PRINCETON UNIVERSITY
STEBBINS, ALBERT J.	CANADIAN INST. FOR THEORETICAL ASTROPHYS.
STEINER, HERBERT	LAWRENCE BERKELEY LABORATORY
STEPHENS, S. ALFRED	NEW MEXICO STATE UNIV.
STODOLSKY, LEO	PRINCETON INST. FOR ADVANCED STUDY
STOEFFL, WOLFGANG	LAWRENCE LIVERMORE NATIONAL LAB.
STRAUSS, MICHAEL A.	UNIV. OF CALIFORNIA, BERKELEY
STRINGFELLOW, GUY S.	UNIV. OF CALIFORNIA, SANTA CRUZ
STUBBS, CHRISTOPHER W.	UNIVERSITY OF WASHINGTON
SULLIVAN, NEIL S.	UNIVERSITY OF FLORIDA
SUR, BHASKAR	LAWRENCE BERKELEY LABORATORY
SVOBODA, ROBERT C.	UNIV. OF CALIFORNIA, IRVINE
SYMONS, JAMES	LAWRENCE BERKELEY LABORATORY
TAKAHASHI, KASUKE	NATL LAB. FOR HIGH ENERGY PHYSICS

NAME	INSTITUTION
TENORIO, LUIS F.	UNIV. OF CALIFORNIA, BERKELEY
THOMPSON, A. CHRISTOPHER	CALIFORNIA INSTITUTE OF TECHNOLOGY
THORNTON, GREGORY J.	UNIVERSITY OF MICHIGAN
TIMBIE, PETER T.	UNIV. OF CALIFORNIA, BERKELEY
TOTSUKA, YOJI	UNIVERSITY OF TOKYO
TRIPP, ROBERT D.	LAWRENCE BERKELEY LABORATORY
TURNER, MICHAEL S.	FERMI NATIONAL ACCELERATOR LABORATORY
VANBIBBER, KARL A.	LAWRENCE LIVERMORE NATIONAL LAB.
VANDERVELDE, JACK	UNIVERSITY OF MICHIGAN
VEERARAGHAVAN, SHOBA	UNIV. OF CALIFORNIA, BERKELEY
VELTMAN, HELENE	LAWRENCE BERKELEY LABORATORY
VILLUMSEN, JENS VERNER	OHIO STATE UNIVERSITY
VITALE, SANDRO	SEZIONE INFN, GENOVA
VOGEL, KURT R.	LAWRENCE BERKELEY LABORATORY
WAGONER, ROBERT V.	STANFORD UNIVERSITY
WANG, LI-PING	LAWRENCE BERKELEY LABORATORY
WANG, NING	UNIV. OF CALIFORNIA, BERKELEY
WEEKES, TREVOR C.	SMITHSONIAN ASTROPHYSICAL OBSERVATORY
WEISS, STEVEN C.	UNIV. OF CALIFORNIA, BERKELEY
WESTPHAL, ANDREW J.	UNIV. OF CALIFORNIA, BERKELEY
WHEELER, J. CRAIG	UNIVERSITY OF TEXAS AT AUSTIN
WIDGOFF, MILDRED	BROWN UNIVERSITY
WILBANKS, THOR M.	UNIV. OF CALIFORNIA, BERKELEY
WILKERSON, JEFFREY A.	UNIV. OF CALIFORNIA, BERKELEY
WILKES, R. JEFFREY	UNIVERSITY OF WASHINGTON
WILKINSON, DAVID	PRINCETON UNIVERSITY
WILLIAMS, JONATHAN	UNIV. OF CALIFORNIA, BERKELEY
WILLIAMS, WINTHROP T.	UNIV. OF CALIFORNIA, BERKELEY
WILSON, JAMES R.	LAWRENCE LIVERMORE NATIONAL LAB.
WITEBSKY, CHRIS	LAWRENCE BERKELEY LABORATORY
WITHERELL, MICHAEL S.	UNIV. OF CALIFORNIA, SANTA BARBARA
WOLFENSTEIN, LINCOLN	CARNEGIE-MELLON UNIVERSITY
WOOSLEY, STANFORD	UNIV. OF CALIFORNIA, SANTA CRUZ
XIAO, QIANG	UNIVERSITY OF NOTRE DAME
YANG, SCOTT	LAWRENCE BERKELEY LABORATORY
YODH, G. B.	UNIV. OF CALIFORNIA, IRVINE
YOUNG, BETTY A.	STANFORD UNIVERSITY
ZHOU, BING	BOSTON UNIVERSITY
ZIOUTAS, KONSTANTIN	ARISTOTLE UNIV. OF THESSALONIKI

AUTHOR INDEX